HINDU CHEMISTRY

HINDU CHEMISTRY

Text with Translation into English

B. N. SEAL

BHARATIYA KALA PRAKASHAN

Delhi : : (India)

Revised Edition : 2008

Published by **C.P. Gautam**
BHARATIYA KALA PRAKASHAN
3421-A, IInd Floor, Narang Colony Tri Nagar,
Delhi-110035. Phone : 27393083, 32967505
E-mail : bkp3421@yahoo.co.in

ISBN : 978-81-8090-186-7

Printed By :
BDH Printer at Salasar Imaging Systems
Delhi

CONTENTS

Chapter IV

Chapter V

Chapter VI

Chapter I

The Age of Nāgārjuna and the Buddhist Alchemical Tantras

Rise, Development and Propagation of Mahāyānism

In this book it has been incidentally mentioned that side by side with the Śivaite. Tantras we have their Buddhist counterparts in which the salient features of the former appear.[1] The question now arises; when did the votaries of the religion of Śākyamuni find themselves called upon to engraft upon their simple and pure creed the *ensemble* of the gross and grotesque superstitions, the hideous incantations, as also the speculative, the metaphysical and esoteric phases of spiritual aspirations of which the Tantras are repositories ? The answer cannot be given off hand in a few words, for it will lead us to a discussion of the origin and development of Mahāyānism. Under ordinary circusmstances we should have considered it beside our purpose to devote much space to this subject in a work dealing with the history of chemistry. But it so happens that the most prominent figure in Indian alchemy, who is knowledged on all hands to be the discoverer of the processes of distillation, sublimation etc., no other than the reputed founder of the Mādhyamika

1. Vol. i, Intro lxx.

system of philosophy, renowned Nāgārjuana. A brief and rapid survey of the cases which led to the origin on Mahāyānism and its intimate connection with the tenets of the Śivaite cult will be of help to us.

It is one of the saddest episodes in the history of the great religions of the world that the purity of life, right conduct, right living—in short, the moral code upon which their founders have always laid particular stress, soon begin to occupy a subordinate position and a dogmatic theology springs up in time, which gives occupation to the sublest intellect. As for the masses of the people it is doubtful if ever they have remained contented with the mere ethical aspect of a religion. At any rate we find that immediately after the tidings of the death of Buddha had spread about there was a squabble among his followers for the possession of the master's corporeal relics and in course of time stūpas were raised over them, which no less than the three jewels (*Ratnatraya*)—the Buddha, the Dharma and the Saṁgha—and the Bodhi-tree became objects of worship of the devotee. The Caityas, Vihāras and other sanctuaries of divers sorts, with which Buddhistic India was dotted over in the first and second centuries after the decease of the great Teacher, afforded asylums for a vast multitude of monks, who, freed from all worldly cares, found ample leisure to formulate and draw up a code of spiritual and disciplinary exercises.

Although their high ideal was "to lead sweet lives of purest chastity"and induce their lay brethren to follow in their footsteps, rituals and ceremonials of an imposing and elaborate character soon obscured the ray of light which shone forth from the founder. Thus, we find that within 150 years after the passing away of Buddha tedious and circumstantial regulations as to the quality of robes to be worn by the Bhikṣus, the mode of bathing and fasting, the taking of the midday meal, the drinking of unchurned milk, probation and penance, dwellings and furniture and so forth were amongst the weighty subjects for discussion in the Second General Council.[1] We need not pursue the subject further. The reader who is interested in the study of comparative religions will find striking parallels in the early history of the Christian Church.[2]

We are not concerned here with tracing the rise, progress or declined of Buddhism if we have at all referred to its excrescences, it is only to prepare the mind of the reader for the proper understanding of the Mahāyānist development.

The India of Aśoka and of the fourth and third centuries B.C. was in the main Buddhist, but it should not for a moment be supposed that the old faith was extinct. The triumph and ascendancy of the teachings of Śākyamuni and his followers were due in a large measure to the fact that they drew upon, and incorporated into, their creed much that was essentially of Hindu origin.[3] As Dr. Bhāṇḍārkara observes—

"But it was not the metaphysical doctrines of Buddhism

1. "On the daily Life of the Bhikṣus." Vinaya Texts (S.B.E. Series). The Chullavagga, Pt, iii, p. 66.
2. The sermon on the Mount was more or less forgotten. Idolatry disappeared, it is true, but only to re-appear in the shape of Mariolatry and the worship of Saints and relics. In the heat of the schismatic strifes, Christian Charity had to make shift for itself as best it could. Theologains were busy with the interpretation on the writings of the fathers. In a word, the dogmatic to a large extent superseded the moral element of religion.
3. Prof. Rhys Davids expresses the same views in several places :— "There is ample evidence even is the books of the orthodox body of Brahman teachers to show that when Buddhism arose there was not only much discussion of the ultimate problems of life, and a keen interest in the result, but also that there was a quite unusually open field for all sorts of speculations."–"Buddhism" American lectures on the History of Religions (1896), p. 26. Again : "But Buddhism is essentially an Indian system. The Buddha himself was, throughout his career, a characteristic Indian.........he was the greatest and wisest and best of Hindus.–*ibid*, p. 117. The same high authority puts it more tersely when he says : "This is partly, no doubt, because we call them Buddhists, and imagine them, therefore, to belong to a separate class, quite distinct from other Indians of that stock. The Buddhist were, as a matter of a fact, characteristically and distinctly Indian." "Buddhist India," p. 165. *Cf.* "Buddhism, in fact, may be regarded as a reformed

that influenced the masses of the people. What proved attractive was its ethical side. the Buddhist preachers dicoursed on Dharma or righteousness of the people. Such discourses on Dharma without the introduction of any theistic idea have their representatives in the Brahmanic literature. In many of the episodes of the Mahābhārata especially in the Śānti and Anuśāsanika books we have simply ethical discourses without any reference to God, of the nature of those we find in the Buddhistic works ; and sometimes the verses in the Mahābhārata are the same as those occurring in the latter. There appears to be at one time a period in which the thoughts of the Hindus were directed to the delineation of the right conduct in itself without any theistic learning. And Buddhism on its ethical scale represents that phase. Right conduct is the last of the four noble truths of Buddhism."

× × ×

"It was this phase of Buddhism that with the strenuous efforts of the missionaries and of the Emperor Aśoka enables it to achieve success amongst the masses of the people ; and what was wanting on the theistic side was supplied by the prefection and marvellous powers attributed to the founder of the religion. Without this faith in the prefection or, what we should call the divine nature of Buddha, a mere ethical religion would probably not have succeeded. Buddhism was not a social revolution as has been thought by some writers. It was a religion established and propagated by persons who had renounced the world and professed not to care for it.

---

phase of Hindu religion and ethical activity. The Buddhist doctrine of the vanity of the world had been thoroughly disseminated by Yājñavalkya (in the Śatapatha Brāhmaṇa) and with it the practice of subsistance upon alms as Pravrājaka or Bhikṣu ; and a fruitful soil had thereby been prepared for Buddhism. "Weber's Hist. Sans. Lit." trans. 3rd Ed. (1892) p. 285. Again "This teaching contains, in itself, abso-lutely nothing new ; on the contrary, it is essentially identical with the corresponding Brahminical doctrine ; only the fashion in which Buddha proclaimed and dissominated it was something altogether novel and unwonted."–*ibid.*, p. 289.

From times of old there existed in the Indian community such persons who were called Śramaṇas and belonged originally to all castes. These gave themselves to contemplation and sometimes propounded doctrines of salvation not in harmony with the prevelent creed. Buddhism was not even a revolt against caste, for though men from all castes were admitted to the monastic order, and though in the discourses of Buddha himself and others the distinction of caste in pronounced to be entirely worthless, still the object of those who elaborated the system was not to level caste distinctions. They even left the domestic ceremonies of their followers to be performed according to the Vedic ritual. This is one of the arguments b ought against Buddhism by Udayanācārya. "There does not exist," he says, "a sect, the followers of which do not perform the Vedic rites beginning with Garbhādhāna and ending with the funeral, even though they regard them as having relative or tentative truth."[1] Buddhism, however, was a revolt against the sacrficial saystem and denied the authority of the Vedas as calculated to point out the path to salvation. And this is the root of the hostility between itself and Brahmanism."[2]

The zealous missionaries of Buddhism captured the heart of the masses by appealing to their moral instincts. Persuation and not persecution was the instrument they chose to wield. King Aśoka proclaimed universal toleration ; he inculcated respect for Brahmins as well as Śramaṇas or ascetics of all sects. It is a relief to find that the mighty monarch had never recourse to brute force for the propagation of his creed. We

---

1. नास्त्येव तद्दर्शनं यत्र सांवृतमेतदित्युक्त्वापि गर्भाधानाद्यन्त्येष्टिपर्यन्तां वैदिकीं क्रियां जनो नानुतिष्ठति। Ātmatattvaviveka, Colc. Ed. of Saṁvat 1906, p. 19. सांवृत relating to संवृति, a Buddhist technical term.
2. A peep into the Early History of Indian from the found action of the Maurya Dynasty of the fall of the imperial Gupta Dynasty (B.C. 322 Circa 500 A.D.) pp. 362-363.
   Journal of the Bombay (Mumbai) branch of the Royal Asiatic Society (1900), Vol. xx. No. LVI.

are spared all the scenes of blood, reapine and violence, which disfigure the pages of religious history in the West.

It has already been pointed out that Buddhism itself may be looked upon as an offspring of Hinduism ; nay, it is a logical outcome of the principles enunciated in the Sāṁkhya system of the philosophy. If India so readily responded to the teachings of the new faith, it is not because she renounced Hinduism but because she found that all that was highest and noblest in the latter was absorbed in the former. The great Teacher who now arose gave only a new shape and direction and a vigorous impetus to the germ of ideas already in existence and turned them to capital account. A good deal of confusion may be avoided by bearing in mind this aspect of the question. When speaking of Buddhist India one is apt to rush to the conclusion that every vestige of Hinduism had disappeared off the face of the vast continent. As a matter of fact even during the zenith of Buddhistic glory. Brahminism was rearing its head and was professed by not a few among the cultured and intellectual classes and was ready to assert itself whenever a favourable occasion arose.

In the second century B.C., we find the Kabul valley the Punjāb and Mālava acknowledging the sway of the Princes of the Turkish race.[1] Wema Kadiphses, the second of the Kuṣāṇa Dynasty is spoken of as a devotee of Maheśvara ; his coins bear the emblem of Nandin on the reverse, which is accompanied by a human figure which holds a trident in its right arm representing Śiva.[2] The last three Kuṣāṇas-Kaniṣaka, Huviṣka and Vāsudeva—have been noticed in the Rājataraṅgiṇī,[3] and the emblems on the reverse of their coins

---

1. Lassen "Indische Alterthumskunde," Vol. ii, pp. 811-13, Ed. 1874.
2. *Ibid.*, pp. 808-811.
3. See next page. Regarding the reliability of the account in the Rājataraṅgiṇī, Dr. Stein observes :- "Kalhaṇa's account of the reign of these kings who are supposed to have ruled simultaneously, is brief enough, but undoubtedly preserves data of genuine historical tradition. It clearly describes them as princes of Turuska, *i.e,* Turkish

are figures of deities borrowed from the Greek, Persian and Brahmanical pantheon and of Buddha. Thus, the Gods of all these four religions shared the adoration of these. Turkish Kings. But the figure of Buddha in the sitting or meditative and the standing posture is to be found on the coins of Kaniṣka alone.[1]

In the Buddhist ecclesiastical history Kaniṣka occupies a prominent place. It was in his reign and under his patronage that the Third Council was held to settle the cannon once again. The church was convulsed by internal dissensions and schisms resulting in its being split up into as many as 18 sects. "The most significant trait of the Third Council is that it closed a period of old quarrels between the sects ; it did not prevent the rise of new aspirations. Mahāyānism, which is an incipient stage was already existing erelong, boldly raised its head."[2]

Hinduism was now very much alive. After the Third Council and distinction between the Hinayānists and the Mahāyānists became more and more accentuated. The apostles of the latter development became convinced that in order to draw in into their folds the bulk of the people some sort of compromise was necessary, the orthodox Buddhism in all its rigidity must be given up. In other words, the neo-Buddhism which now sprang into existence began to absorb and assimilate the popular form of Brahminism and thus swell the ranks of its followers. A pourely ethical creed has

---

nationality, as powerful sovereigns and as faithful patrons of the Buddhist Church. On these points the statements of the Chronicle are fully supported by the evidence of our most authentic records. The continued existence of the three places, Kaniṣkapura, Huṣkpura, and Juṣkapura, which are described as foundations of these kings and which still survive to the present day is likely to have assisted in preserving a recollection of their founders.

1. Percy Gardner : "Coins of Greek and Scythic kings" pp. 129 ff. Cunningham : "Num. Chr." 1892, pp. 63 ff.
2. Kern : Manual of Ind. Bud. (Grundiss d. Indo.-Arischen Phil). p. 122.

never secured a following except perhaps among a chosen few. To quote the eloquent words of the historian of "Rationalism in Europe"—

"There arise from time to time men who bear to the moral condition of their age much the same relations as men of genius bear to its intellectual condition. They anticipate the moral standard of a later age, cast abroad conceptions of disinterested virtue, of philanthropy, or of self-denial that seem to bear to relation to the spirit of their inculcate duties and suggest motives of action that appear to most men altogether chimerical. Yet the magnetism of their prefection tells powerfully upon their contemporaries. An enthusiasm is kindled, a group of adherents is formed, and many are emanciapated from the moral condition of their age. Yet the full effects of such a movement are but transient. The first enthusiasm dies away, surrounding circumstances resume their ascendency, the pure faith is materialised, encrusted with conceptions that are alien to its nature, dislocated and distorted till its first features have almost disappeared. The moral teaching being unsuited to the time becomes inoperative until its appropriate civilization has dawned, or at most it faintly and imperfectly filters through an accumulation of dogmas, and thus accelerates in some measure the arrival of the condition it requires." Vol. i, p. 305, Ed. 1900.

The same historian in explaining the rapid spread of Christainity in Europe observes in another place :—" It triumphed not so much by superseding rival faiths as by absorbing and transforming them. Old systems, old rites, old images were grafted into the new belief, retaining much of their ancient character, but assuming new names and a new complexion."

Nāgārjuna : a Leading Representative of Mahāyānism

Among the bold spirits who took a leading part in this

renovation the name of Nāgārjuna stands conspicous. The Mādhyamika system with its axiom—*sarvaṁ Śūnyam*—a form of Pyrrhonism pushed to its extreme limits, which is an essential part of Mahāyānism is generally ascribed to him. Northern Buddhistic literature is replete with the marvels and miracles performed by him, and tradition has invested him with superhuman powers. Hiouen 'Thsang calls him along with Deva, Aśvaghoṣa and Kumāralabdha, "as the four suns which illumine the world."[1] As early as A.D. 401-409 we find a life of Nāgārjuna Bodhisattva translated into Chinese.[2] Tāranātha has committed to writing all the floating mass of legends connected with this venerable name, but as the Tibetan monk wrote so late as the beginning of the 17th century A.D., and as nothing was too astounding for his pious credulity, we have here only nuclei of facts round which have gathered accretions of vast proportions. It is now almost a hopeless task to separate the grain from the chaff. Nevertheless we can glean certain historical data from all that have been handed down. Before proceeding further we think it proper to treat that reader to a specimen of the materials with which one has to deal. We cull the following bits from the "Life and Legends of Nāgārjuna" gathered chiefly from Tibetan records including Tāranātha's "History of Buddhism."

"A rich Brahmin of the Vidarbha country to whom no son had been born for many years, once saw in a vision, that if he gave alms to, and entertained one hundred Brahmins, he could get a son. Accordingly he made offerings and prayers to the God the entertained on hundred Brahmins. After ten months his wife gave birth to a son. The rich man invited learned astrologers to predict the fortunes of his child; but they found that it could not live more than a week. In all other respects the child was calculated to be fortunate.....The astrologers assured them (his parents) that if they entertained one hundred Bhikṣus, it would live seven years,

1. Julien's texts, Vol. ii, p. 214.
2. Bun. Nanjilo's Catalogue, Ap. i, No. 3

beyond which its life could not be prolonged by any means whatever. When the seventh year about to expire the parents were overwhelmed with grief. To avoid the painful sight of the their son's predicted death, they caused him to be removed to certain solitary place in company with a few retainers.

"As the boy (Nāgārjuna) was passing him mournful days, one day the Mahābodhisattva. Avalokiteśvara Khasarpaṇa visited him in disguise and advised him to go to the great monastery of Nālendra in Magadha as the surest means of escaping from the hands of death. He accordingly repaired to that famous Vihāra and arriving at the gate recited some gāthās. During that time the great sage Śrī Sarah Bhadra was the high priest of Nālendra,......Who ordained Nāgārjuna a Bhikṣu of the Vihāra. During the latter part of this office the country was visited by a famine in consequence of which the monks fell into great distress. The Manager became very thoughtful about the terrible effects of the natural calamity. Distress and scarcity compelled the congregation more keenly to feel the necessity of money. The monks now determined to devise some means of acquiring treasures for the support of the famished congregation, and Nāgārjuna accordingly started on an expedition to visit an island in the great ocean where lived a great saint well versed in the art of alchemy. As the sea could not be crossed by any earthly means, he, by dint of his divine learning, got two leaves of an enchanted tree, by means of which he crossed the ocean and miraculously visited the island and presented himself before the sage who was greatly suprised to see a human being arrived at his abode deemed inaccessible to mortal beings. The sage earnestly inquired how he succeeded in achieving the wonder. Nāgārjuna replied respecfully stating to him the reasons of his visit and circumstances that brought him thither. He also showed him one of the enchanted leaves, concealing the other in his medicant's platter. He begged him to teach him the art of turning metals into gold. The sage consented to the proposal, but not liking to let the wonderful art be known in Jambudvīpa he determined to detain him for ever in the island by depriving him of the enchanted leaf. To effect this, he said that he could not teach the art of alchemy unless Nāgārjuna consented to part with the leaf. Nāgārjuna consented, and was taught the art. When it was fully mastered he flew towards the Indian Continent by the help of the remaining leaf. Returning to Nālendra, by means of his easily acquired wealth he supported the whole body of monks. By his religious practices he obtained Siddhi (perfection). He refuted the theories of Śaṅkarācārya[1] and imparted religious instruction to the monks of Nālendra.

"Nāgārjuna returned to his country after a visit to Uttarakuru and erected many caityas and temples, composed many works on science, medicine, astronomy and alchemy. After the death of Saraha Bhadra, the office of high priest fell upon Nāgārjuna, which he managed with great ability and indefatigable zeal. He matured the Mādhyamika philosophy which was only conceived by his illustrious teacher Saraha."

"Nāgārjuna is said to have been a great friend of king De-Chye (Śaṁkara) of S. India, whom he convented to Buddhism."[2]

One thing seems to be clear from the above, namely the Nāgārjuna was born and brought up in the Brahminical faith but was afterwards converted to Buddhism and was celebrated as an alchemist.

Nāgārjuna as an Adept in Magic and Alchemy.

Tāranātha, it is true, completed his History in 1608 A.D.,[3] but he derived his materials from Tibetan sources and the analysis of Dulva by Csoma goes to confirm his account. We

1. An instance of glaring anachronism. Śaṁkara flourished in the 8th-9th century A.D. See below, foot-note to p. 13.
2. S.C. Dās : Journ, As. Soc. ii, Pt. i, pp. 115-120.
3. Schiefner : Gesch. d, Bud. Pref. vi.

have already made use of the record left us by Hiouen Thsang.[1] It is thus clear that all the testimonies concur in ascerting Nāgārjuna not only as the originator of the Mādhyamika Philosophy but also as an adept in magic, conjuration and alchemy, and that even so early as the 7th century A.D. The exact time during which he flourished is a matter of controversy. He is generally regarded as a contemporary of Kaniṣka. One cannot go far wrong in assigning circa 150 A. D. as the date fo his succeeding to the Patriarchate.[2]

Nāgārjuna as a Contemporary and Friend of king Sadvāhana.

An important document has been preserved for us in its Tibetan and Chinese versions, which seems to have an historical basis. It is in the shape of a "Friendly Epistle of Nāgārjuna to king Udayana."[3] The original in Sanskrit,

1. *Vide* Vol. i, intro. xciii.
2. According to Lassen Nāgārjuna live about A.D. 23 during the reign of Kaniṣka. The Rājataraṅgiṇi says :
"Then there were in his land kinds called Huṣka, Juṣka and Kaniṣka, who built three towns named after them (Huṣkapura, Juṣkapura and Kaniṣkaura).........During the powerful reign of these (kings) the land of Kāśmīr was to a great extent, in the possession of the Buddhas. At that time 150 years had passed in this terrestrial world since the blessed Śākya Siṁha (Buddha) had obtained completed beatitude (Nirvāṇa)......And a Bodhisattva lived (then) in this country as the lord of the land, namely the glorious Nāgārjuna.....Stein's trans. Vol. I, pp. 30-31. As the tradition of the Northern Buddhists as recorded by Hiouen' Thsang (Si-yu-ki, I, pp.l 99, 151, the Tibetan Dulva (Csoma, As. Res. XX, pp. 92, 297) as also Schiefner (Tār. Gesch. D. Bud. p. 301) concur in placing the commencement of Kaniṣka's rule 400 years after Buddha's Nirvāṇa, the date assigned here to Nāgārjuna is rather curious. Lassen sums up his conclusion on this knotty point in these words:—"Wegen der grossen Anzahl seiner Mūnzen mus dem Kaniṣka eine ziemlich lange Regierung Zugeschrieben werden ; ich glaude daher annehmen zu durfen, as er etwa bis 40 nach Chr. G. regiert habe." Fleet maintains B.C. 57. V. Smith c. 120 A.D. While Messrs. Bhāndārkar c. 278 A.D. (Journ Mumbai Br. Roy. As. Soc. Vol. No. lvi, pp. 269-396) as the date of Kaniṣka.
3. *Vide* Wenzel ; Journ. Pāli Text Soc. (1886), p. 1.

entitled *Suhrillekha,* has not yet been recovered and probably been lost. Udyana in the Tibetan Subscription is Bdye-Spyod=Sadvāhana.[1] This Sadvāhana is a prominent figure in the history of S. India. In ancient Sanskrit literature he is frequently spoken of as a patron of learning and there are several literary reminiscences associated with this name.[2]

Now "the Andhrabhṛtyas or Śātavāhanas ruled over the Deccan from B.C. 73 to about A.D. 218, *i.e.,* for about 3 centuries.

"The period during which they ruled over Mahārāṣṭra must have been a prosperous one in the history of the country. Hence, several traditions which regard to different kings have been preserved. But that Śālivāhana or Śātavāhana was a family name must have been forgotten and different princes of the dynasty have been confounded and indentified. Thus Hemachandra in this Desikoṣa gives Śālivāhana, Śālana, Hāla, and Kuntala as the names of one individual."[3]

The So-to-p'o-ho of the Chinese version of the "Friendly Epistle" has been indentified with one of the Śātavāhanas or Andhra kings, possibly Yajña-Śrī-Śātakarṇi, who seems to have reigned about 172-202 A.D. There is, therefore nothing improbable in this particular, Śātavāhana being a contemporary of Nāgārjuna. In the alchemical Tantra "Rasaratnākara" ascribed to Nāgārjuna, there is a dialogue

1. S.C. Dās (l.c.) says : "Nāgārjuna is said to have been great friend of king De-Chye (*i.e.* Śaṅkara) of S. India, whom he converted to Buddhism. Now De-Chye=Bdye-byed according to modern prono-unciation. Dās is evidently in error in rendering De-Chye as Śaṅkara (Wenzel).
2. *Cf.* Kāmasūtra of Vātsyāyana. कर्तर्या कुन्तलः शातकर्णिः शातवाहनी महादेवीं मलयवतीं [जघान] । Referring to a "new" MS. of *Harṣacaritaṁ* his possession Hall remarks—"for Śālivāhana I there find Śātavāhana "Intro. to "Vāsavadattā," p. 54. This by implication would suggest that the another MS. had the reading Śālivāhana.
3. Bhāṇḍārkara's "Early History of the Deccan" p. 36.

between this sage and Śālivāhana. We shall revert to this subject later on.

Nāgārjuna as a comprehensive name of the activity of Mahāyānism

Numerous works have been fathered on Nāgārjuna and it is an open question if any of them be genuine.[1] As is well known Vyāsa or more properly named Vedavyāsa has been taken to be compiler of not only the four Vedas but also of all the Purāṇas put together. Piety and credulity go hand in hand and are seldom troubled with questions of anachronism. Plato, Democritus and Geber have been help responsible for writings which appeared several centuries later Names, venerable and illustrious, have often been passed into service to lend weight and dignity it productions which otherwise would not have commanded a respectful hearing. On the whole we are inclined to agree with Kern when he says, "The figure of Nāgārjuna, so prominent in the history of the rise of Mahāyānism, shows a double character. It is, on the one side, the name of an influential person, the first eminent leader of a school imbued with Hinduism and the methods of Indian scholastic philosophy. On the other hand, Nāgārjuna is simply a comprehensive name of the activity of Mahāyānism in the first phase of its onward course."[2]

1. In Bunyin Nanjio's Catalogue of the Bud. Tripiṭaka there is a list of 24 works ascribed to Nāgārjuna. The Suhṛllekha was translated into Chinese in 434 A.D. App. pt. I, p. 368.
2. In Vol. I of this work (Intro. xciii), we quoted Albérūnī as to the date of Nāgārjuna. This cultured Arab, ordinarily a very trustworthy guide, derived his information from the Brahmins of the part of Indian from which every vestige of Buddhism how disappeared in the 11th century and he was evidently misled on this point as the treaditions relating to Nāgārjuna had at that distance of time become very vague. This will be clearly seen from what Prof. Sachau says in the preface to the Arabic edition to Albérūnī.

Asaṁga

From the time of Nāgārjuna onwards Mahāyānism began to be tinged more and more with Brahmanical bias. A notable and decided step in this direction was taken by Asamga, a monk of Gāndhāra, who composed the Yogācāra-bhūmi-śāstra, in which by assimilating the doctrines of Patañjali he paved the way for the growth of the Tantras. He seems to have lived about 400 A.D.[1] Vasubandhu, Asaṁga's younger brother, was another zealous adherent of this school and is said to have been a teacher in the college attached to the monastery of Nālendra—the "Oxford of Indian Budd-hism." From Tibetan sources we learn that the celebrated logician Dignāga was a disciple of Vasubandhu, a contemporary to Lha-tho-ri, king of Tibet, who lived up to 371 A.D.[2]

"Its civilization was then essentially Brahmanical as it had come to be in a protracted struggle with Buddhism. Albérūnī does not know Indian Buddhism from personal experience, though it had not yet entirely withdrawn from India and in some part was still a political power. (p.v.)

"The valley of the Kābul river and the Punjab are all that Albérūnī has seen of India. (p. xiii).

"The high schools of Hindu science and learning, Kasmir and Benares, were in Albérūnī's time unapproachable for Muslems. (p. xiv)."

1. *Cf.*–Takakasu: Journ. Roy. As. Soc. Jan. 1905.
   A life of Vasubandhu was translated into Chinese by Paramārtha. A.D. 557-569–Bun. Nunj. Cata. p. 371. Csoma Korosi following the Tibetan Chronology says :–"I know that Ārya Asaṁga lived in the sixth or seventh century after Jusus Christ." As. Res. XX, p. 513. This date has been accepted by Lassen (Ind. Alt. ii, p. 460) as also by R. Davids (Buddhism. p. 207). But this Chronology has now become untenable. The date of Asaṁga and his brother Vasubandhu should be put back by about two centuries, as some of their works were translated into Chinese in the beginning of the fifth century and perhaps earlier. (Bun. Nanj. Cata. App. i, p. 371). It must have taken a century or two to have works sufficiently recognised in India before they would deserve a place in the Chinese Tripiṭaka.
2. S.C. Vidyābhūṣaṇa : Journ. As. Soc., Vol. i, (1905) p. 227.

Gradual and Imperceptible fusion of Mahāyānism with Śaivaism

Mahāyānism now began to adept itself to its environments. The absorption of the Yoga ideas made the transition into the Tāntric cult easy and Northern Buddhism began to develop and expand by entering into an alliance with Śaivaism, which favoured the growth of Buddhist Tantras. The origin and development of this class of literature have been the subject of a masterly exposition by Burnouf. The Mahāyānism not only set up their own deities but borrowed copiously from the pantheon of the Hindus. Thus in the Śivaite Tantras while the God Śiva is the fountain of all knowledge and sciences, in their Buddhist counterparts we have the celestial, metaphysical and potential Buddhas[1] occupying the same position The worship of the female energies (Śakti) which plays such an important part in the Tantras was encouraged in the person of Tārā. The Hindu Gods and Goddesses were also objects of adoration, only they were assigned a subordinate position. The question has often been asked : why did Buddhism seek alliance with Śaivaism in preference to the Viṣṇuvite cult? The answer seems to lie in the fact that it was precisely in those parts of India where the worship of Śiva, especially in its Tāntric form, had struck root that Mahāyānism asserted its sway;[2] and thus the apostles and propagators of the latter made most of the former. A few centuries later when Vaiṣṇavism gained as ascendancy in Bengal the tables seem to have been turned. A gradual and imperceptible fusion took place between the rival creeds. Thus in the celebrated adoration in the opening lines of Gītagovinda we find Buddha freely acknowledged as an incarnation of Viṣṇu and extolled for his abhorrence of sacrificial rites.[3]

---

1. *i.e.* the Bodhisattvas.
2. *Vide ante* p. xi (under Wema kadiphses).
3. निन्दसि यज्ञविधेरहह श्रुतिजातम् सदयहृदयदर्शितपशुघातम्।
   केशव धृतबुद्धशरीर जय जगदीश हरे।।

From Fah-hian's travels we gather that in the beginning of the 5th century A.D. the Mahāyānism were gaining the upper hand, though their rivals, the Hīnayānists, were still holding their own in several localities. Thus at Mathurā and Pāṭaliputra he found the members of both the sects living side by side and having monasteries of their own. From the fact, however, that the pious Chinese pilgrim repeats the Suraṅgama Sūtra for his protection, we may conclude, since he is by no means a pronounced Mahāyānist, that the predisposing cause for the origin of the Tantras were already in existence. "In this Sūtra is contained the most complete list of Dhāraṇīs (invocations) found in any Chinese compilation. There are 426 distinct sections containing the names of the different Buddhas and Hindu deities worshipped at the time of the composition of the Sūtra. Considering that Fah-hian in the early part of the 5th century regarded this book with reverence... We may reasonably assign it to a period not later than the end of the 1st century. Now amongst the invocations, we find distinct reference to Dhyānī Buddha, Vairocana, Akṣobhya, Amitābha and others shewing that they were coming to be recognized and worshipped even at that early date."[1]

Dhāraṇīs as the Precursors of the Tantra

These Dhāraṇīs may be looked upon as the precursors of the Tantras and they fully support the views that when they were composed Buddhism had turned a new leaf. The repeating of certain magic formulas along with the names of Buddha Amitābha, etc., was to secure, salvation, in other

---

It is scarcely correct to assert that Buddhism was exterpated in the land of its birth by cruel persecution. *Cf.* Vol. i, Intro. lxvii. The Bengali poet Ramcandra Kavibharatī, author to *Bauddha Śataka,* though a devoted Buddha, belonged to the same class of thinkers as Jayadeva. Pandit S.P. Śāstrī is inclined to place him in the latter end of the 13th century.—Journal of Buddhist Text Society, Vol. I, Pt. iii. So late as 1441 A.D. MSS. Buddhist works used to be copied from in Magadha—Bendall's Camb. Cata of Bud. Sans. MSS. (1883), Pref. iv.

1. Intro. to Beal's "Fah-hian." (Lxxii).

words, "instead of the old doctrine [of *Karman]*—as a man sows, so he shall reap—a new and easier way of salvation is here preached, *viz.*, as a man prays, so he shall be saved. It is what is known to us a salvation by faith rather than by works....It would almost seem as if this popular and easy doctrine had secured to itself the name of Mahāyānism as meaning the Broad Way in opposition to the Narrow Way, the Hīnayāna.[1]

Vaipulya Sūtras

The new class of literature which now arose in order to meet the demands of the Mahāyānism revival is collectively known as the Vaipulya Sutras on the Sūtras of the developed school, of which the Dhāraṇīs are an integral part. It found expression in such works as the "Saddharmapuṇḍarīka," "Lalitavistara," "Tathāgataguhyaka" "Prajñāpāramitā," etc., all of which soon acquired almost canonical sanctity in the estimation of the N. Buddhists. Burnouf has tried hard to differentiate between the simple Sūtras as represented by the literature of the primitive orthodox Buddhism and the expanded Sūtras of which the distinguishing mark is the occurrence of Bodhisattvas.[2] This drawing of hard and fast

---

1. Max Müller, Buddhist Mahāyāna Sūtras. Intro. to larger Sukhāvatī-Vyūha, ix, S.B.E. Series, Vol. xlix.
2. La presence des Bodhisattvas ou leur absence intéresse donc le founds mems des livres oū on la remarque, et il est bien evident que ce seul point trace une ligne de démarcation profonde enter les Sūtras ordinaires et les Sūtras développés–Burnouf, Intro. p. 112 ; *ibid*; p. 120. Ed. 1844. *Vide* Max Müller's remarks at the end of the Smaller Sukhāvatī-Vyūha, p. 102, S.B.E. Series, Vol. xlix.
   Cf. also "Some Vaipulyas are, materially, much like the old Sūtras, whole passage *e.g.* of Lalita-Vistara recur almost word for word in the Pāli Scriptures." Kern (Ind. Bud., p. 5). This eminent Buddhist scholars also very appositely remarks elsewhere. "The results arrived at by Burnouf may be right so far as Mahāvaipulya Sūtra as a whole is concerned, they cannot be applied to all the component parts of such a work. Not to go further than the Saddh. and the L. Vistara, it can hardly be questioned that these works contain parts of very different dates, and from various sources." Kern—Intro. to Saddhar,

lines has been found to be well nigh untenable.[1]

By the second century A.D. we find the leading beliefs and ideas as crystallised in the Mahāyāna literature fully in vogue. Thus we come across in the "Buddhacarita" of Aśvaghoṣa, who is admitted on all hands to be a centemporary of Kaniṣka, such a passage as this; "this, Sirs, is the Mahāyāna, the instrument of the law of the perfect Buddha, which is the establisher of the welfare of all beings, set forth by all the Buddhas."[2] The "Lotus of good Law" was translated into Chinese at the close of the second century A.D.[3]

Buddhistic Tantras

Side by side with the growth of the scriptures another class of literature was called into being by the exigencies of the times—we mean the Tantras, the necessity for which has been hinted at in the first volume.[4] However, esoteric may be the doctrines sought to be conveyed through the medium of these productions, it would be idle to deny that there is much in them which is calculated to pander to the baser

---

pp. x-xi. Passages in L.V. are literally indentical with those in Mahāvagga. Again :—"These few examples I have chosen will suffice to prove that the material of a Mahāvaipulya Sūtra is partly as old as that of any other sacred books of the Buddhists"—*ibid*, xiv.

1. The truth seems to that "the historical relation between the Hīnayāna and the Mahāyāna schools of Buddhism is to me a great a puzzle as ever.—Max Müller, Intro. to L. Sukh. Vyūha, ix. Even the very distinction between the N. and the S. Buddhism has been taken strong exception to by R. Davids, who says:—"There is not now and never has been, any unity either of opinion or of language in what is called northern, or in what is called southern Buddhism. There is a distinct disadvantage in continually suggesting a unity which has no existence in fact. In a word the current division of Buddhist literature into northern and southern is entirely unscientific, and misleading."—Bud. India, p. 173.
2. Cowell's trans. p. 184. S.B.E. Series, Vol. xlix.
3. Edkin's Chinese Buddhism.—p. 89. There were sevral versions of this sacred book.–*Vide* Kern's Intro. p. xxiii (S.B.E. series).
4. Vol. i, Intro. lxx.

instincts of frail humanity. At what particular time these Tantras appeared on the sense is yet a matter of conjecture, but the views of Wilson which held the ground for more than 30 years have now been found to be erroneous. Evidence is now forthcoming from quite an unexpected source, which goes to prove that Buddhist Tantras existed as early as the 5th or 6th Century A.D., if not earlier.[1] If Buddhist Tantras, again, pre-suppose the existence of Hindus Tantras, one need not be considered as rash in assigning an earlier age to the latter.[2]

Tantras imported into Tibet

The Tantras found a congenial home in China. Amoghavajra, a Śramaṇa of northern India and a Brahmin by caste resided in the Celestial Empire for several years between 746 and 771 A.D., and under his influence the Tāntric doctrines dealing with talismanic forms and professions of supernatural power first gained currency there.[3] Hence long before his time Tantras must have been popular in India. From the eighth to the eleventh century A.D. we are in possession of authentic records as regards the prevalence to Tantras in Northern India; as it was about this period that

1. *Vide* the palm-leaf MSS. discovered in the monastery of Horiuzi in Japan and carried from Central India. One of these is in the handwriting of a famous Chinese priest, named Kanshin, who came to Japan in A.D. 753. The MS. contains besides a Dhāraṇī, five Tantras. "Bud. Texts from Japan," ed. by Max Müller. Vol. i, pt. i, Intro. pp. 6-8. It is to conclude that these Tantras existed in the land of their birth at least two centuries earlier.
2. Waddell says:—"No one has yet realised the vast extent to which Mahāyāna and Tāntric Buddhist remains cover India ; nor sufficiently realised the leading part played by the Mahāyāna in Indian Buddhism during its popular period." Journal of the Roy. As. Soc. (1894).
3. Bun. Nanjio's Cata. App. ii, p. 445. Amoghavajra translated 77 works including Uṣṇiṣa-Cakravarti-Tantra, Garuḍagarbhaja-tantra and Vājrakumāra-tantra.
   Cf.—Also "The existence of the Tantra Śāstras may thus apparently be traced at least as far back as the 6th century A.D. *Vide* "Annual Report", As. Soc. Beng. 1906.

several of these were gradually imported into Tibet by Indian Paṇḍits, but they must have been composed much earlier.[1]

Atisa gave a fresh inpetus to Tantrism in the land of snow. From the analysis of Mdo by Csoma we also come to know in detail the names with the dates of the Indian scholars who with the aid of the local interpreters (*lochavas*) rendered into Tibetan the various Sanskrit Tantras. The faithful accuracy with which these translations were made and their fidelity to the original enable us often to reproduce of Sanskrit. Texts and thus we are in possession of valuable historical data.[2] Some of these Tantras deal with alchemy and their contents reveal to us the knowledge of chemical processes in India from about the 6th to the 8th century A.D.[3]

1. In the first half of the eighth century two eminent Paṇḍits of Bengal visited. Tibet at the invitation of its king and formally introduced there Buddhism ; these Paṇḍits were Śānti Rakṣita, high priest of the monastery of Nālenda and his co-adjutor Padma Sambhava, a native of Udyana, who took charge of the Tāntric part of the Buddhist liturgy. They were followed by the sage Dipaṁkara-Srī jñāna (Atisa), b. A.D. 980, d. 1053. He acquired proficiency in three Piṭakas of the four classes of the Hīnayāna Śrāvakas, in the Vaiśeṣika philosophy, in the three piṭakas of the Mahāyāna doctrine, the metaphysics of the Mādhyamika and the Yogācārya schools and the four classes of the Tantras. At the request of king Naya Pāla he accepted the post of high priest of the monastery of Vikramaśilā.—*Vide* S.C. Dās,—Journal of the Buddhist Text Soc. Pt. i.
2. Speaking of the Tibetan translation of Aśvaghoṣa's *Buddha-Carita*, Cowell remarks ; "The Tibetan version appears to the much closer to the original Sanskrit then the Chinese; in fact from its verbal accuracy we can often reproduce the exact words of the original, since Sanskrit words are always represented by the same Tibetan equivalents, as for instance the prepositions prefixed to verbal roots." l. c. Intro. p. x. Waddell is equally impressed with the "profoundly accurate and scholarly nature of the Lāmaist translations of Sanskrit Buddhist books;" and he again observes : "It is clear that the Tāntric and Mahāyāna features of Lāmaism were imported *en bloc* from Indian Buddhism." Journal of the Royal As. Soc. for 1894, p. 15.
3. "Analysis of the contents of the Mdo (Sūtra) by Csoma de Korosi Asiatic Researches, XX (1836), p. 583. "A work on preparing quicksilver, the most powerful for subduing every sickness and for improving the vigor of the body."—"A work on turning base metals into gold." We are thus reminded of the contents of the Rasārṇava and the Rasaratnākara. *Vide* p. 2

Rasaratnākara

Fortunately we are not dependent upon the Tibetan Tantras alone for gaining an insight into this dark recess in the history of Hindu intellectual activity. In the course of our search for MSS. of alchemical Tantras we have come upon a precious find, in the shape of a Buddhist Tantras, with Nāgārjuna as its reputed author. Of alchemical Tantras we have had enough had to spare; but there is great difficulty in assigning dates to them, as they one and all pretend to emanate from the mouth of the God Śiva himself. We are only left to internal evidences—evidences based upon the gradual evolution of chemical processes, which we have not been slow to take advantage of.

Its Historical Importance.

The MS. in question is a mare fragment, but it is calculated to evoke all the zeal and enthusiasm of a palcon-tologist— of an Owen or a Marsh—in his efforts to restore an animal and assign to it its proper place in the economy of the laws of evolution, when he luckily chances upon a fossil impression of it tooth or claws. From this point of view Rasaratnākara, for such is the name of our MS., is of uncommon interest. It is a Tantra of the Mahāyānist school and as such its invocations are addressed to all the Buddhas;[1] and in one place there is a pointed reference to Prajñā-pāramitā[2] (perfection of wisdom) appearing before Nāgār-juna in a dream and revealing to him chemical knowledge.[3]

Also a typical production of the Mahāyānist period of intellectual activity

A noteworthy feature of this work is that some chemical

1. प्रणिपत्य सर्वबुद्धान्। *Cf.* The opening invocation in the Sukhāvatī Vyūha: ओं नमः श्रीसर्वबुद्धबोधिसत्त्वेभ्यः। also नमः प्रत्येकबुद्ध आर्यश्रावकाणाम्, नमो बोधिसत्वानाम् in Ārya Mañjuśrīmūla Tantra.—*Vide* As. Res. xx, p. 512.
2. *Cf.*—The invocation in Vajracchedikā: नमो भगवत्या आर्यप्रज्ञापारमितायैः.
3. *Vide* Sans. Texts, p. 10, also Trans. p. 5.

processes are discussed in the form of a dialogue between Śālivāhana and Nāgārjuna, and Ratnaghoṣa and Māṇḍavya.[1] These last two names are held equally in veneration with Nāgārjuna and grateful acknowledgments to their services occur in some later chemical treatises.[2] The bringing together of these four *dramatis personac,* especially of the first two has a significance of its own. We have already seen that references to Nāgārjuna and his contemporary king Śālivāhana are only to be met with in ancient classical literature. It seems probable that Rasaratnākara was written at a time when the memory of these personages was still fragrant.

Judging from internal evidences also we come to the conclusion that it is one of the earliest works extant on the subject. In our attempt at throwing light on the text of Rasārṇava, we had to quote several parallel passages from it,[3] and from a careful perusal of both we are of opinion that the latter is the inspirer rather than the borrower. Rasaratnākara, in short, seems to us to be a typical production representing the Mahāyānist period of intellectual activity and we may not be wide of the mark if we put down 7th or 8th century A.D. as its latest date.

From the 5th to the 11th century. A.D. the colleges in connection with monasteries of Pāṭaliputra, Nālandā, Vikramaśīlā, Udaṇḍapura, etc., were the great seats of learning as the temples attachted to the pyramids in ancient Egypt ; and alchemy was included in the curricula of studies.[4]

---

1. *Vide* Sans. Texts, pp. 12-14, also Trans, pp. 6-8.
2. *Cf.* Vol. i, p. 77.
3. *Vide* Vol. i, Sans. Texts, pp. 7, 12,13 and 18.
4. *Cf.* "By the side of the tower of king Aśoka in built a Saṁghāvāsa belonging to the great vehicle, very imposing and elegant. There is a temple belonging to the little vehicle......In the college attached to the temple one may see eminent Śramaṇas from every quarter of the globe."–Beal's "Fah-hian," ed. 1869, p. 105.

Antiquity of Alchemical Tantras

The existence of a vast ancient Tāntric literature with alchemy as a component part has now been placed beyond, doubt, thanks to the searching examination of the Mahārājā's collection of valuable MSS. of Nepal by the eminent scholars Bendall, H.P. Śāstrī and S. Lévi. Of surpassing interest is the discovery of a Tantra belonging to the extinct school Kubjikāmata, written in Gupta Character and copied about the sixth cetury A.D. This school, though itself very ancient, presupposes the existence of other schools and we have distinct mention of the Mahāyāna.[1] We now learn that the gradual fusion and amalgamation of Śivaite and Buddhist Tantras had begun even anterior to the sixth century. The Kubjikā-tantra was evidently in Nepal. The Goddess Pārvatī encourages her consort to proceed to the continent for the spread of the tenets propounded by it.[2] In one place we come across a passage[3] in which Śiva himself speaks of pārada (mercury) as his generative principle and eulogises its efficacy when it has been killed six times.[4] We also find allusions to the transmutation of copper into gold with the aid of mercury. In short, we have ample references to alchemical processes described in the very technical terms

1. Catalogue of Palm-leaf and selected paper MSS. belonging to the Durbār Library, Nepal, by H. P. Śāstrī (1905). lxxviii, *et. seq.*
दक्षिणे देवयानी तु पितृयानस्तथोत्तरे।
मध्यमे तु महायानं शिवसंज्ञा प्रजीयते।।
2. गच्छ त्वं भारते वर्षेऽधिकाराय सर्वतः।
3. मद्वीर्य्यः पारदो यद्वं पतितः स्फुटितं मणिः।।
मद्वीर्य्येण प्रसूतास्ते तावार्य्या सूनके वहि।
तिष्ठन्ति संस्कृताः सन्तः भस्मा षड्विप्रजारणाम्।
For the passages cited here we are indebted to the industry and courtesy of Paṇḍit H.P. Śāstrī, who has wended his way through the bulky MS.
4. Even at the present day mercury, which has been *killed* six times in succession, each time with an equal weight of sulphur षड्गुणबलिजारित, is reputed to be one of the most potent of remedies.

in which Rasārṇava, Rasaratnākara and other typical works of the Tāntric period abound.[1]

Alchemy as Integral part of Mahāyānist Activity

It is now only necessary to wind up this chapter with an extract from the work of a eminent Buddhologist with this reservation that the growing influence to Tantrism began as we have seen long before the 8th century.

"The decline of Buddhism in India from the 8th century downwards nearly coincides with the growing influence of Tantrism and sorcery, which stand to each other in the relation of theory to practice. The development of Tantrism is a feature that Buddhism and Hinduism in their later phases have in common. The object of Hindu Tantrism is the acquisition of wealth, mundane enjoyments, rewards for moral actions, deliverance, by worshipping Durgā, the Śakti of Śiva—Prajñā in the teminology of the Mahāyāna—through means of spells, muttered prayers, Samādhi, offerings etc. Similarly, the Buddhist Tantras purpose to teach the adepts how by a supernatural way to acquire desired objects, either of a meterial nature, as the elixir of longevity, invulnerability, invisibility, alchemy; or of a more spiritual character as the power of envoking a Buddha or a Bodhisattva to solve a doubt, or the power of achieving in this life the union with some divinity. There is an unmistakablle affinity between Tantrism on one side, and the system of Yoga and Kammaṭṭhāna on the other. Tantrism is, so to say, a popularised and, at the same time, degraded form of Yoga, because the objects are commonly of a coarser character, and the practices partly more childish partly more revolting.

"Tāranātha informs us that Tantrism existed and was

1. *e.g.* पलेन विहितो वेधः किं व्यञ्जता न विध्यते।
रसविद्धं यया ताम्रं न भूयस्ताम्रतां व्रजेत्।।
For the meaning of the word वेध (Vedha) see this Vol. p. 18, *footnote*, also Vol. i, p. 120.

transmitted in an occult manner in the period between Asaṅga and Dharmakīrti, but that after Dharmakīrti's times the Anuttarayoga became more and more general and influential. Substantially his statement is certainly right. He adds that during the reign of the Pāla dynasty, there were many maṣters of magic, Mantra-Vajrācāryas, who, being possessed of various Siddhis, performed the most prodigious feats.

"The king of the Pāla dynasty, whose sway over Gauḍa and the adjacent regions lasted from about A.D. 800 and 1050, are known both from the annals and their inscriptions as protectors of the Faith. It was during that period that the monastery of Vikramaśīlā was a renowned centre of Tantrist learning.

"The Sena kings, who followed the Pālas in the dominion over Eastern India, though belonging to a Hindu persuasion, were not hostile to the Faith. Still Buddhism decline during their reign and more so after the invasion of country by the Muhammedans in A.D. 1200. The monasteries of Udaṇḍapura and Vikramaśīlā were destroyed; the monks were killed to fled to other countries. The learned Śākyśrī went to Orissa, and afterwards to Tibet ; Ratnaraskṣita to Nepāl ; Buddhamitra and others sought a refuge in S. India, Whilst Saṅgama-Śrījñāna with several of his followers betook themselves of Burma, Camboja, etc. And thus the Law of Buddha became extinct in Magadha.

"Many emigrants from Magadha rejoined their brethren in the South and founded colleges on a modest scale in Vijayanagara, Kaliṅga, and Koṅkana. The Comparatively satisfactory condition of Buddhism in Deccan about that time is attested by the rich donations to the monastery at Dambal."[1]

Spread of Alchemy

It will be noted that the monks of the monasteries of

1. Kern—Manual of Ind. Bud., pp. 133-134.

Udaṇḍapura and Vikramaśīlā on their dispersion carried with them their learning and arts in the same manner of the Byazantine Greeks on their expulsion from Constantinople bore with them their intellectual treasuries to the Italian cities. In the kingdom of the Deccan and in Tibet the Buddhist refugees found hospitable asylums,[1] just as the Greek Philosophers did in the Florentine Republic under the Medicii.

Some eminent orientalists, whose opinions naturally carry weight, have hitherto taken for granted the knowledge of Chemistry. Such as we find reflected in the Tantras quoted by Mādhava in Reseśvaradarśana[2] was derived by intercourse with the Arabs.[3]

---

1. *Cf.* "The Deccan, which from the eleventh century was the refuge and centre of literary activity generally. In Hindustan it had been substantially arressted by the inroads and ravages of the Mahmmadans."—Weber. Hist. Ind. lit. p. 283.
2. Vol. i, Intro. lxxviii, *et seq*.
3. *E.g.* Barth : "In regard to alchemy, anyhow in which the Sittars are zealous adepts, they were disciples of the Arabians, although other Śivaites has preceded them in the pursuit of the philosopher's stone. Already, in his exposition of the different doctrines of the Śaivas, Sāyaṇa thought he ought to dedicate a special chapter to the *Raśesvaradarśana* or "system of mercury," a strange amalgamation of Vedāntism and alchemy. The object contemplated in this system is the transmutation of the body into an incorruptible substance by means of *rasapāna i.e.*, the absorption into it of elixirs compounded principally of mercury and mica, that is to say, of the very essential qualities of Śiva and Gaurī, with whom the subject of the operation is thus at length indentified. This species of transubstantiation consitituës the *jīvanmukti*, or state of deliverance commencing with this present life, the sole and indispensable condition of salvation. It is clear that the devotional fromulac of the Vedānta are here only a sort of jargon, under which there lies hid a radically impious doctrine ; and it is not less clear that in this doctrine, which had from the fourteenth century produced a rather considerable literature, there is an infusion of Mohammedan ideas.....The Arabs of the Khalifat had arrived on these shores in the character of travellers had arrived on these shores in the character of travellers of merchants, and had established commercial relations and intercourse with these parts long before the Afghans, Turks, or Mongols. their coreligionists, came as conquerors."–"Religions of India," ed. 1891, pp. 210-211.

In Tibet and the Deccan

The attentive reader who has followed us all along could not have failed to notice that it was in the Universities of Nālandā, Udaṇḍapura and Vikramaśīlā, in Central India and Magadha, precisely the regions which were cut off from communication with the external world, that Tāntric mysticism with alchemy as an integral part was cultivated and from thence spread to Bhot (Tibet) and the regions lying to the south-east of it,[1] and South India.[2]

1. See under colophons to Rasahṛdaya and Rasasāra respectively.
2. The last rallying point of Hindu learning and sciences was the kingdom of Vijayanagara, which was in the heyday of its glory under Bukka I (A.D. 1354-1371) with Mādhava as his prime Minister. Wilson says:—"The history of Vijayanagara is a subject of considerable interest in the annals of India, as the last barrier that was opposed to Muhammedan invasions, and that preserved the sourthern part of the Peninsula from foreign rule until a very modern period. *As. Res.* Vol. xx. (1830), p, 1. See also Sewell's "A Forgotten Empire." (Vijayanagara).

Chapter II

Further Cultivation of Alchemy

Having thus far prepared the ground for believing that alchemy along with other cognate branches of learning was the outcome of Indian intellectual activity, we shall now proceed to follow it up with its further development. At the outset it is, however, necessary to remember that on the decline of Buddhism the vigorous impetus which its followers gave to literature and science was not lost to India. Nay, during and after the revival of Brahmansim under the Gupta dynasty and its successions we have some mighty intelletual gaints whose productions will continue to shed lustre as long as the Hindu nation exists. Kālidāsa and Bhavabhūti, Brahma Gupta and Āryabhaṭṭa, Śaṁkara and Rāmānuja are names which may be regarded as the heritage not of India alone but the entire civilized world. At the time of the Brahmanic revival Buddhist works at acknowledge merit far from being cast aside were held in veneration. Amarasiṁha in his Lexicon and Vāgbhaṭa in his Aṣṭāṅgahṛdaya[1] as also the psuedo Vāgbhaṭa commence with an invocation to Buddha, which has never shocked the tolerant spirit of the Hindu. Caraka and Suśruta also bear distinct impress of Buddhist retouching.

Buddhist Alchemical Tantras absorbed into Hindu Tantras

The Buddhist Tantras became likewise a part and parcel

1. Vol. i, p. 76.

of Hindu religio-philosophical literature,[1] the subject-matter of the fomer was incorporated into the latter, the names of Tārā, Prajñāpāramitā and Buddha being simply changed into those of Pārvatī and Śiva. In Rasaratnākara itself we have distinct indications that it is an admixture of both.

It should thus be understood that the works under review in this chapter are based upon their Buddhist counterparts ; in short, there is no disposition to ignore the debt of obligation the authors are under to Nāgārjuna, Ratnaghoṣa, Māṇḍavya and others.

Rasahṛdaya

In the introduction to the first volume we stated ; It is to be regretted that of the several works quoted by Mādhava [on the science of mercury] Rasārṇava alone seems to have survived to our days" (l. c. lxxxiii). Since then we have been fortunate enough to procure as many as three transcripts of MSS. to Rasahṛdaya by Govidabhagavat from different quarters. As Mādhava speaks of this author in terms of the deepest reverence and regards him as "ancient,"[2] we think we should not err on the wrong side if we place the author some three centuries before his time. If our surmise be correct he should have lived about the eleventh century A.D. The only personal allusion which the author has condescended to offer for the edification of posterity is that he wrote his book at the request of the king of the Kirātaland, *i.e.*, the region adjoining modern Bhotan. Portions of this remarkable treatise after a careful collation of three MSS. have been

---

1. Speaking of a typical Buddhist production belonging to this category Burnouf observes : "Ce morceau est exclusivement spéculatif, et il nous offre une nouvelle preuve de l'alliance intime quele systeme des Tantras a contractee avec la philsophie buddhique la plus elevee." (l. c.p. 543 ed. 1844).
2. *Cf.* Vol. i, Intro lxxx.
तत्प्रपञ्चस्तु गोविन्दभगवत्पादाचार्य्यसर्वज्ञरामेश्वरभट्टारकप्रभृतिभिः प्राचीनैराचार्य्यैर्निरूपितः।
The qualifying epithet भगवत् is only applied to venerable *Ṛṣis* of old.

reproduced in the proper place. The worm-eaten fragmentary MS. procured from Benares is 386 years old. Internal evidence equally goes to establish the antiquity of this work. The author was evidently a Buddhist as we learn from the colophon to the Benares MS. (*vide* trans. p. 12). It is, however, not to be found either in the India office or the Nepāl MS. This important omission is easily accounted for. It is the reluctance of the latter day Hindus to acknowledge their obligations to a Buddhist author.

Rasārṇava which has been noticed at length in the first volume and to which was assigned the 12th century. A.D, as also Rasaratnasamuccaya, (13th to 14th century), need not further arrest our attention.

The presentation of the various treatises of the Tāntric and iatro-chemical periods under discussion in their exact chronological sequence is not an easy task ; in the absence of more definite information, we are afraid, we shall now and then have to hazard a conjecture.

Resendra Cūḍāmaṇi

The introductory line of Books viii and ix of Rasaratnasamuccaya, in which Somadeva is mentioned as the author of the descriptive part, led us to supspect that they are merely reproductions from "a standard work on the subject by Somadeva, no longer extant.[1] Our surmise has proved to be partially correct. We have at last been able to recover the supposed lost work. A transcript of a MS. of Rasendracūḍāmaṇi by Somedeva, preserved in the Library of the Deccan College Poona, reveals to us the fact that practically all the important portions of this book have been woven into the text of Rasaratnasamuccaya ; and this gives us an additional opportunity of collating many dubtful passages in both. But the present work itself does not lay claim to originality. It makes a candid admission that it culls all its materials from

1. Vol. i, p. 118.

pre-existing chemical treatises.[1] It further puts forward the alchemist Nandi as the inventor of the process of sublimation and of the Koṣṭhī apparatus (Vol. i, pp. 69 and 89).[2] Nāgārjuna, Daṇḍī, Śambhu and the sage Brahmajyoti are also spoken of as sources of his inspiration. We get very little inkling into the personal history of Somadeva, except that he was the ruler of a city named Karavālabhairava. The fact that R.R.S. lays Rasendracūḍāmaṇi so amply under contribution would suggest that the latter had become somewhat rare or that its contents had been well nigh forgotten when the former was compiled. Its date may therefore be put down between the 12th and 13th centuries A.D.

Rasaprakāśasudhākara

Another important work of this period is Rasaprakāśasudhākara by Yaśodhara. The author of Rasratnasamuccaya in the opening lines mentions the names of 27 alchemists to whom he is beholden, among which occurs that of Yaśodhana. We have little doubt the correct reading is Yaśodhara. We now find that there is very little original matter in Rasaratnasamuccaya, it being made up to citations from Rasārṇava and the works of Somadeva and Yaśodhara. We should not, however, be understood to retract the high encomiums we thought fit to bestow upon Rasaratnasamuccaya for "its methodical and scientific arrangements of the subject-matter, which would do credit to any modern work."[3] Its author has never laid claims to originality, but on the contrary has freely acknowledged the sources he drew upon. The student of Hindu chemistry, however, like a weary but devout pilgrim must wind his way through intricate and

1. अथ यन्त्राणि वक्ष्यन्ते रसतन्त्राण्यनेकशः।
   समालोक्य समासेन सोमदेवेन साम्प्रतम्।।
2. उद्ध्र्वपातनयन्त्रं हि नन्दिना परिकीर्त्तितम्।
   कोष्ठिकायन्त्रमेतद्धि नन्दिना परिकीर्त्तितम्।
3. Vol. i, Intro. lxxxvi.

thorny paths and feel ineffable joy at being able to approach the original fountain undefiled. It now transpires that the credit for the accurate observations on the metallurgy of zinc (Vol. i, pp. 88 & 156) really belongs to Yaśodhara. While the author of Rasaratnasamuccaya always scrupulously admist that his work is a mere compilation, Yaśodhara, on the other hand, is anxious to let the world know that the processes he describes have been verfied by experiments perfomed with his own hands.[1] Among the authorities he cites are Nāgārjuna, Devīśāstra (probably Rasārṇava). Nandi Somadeva, Svacchandabhairava and Manthanabhairava. As Yaśodhara quotes Somadeva, he must have been posterior to the latter by at least a hundred years and he should therefore be placed in the 13th century. A.D.[2]

Rasakalpa

Rasakalpa likewise seems to belong to this period. It is in the shape of a Tantra, but curiously enough it is not much troubled with consistency. In the opening salutation Śiva, "the king of mercurial lore" and his consort Caṇḍikā are reverentially bowed to—a procedure scarcely reconcilable with the position of a revealed document. The colophon at the end of each Chapter (उल्लास) claims the work to be a part and parcel of Rudrayāmala Tantra ; at the same time due acknowledgment is made to the contributions of Govinda, the author of Rasahṛdaya, Svacchandabhairava and other adepts.[3] Towards the end the author says : "The processes described have all been verified by me and not borrowed at secondhand from my teachers." Needless to add here that the God Śiva would not have been at pains in justifying the

1. *Cf.* स्वहस्तेन कृतं सम्यक् जारणं न श्रुतं मया।
   स्वहस्तेन भवयोगेन कृतं सम्यक् श्रुतेन हि।।
   धातुबन्धस्तृतीयोऽसौ स्वहस्तेन कृतो मया।
   दृष्टप्रत्यययोगोऽयं कथितो नात्र संशयः।।
2. We are using a copy of the MS. preserved in the Runbir Library, Kāśmīr. The readings are on the whole accurate.
3. *E.g.* सर्वाचार्य्यैरुदीरिताः।

necessity of his production with such a timorous apology. From internal evidence it is abundantly clear that it could not have been composed earlier than the 13th century A.D.

As the author has confined himself solely to a description of the metals, minerals and the processes of "killing" them with the agency of various apparatus and has not gone into the treatment of diseases with the aid of mineral preparations, he has been able to condense much important information within the narrow compass of a few short chapters.

Rasarājalakṣmī

Rasarājalakṣmī also deserves some prominence here not because of its intrinsic worth but because of the references to previous Tantras and alchemists, amongst which are Rasār-ṇava, Kākacaṇḍīśvara, Nāgārjuna Vyāḍi, Svachanda, Dāmodara, Vāsudeva and Bhagabat Govinda. We have also categorical mention of the purely Ayurvedic works, such as Caraka, Suśruta, Hārīta and Vāgbhaṭa, to which our author is equally indebted.[1] From the colophon to the MS. procured from Benares we learn that the author Viṣṇudeva was Court Physician to king Bukka.[2] If this statement be accepted—and *prima facie* there is no reason to question it—this treatise must have been complied in the latter part of the 14th century. Internal evidence equally supports this date.

1. *Vide* Sanskrit Texts, pp. 80-81.
2. *Vide ante.*, footnote, xlix.

Chapter III

Iatro-Chemical Period–C. 1350 A.D.

We now proceed to examine the contents of another series of works belonging also to the Iatro-Chemical period ; the one characteristic feature is of which is that opium is recongnised as an official drug in the materia medica portion. These may be taken to date from the middle of the 14th century A. D.

Rasanakṣatramālikā by Mathanasiṁha, physician to the king of Malwa, should find a place in this chapter. For particulars as the centents the reader may consult p. 22. The date given by the copyist of the MS. is Samvat 1557 *i.e.* 1500 A.D. The work itself is therefore older.

Rasaratnākara by Nityanātha

Rasaratnākara by Siddha Nityanātha, son of Pārvatī, comes under this category.[1] The author gives the sources of his information and explains the object of his compilation in the following words : "Whatever has been revealed by Śiva in Rasārṇava under the preparations of mercury ; the Dīpikā of Rasamaṅgala on mercury; all that has been said by Nāgārjuna for the benefit of people afflicted with diseases, as also by Siddha Charpaṭi, by Vāgbhaṭa and Suśruta—all

1. Two editions of this somewhat voluminous work have been published—one by Gaṅgeśa Chandra Ghosa (Kolkota), the other at the "Venkatesvara press" (Mumbai). We have compared the reading of these two editions with the MS. in the Library of the Sanskrit College, Kolkata. They agree fairly well.

these and many other treatises on mercury and mineral being consulted, I have, after rejecting the drugs and medicaments which have become rare and difficult to procure, put together [in my work] the essential features thereof. All that I have learned from my teachers and have been in a position to subject to practical tests has been incorporated into my work for the benefit of manking." In addition to the authorities cited above Nityanātha quotes from Cakrapāṇi and Rasendracūḍāmaṇi.

Rasendracintāmaṇi

Rasendracintāmaṇi will next claim our attention., We have before us the printed copy, edited by Umesa Chandra Sena Gupta, late librarian, Sanskrit College, Kolkata, in which the text has been adopted, as the editor informs us the preface, after collation of 3 or 4 MSS. We have carefully gone over this edition and compared from Benares and another from the Runbīr Library, Kāśmīr. All these agree in the main, but there happen to be certain omissions and additions, which form an interesting chapter in the history of interpolations in the literature of his description. To begin with : there is a dispute over its very authorship. Some MSS. ascribe it to Ḍhuṇḍhukanātha, disciple of Kālanātha, others to Rāmacandra. After the customary salutation there occur two couplets in the Kolakata edition which are not met with in the Benares and Kāśmīr exemplars. In these the author is made to justify his *raison d'etre* for undertaking his task in these words : "I shall give publicity only a such processes as I have been able to verify by my own experiments." Such protestations have already been noticed more than once.[1] In another placd we have ; "Those mercurial operations alone

1. Such pretentions put forth on behalf of the author, especially in the first person singular, lead one to suspect that they are later interpolations. In the Arabians alchemy Geber is often made to decalre to same effect : *e.g.* "J'ai expérimenté moiméme tout ce que je repporte."—Berthelot, "La Chimie au moyen Age," pt. 1, p. 338. The well known egoistic couplet in Suśruta, Ch. i, अहं हि धन्वन्तरिरादिदेवो is generally regarded as an interpolation.

have found a place in my book, which I have been able to put to tests. Those who teach without being able to perform experiments labour in vain." This last couplet is quite out of place as it hangs like a tail to a long extract from Rasārṇava and is in fact conpicuous by its absence in the Benares and Kāśmīr MSS.[1] For the second time we find calomel designated as Rasakarpūra (lit. camphor of mercury;). This preparation no doubt occurs in the older works, *e.g.* Rasārṇava ; but it is described there as the while "ash" of mercury (*śvetabhasma*). The recipe for calomel as given in the Kolkata edition, however, is quite different from that in the Benares and Kāśmīr MSS. and it is altogether wanting in one of the MSS. in the Kolkata Sanskrit college Library. We are therefore not without suspicion that it has been foisted into the original composition. The author refers Rasārṇava, Nāgārjuna, Govinda, Nityanātha, Siddha Lakṣmīśvara, Trivikramabhaṭṭa and Cakrapāṇi. There is a lengthy colophon but no direct information can be derived from it as to the sage of the author.

Rasasāra— It is a comprehensive but purely chemical work dealing with 18 operations on mercury; various chemical processes are incidentally described, a good many of which, however, overlap each other. There does not seem to be any justification for coining a multiplicity of technical jargons to cannote each of these. But prolixity was never regarded in the light of a vice by the alchemical writers of the middle ages either in India or in Europe. The contents of Rasasāra often remind us of a poem on alchemy by Georga Ripley, canon of Bridlington in Yorkshire (b. about 1460), of which Rodwell says :—

"But we cannot point to a new fact which he eluciadated. He divided all chemical operations into twelve processes—calcination, dissolution, separation, conjunction, putrefaction, congelation, cibation, sublimation, fermentation, exaltation, multiplication and projection. Several MS. copies

1. The extract commences with त्वं माता सर्वभूतानां.

of his poem exist in the British Museum, bound up with copies of the works of Roger Bacon and earlier writers. Here is a specimen of his rugged rhymes :—

The first chapter shall be a natural *Calcination;*
The second of *Dyssolution*, secret and phylosophycall ;
The third of our elementall *Separation* ;
The fourth of *Conjunction* matrimoniall;
The fifth of *Putrefaction* then followed shall;
Of *Congelation Albyficative* shall be the Sixt;
Then of *Cybation*, the seaventh shall follow next."[1]

The author, Govindācārya, as a devout Hindu begins with his adoration with his adoratoin of Śiva and Viṣṇu and tells us at the outset that his treastise is a compilation and epitome based upon standard works on the subject and the contributions of the adepts. We have every reason to be thankful to him for the many hints he throws on the sources of his information. Of surpassing interest is his declaration that for a knowledge of certain processes he is idebted to the Buddhists of Tibet.[2] This must be taken to mean that the cultivation of alchemy had become neglected and almost forgotten in India proper and the earnest searchers for his lore had to repair during the time of our author to Tibet. Confirmatory evidence to this effect is equally frunished in the colophon to Rasahṛdaya. During the decadance of Buddhism and the corresponding ascendancy of Brahminism in Bengal, say about the 11th to 13th centuries A.D. this

1. *Vide* "Birth of Chemistry."
2. एवं बौद्धा विजानन्ति भाटदेशनिवसिन: ।

Towards the close of his work the author again acknowledges his obligation to the Buddhist.

. बौद्धमतं तथा ज्ञात्वा रससार: कृतो मया ।

We have in our possession five transcripts of the MSS. of Rasasāra procured from the Libraries of Kāśmīr, Tanjore, Ulwar, Chennai and the Temple of the Goddess Kālī, at Ramnā, near Decca, Bengal.

branch of science also found a sage retreat in the land of the Kirātas as we have already seen. Burnell, under the influence of his preconceived notion, namely that Indian alchemy owed its origin to the Arabs, very naively remarks: "By *Bauddhas* he (the author) probably means the Muhammedans (*cf.* the vernacular Malayālam use of the word), though studies of this nature were much pursued by the later Buddhists."[1]

Some idea of the date of Rasasāra may be formed from the fact that there is mention of opium in it. But our author was evidently quite in the dark as to its origin ; for, says he: "There are four different kinds of poisonous sea-fish, and it is from their foam that four different kinds of opium are derived, namely, white, red, black and yellow ; while others maintain that it is derived from the foam of the snake and it is very properly used for chemical operations,"[2] *e.g.*,—killing and fixation of mercury. From the context it is, however, doubtful if the narcotic drug is really alluded to here. But independently of the use of this term we may safely place this work in the 13th century A.D.

---

1. Catalogue of the Tanjore Palace MSS. pt. 1, p. 70. Burnell draws his conclusions from the reference to Buddhists in the colophon ; but we have quoted above another passage in which distinct mention is made of Buddhists of Tibet.
2. समुद्रे चैव जायन्ते विषमस्त्याश्चतुविधाः।
   तेभ्यः फेनं समुत्पन्नम् अहिफेनो विषं स चतुर्विधं।
   केचिद्वदन्ति सर्पाणां फेनं स्यादहिफेनकं।
   यथा–धारणं श्वेतवर्णं च रक्तवर्णं च जारकं।
   सारणं पीतवर्णं च कृष्णवर्णं च मारणं।
   विषविदुत्तमं फेनं युज्यते रसकर्मणि।
   The Sanskrit word for opium, "ahiphena" lit; foam of the snake, is of modern date. It is generally held that this word has been Sanskritised from the Arabic "aphiyūn."
   These indential distichs also occur in the commentary on Śārṅgadhara by Āḍhamalla with the additional line, षाषजः (खाखजः) क्षीरविशेषः, namely, it is the milky juice of poppy ; proving that during the time of the latter is real origin was known, at the same time its supposed origin was not disbelieved.

Śārṅgadhara-saṁgraha—a compilation by Śārṅgadhara : its peculiarity is that it is based upon the Āyurvedas (Caraka, etc.) on the one hand and the Tāntric chemical treatises on the other. In the chapter on the purification and incineration of metals, seven of these are recongnised, but strange to say there does not occur any mention of zinc. Later on nine metals are named including two alloys, brass and bell-metal, after the nine planents, the significance of which will be discussed in the proper place.

The date of Śārṅgadhara can be ascertained with accuracy, as besides this medical compendium he is the author of the Paddhati, a voluminous miscellaney, containing, with other matters, a poetical anthology. The author dates his works in *Samvat* 1420 or A.D.1363. The name of the author's father is Damodara and that of the grand-father Rāghavadeva. That latter was highly considered by Rājā Hammīra, the Chauhan.[1] There is a big commentary on this work by Āḍhamalla, whose ancestors settled in the city founded by Hammīra.[2]

1. After the sack of Chitor by Alla-ud-din in A.D. 1303, Mewar was occupied by the garrisons of Delhi. Hammīra recovered his ancestral throne and during his prosperous rule of sixty-years (1301-1365) and that of his succes sors arts and literature found liberal patros (*vide* Tod's "Annals of Rājastan.") The court of Hammīra attracted another distinguished physician named Saugata Siṁha as gather from the colophon to a noted recipe,

   एषा सौगतसिंहनामभिषजा लोके प्रकाशीकृता
   हम्मीराय महीभुजे ** संभोजभाजे मृशम्।

2. Āḍhamalla gives an account of his family. His Grandfather was one Cakrapāṇi, not to be confounded with the celebrated medical author of Bengal (Vol. i, Intro. liv) and he himself was attached to the court of a ruler named. Jaitrasiṁha, whose capital Hāstīkāntapurī was situated on the river Carmaṇvatī. We reproduce in the next page the text which is somewhat corrupt from the copy preserved in the Library of the Sanskrit College, Benares.

   श्रीवास्तव्यकुलप्रकाशमणयो ये पूर्वजा विश्रुताः
   श्रीहम्मीरपुरे * * * *
   तदन्वये किं बहुना प्रसिद्धश्रीचक्रपाणिः कुलकैरवेयुः (न्दुः)

Rasendrasārasaṁgraha : The author, Gopālakṛṣṇa begins by admitting that his treatise is a mere compilation based upon many Tantras, though he mentions by name only two, namely Rasamañjarī and Candrikā. Special stress is laid on the therapeutic efficacy of mineral preparations.[1] Like Rasendracintāmaṇi it assigns a minor place to the ancient Ayurvedic method of treatment by Kṣāyayoga, *i.e.*, by herbs and simples ; as far as the knowledge of chemical processes goes, it must be held inferior to the former. Numerous medicinal recipes agree word for word with those in Rasendracintāmaṇi, which only proves that both have drawn on a common stock. There is no question of the one borrowing from the other, as internal evidence proves that both belong to the same period. Abhayānanda Gupta in the preface to his edition says:— "There is a tradition to the effect that Gopāla Kavibhūṣaṇa of Dakṣiṇa Baṅga is the complier of this book ; but this can scarcely be accepted as he lived not long ago." Again. "Its commentor, Rāmasena Kavīndramaṇi (author of Arthabodhikā) was court physician to Nawāb Jaffer Ali Khan of Bengal (Mir Jaffer) some 200 years ago." Rāmasena himself quotes from three commentaries of this book; hence, it was recongnised as of some degree of importance at that time and must have beem composed long before. This work is very popular in Bengal and is a *vade mecum* of the local Kavirājas.

तस्य बुद्धिसुभगस्य तनूजो भावसिंह इति भूपतिमान्यः।
* * * * *
तस्यात्मजो दीर्घमतिः प्रहृष्टो नाम्नाढ़मल्लो * *।
तेन * * वीनाश्यते (?) (विन्यस्यते) शार्ङ्गधरप्रकाशः।
हास्तीकान्तपुरी पुरा पुरजिता काशीव विद्वज्जनै–
र्व्याप्ता यत्र सरः सरिद्गणवरा चर्मणवती पापहा।
यस्यां हृद्गतवासुदेवचरणद्वन्द्वाम्बुजः क्ष्मापतिः
ख्यातो धर्म्मइवास्ति धर्म्मगतिषु श्रीजैत्रसिंहः प्रभुः।।
तस्यां शार्ङ्गधरोक्तानां योगानां च प्रकाशिनी।
क्रियते साढ़मल्लेन संहितायाः प्रदीपिका।।

1. For a description of some of these *vide* "Journal des Svants." April 1898.

Rasendrakalpadurma is another work of this period. It deals chiefty with mineral preparations and is a more compliation from Rasārṇava, Rasamaṅgala, Ratnākara, Rasāmṛta and Rasaratnasmuccaya ; the numerous citations with which it is replete are of some use in correcting many doubtful readings in these latter. Our MS. is incomplete, the first few couplets and the sequel being lost ; for the opening lines of the work see Aufrecht's Catalogue of the Bodleian Library, MSS., pt. 1., p. 321.

Dhāturatnamālā : This epitom is dovoted exclusively to short processes of killing metals and minerals. Six metals are recognised at the outset, namely; gold, silver, copper, lead, tin and iron as in the ancient works; but strange to say later on *kharpara,*[1] which is the mineral calamine, is taken as synonymous with jasada or zinc. We have before us a MS. procured from Benares. It is full of corrupt readings. The copy noticed in the Bodleian Library Catelogue seems to be more correct. From the colophon in the latter we learn that the author is one Devadatta of Gujarāta, but in the Benares exemplar this portion has been coolly tampered with, so that this poor compilation may be palmed off as an integral part of a Saṁhitā of the Aśvinīkumāras, the divine physicians, and thus entitled to the respect due to revelation. This work cannot be placed earlier than in the 14th century. Here we have a serious side light into the history of literary forgeries.

1. Vol. I. p, 158.

Chapter IV

Modern Period. 1500-1600 A.D.

We have now arrived at what may be termed the modern period in Hindu chemistry and materia medica. The characteristic of this period is that over and above opium, we have some other foreign drugs incorporated into the materia medica. By the beginning of the 16th century A.D. the Portuguese had fairly established themselves at Goa and some other parts of India[1] and as a result of intercourse with them, that dreadful scourge—the venereal disease—had made its appearance. Sanskrit medical treastises from the Caraka and Suśruta downwards and ending with Śārṅgadhara are silent about his malady though they give fairly accurate descriptions of diseases of the genital organs (उपदंश). But syphilis had now to be reckoned with and a new name had to be coined for it. Accordingly we find Rasapradīpa one of the standard works of this period given to Tāntric method of treatment, prescribing calomel and *chobchini* (China root, Smilax China Linn.) for what is now termed for the first time Phiraṅgarosa or the disease of the Portuguese. According to Flückiger and Hanbury "the use of this drug as a remedy for syphilis was made known to the Portuguese at Goa by Chinese traders about A.D. 1535." Thus from an independent source we can arrive at the approximate date of Rasapradīpa. This work also given us a detailed

1. The political-influence of the Portuguese as early as 1533 A.D. is evident from the fact that Sultan Bahadur, king of Gujarāta, entered into a treaty with them and through their assistance raised a force of 6,000 Abyssinians—"Memoirs of Humayun".

process for the preparation of mineral acids by distillation which is here termed saṁkhadrāvaka (lit. a solvent for counch shells) and is described as "endowed with the property of dissolving metals." It is thus evident that the use of mineral acids as a solvent for metals was unknown in India before this time.[1] A very noteworthy circumstance inconnection with this is that the Bhāvaprakāśa, which is posterior to Rasapradīpa, is silent about saṁkhadrāvaka, which had come to be prescribed by this time as a favourite remedy for indigestion and derangment of liver and spleen; the solvents beings borax, treacle, etc.—(*cf.* vol. i, p. 130). In the Bhāvaprakāśa among the acids we have only the juices of acid fruits. This is only another instance of the danger one is liable to incur by drawing inferences from *argumentum ex silentiio.*

Rasakaumudī is another compilation belonging to this period and as far as its contents go it is comparable to Rasapradīpa in many respect; both apium and mineral acids are prescribed. From the colophon in the Kolkata Sanskrit College Library MS. we gather that the author is a physician named Mādhava. Curiously enough he is often confounded with the celebrated Mādhavakara, the author of the Nidāna-saṁgraha, who preceded the former by some 8 or 9 hundred years.

The next work which should now claim our attention is the well known Bhāvaprakāśa of Bhāvamiśra. It is a voluminous compilation in which the Ayurvedic, method or treatment has been mainly adopted and as such ample citations are given from the Caraka, Suśruta, Vāgbhaṭa, Hārīta, Vṛnda and Cakrapāṇi. But the Tāntric remedies could not altogether be given a wide berth and accordingly one or two chapters have been devoted to mineral preparations; but these have been borrowed chiefly from Rasapradīpa, Rasendracintāmaṇi, Śārṅgadhara and other standard works.

1. *Vide* Note on mineral acids, vol. i, p. 185.

Bhāvamiśra has been scrupulously can did in his acknowledgments to the various authorities quoted by him. *Phiraṅgaroga* is mentioned as also its treatment with the help of calomel and *chobchini*. the author lived about the time of the Emperor Akbar and evidently in that part of India which is now known as the United Provinces and it is not to be wondered at that Mussulman influence is discernible in his book.

We conclude our survey of this period with the notice of a remarkable production entitled Dhātukriyā or "operations with metals" ; it is in the shape of a dialogue between Śiva and Pārvatī ; in short, it pretends to be a part and parcel of the Rudrayāmala Tantra.[1] The work cannot be placed earlier than in the 16th century A.D. as it contains reference to the country of the *Phiraṅgas* and Rūma, the Arabic name for Constantinople. For the first time we come across the very appropriate term dāhajala (lit. burning water) coined to denote sulphuric acid. The information about the metals is of a very meagre and poor description and it has been needlessly spun into an inordinate length. The author evidently considers it a pious fraud to make such statements as follow emanate from the mouth of Śiva himself :—

> "By using coins made of imitation gold one can carry on exchange ...... and thus enrich oneself."

The curious reader who is referred to the translation of the selected portions of the text given elsewhere will not fail to recollect similar passages in the Leyden Parchment where the dishonest goldsmith gives numerous recipes for "augmenting gold."[2] Fraudulent attempts in the direction of palming off alloys of base metals possessing bright, yellow

1. We have in hand transcripts of two MSS.—one from the Library of the Mahārājā of Ulwar ; the other named somewhat differently, Dhātumañjarī, from Benares. They agree very closely.
2. Berthelot : "Collection des Alchimistes grecs" (trad, p. 287)."La Chimie an Moyen Age," t. 1. p. 31.

lustre for gold, have been made in all ages and in all climes (*cf.* extracts from Rasārṇava, Vol. i, p. 74).

Arkaprakāśa or a treatise on the preparation of medicinal tinctures and essences should also find a place here;[1] its authorship is ascribed to Rāvaṇa, the mythical king of Ceylon. The contents of the work, however, reveal its date. For instance, mercury, treated with Saṁkhadrāvaka, is prescribed as the remedy for "Phiraṅgaroga" (syphilis). Opium also is prescribed ; while the term yasada is used to indicate zinc. As a distilling apparatus a tinned copper vessel is recommended. Mahommedan influence is distinctly discernible in this hybrid production. The very title of the book is no doubt Sanskritised from the Persian *arrak* (essence).

It is unnecessary to proceed further. We have before us the MSS. of Rasamañjarī by Sālinātha, of Rasarañjana, of Gandhaka kalpa (a Tantra), of Rasārṇava (quite distinct from the standard work bearing the same title,) of Rasaratnākara (altogether different from that of Nityanātha) and several others. As all of these repeat *ad nauseam* the processes already described, we refain from examining their contents here.

1. There is a Kolkata edition published by Gaṅgeśa Chandra Ghosha ; we have comapred it with a MS. procured from the Sanskrit College, Benares. They agree remarkably well.

Chapter V
Indigenous Origin of Indian Alchemy

The reader who has been able to follow us throughout with patience will not have been slow to recognise that alchemy in India has been developed all along on independent lines. Its origin and growth are interwoven with a phase of religious activity—the outcome of purely indigenous traits—which has been discussed at length in chapter I.

The only instance in which we find some allusion to adepts outside the pale of Hindu institutions occurs in Rasaratnasamuccaya.[1] But the term Mleccha used there is a generic one applied indiscriminately to the Śakas, the Yavanas (Greeks or Bactrians), the Chīnas (Chinese), etc. The Hindus have never been shy of acknowledging their obligations to the Yavanas ; Varāhamihira for instance expresses his admiration for the proficiency of the latter in the field of astronomy and would gladly learn at their feet.[2] In the Mahābhārata again we find the services of the Yavana architect, Purochana, requisitioned. We have not, however, come across a single passage which may be construed into implying a hint to alchemists in the West. That the term Mleccha in Rasaratnasamuccaya really refers to the Kirātas is evident from the colophon to Rasahṛdaya.[3]

1. म्लेच्छकसाधकान्. *Vide* Vol. i, Sans. texts, p. 56 ; a variant in the Poona edition has मेलकसाधकान्, which must be rejected as both the Benares and Kāśmīr MSS. accept the former.
2. म्लेच्छा हि यवानास्तेषु सम्यक् शास्त्रमिदं स्थितं।
ऋषिवत्तेऽपि पूज्यन्ते किं पुनर्दैवविद्द्विजः।।
3. Sans. texts. p. 39 ; there the four castes (चतुर्वर्ण) are distinctly mentioned and the Kirātas evidently included under the Mlecchas.

Those scholars who have hitherto maintained that Indian alchemy is of exotic origin, have, we are afraid, done so on *a priori* grounds. Indeed the frame of mind which they have brought to bear upon this subject is not calculated to lead to a judicial and dispassionate verdict. The fact that the Hindus are indebted to the Greeks for substantial imporvements in their astronomy is too often urged whenever it is necesary to support such a view.[1] There is, however, very little affinity in the mode of growth of alchemy in Indian and of its *congener* in the West. The progress of magic, witchcraft and alchemy can be traced from the Atharvavedas onwards to the later Tantras according to the laws of evolution without any breach of continuity.

One of the most favourite doctrines of the neo-platonists was that the metals were engendered in the womb of the earth under the influence of the planets. According to olympiodorus (5th century A.D.) lead is begotten by Saturn, "electrum"[2] by Jupiter, iron by Mars, copper by Venus, tin by Hermes, gold by the Sun and silver by the Moon.[3] Now, such a fascinating astrological belief was pre-eminently

1. *Cf.*–"It was, however, Greek influence that first infused a real life into Indian astronomy. This occupies a much more important position in relation to it than has hitherto been supposed ; and the fact that this is so, *eo ipso* implies that Greek influence affected other branches of literature as well, even though we may be unable at present directly to trace it elsewhere." Weber : Hist. Sans. Lit. ed. 1892, p. 251. In jusitice to Weber, however, we should quote here his views on the independent origin of Hindu medicine ; namely, "the weights and measures to be used by the physician are expressly enjoined to be either those employed in Magadha or those current in Kaliṅga; whence we may fairly presume that it was in those eastern provinces, which never come into close contact with the Greeks, that medicine received it special cultivation—" p. 269.
2. An alloy of gold and silver was known by this name. Later on it disappeared from the list of metals, tin being assigned to Jupiter, and mercury to Hermes.
3. The influence of the astral bodies on the production of the metals can be traced to Proclus' commentary on the Timus. Berthelot : "Les origines del' Alchimie : p. 49 ; also Coll. des. Alch. Grecs. trad. p. 100.

calculated to appeal to the Hindu instinct, especially as there was a "commercial and lively intellectual intercourse," to quote Weber's own words, between Alexandria and the West Coast of India. Varāhamihira's Vṛhatsaṁhitā, if anything, is a repertory of astrological notions. In vain do we look even there for any such chance and solitary reference.[1] The alchemical literature of India a equally silent about it.[2] True it is there are two metals which are often designated as the sun and the moon respectively, namely copper and silver.[3] But this has nothing to do with the planetary generation of these metals. In the imagery of the Sanskrit poet the crimson tint of the rising sun and the pale white lustre of the moon are comparable to the colour of copper and silver respectively.

The old Greek writers repeatedly refer to the mystic relationship between the seven planets and the seven metals. In the Sanskrit literature, however, beginning with the Suśruta and coming onwards to Rasaratnasamuccaya, we invariably find six metals recognised, alloys being rigorously excluded from the list and regarded simply as "artificially made."[4] It is only in the Sārṅgadhara and other later works

1. Chapter XVI of this work treats of "countries people and things belonging to the domain of each planet." Among sundry other things "the Sun is the lord of gold and fire." Thousand and one articles are said to be presided over by the Moon, Mars, Mercury, Jupiter, Saturn, Venus, Rāhu and Ketu ; but nowhere is there any connection between the planets and the metals. *Vide* Kern's trans. pp. 97-101.
2. The mythical origin of some metals is given in the Purāṇas ; but no reference to the astral bodies is to be found. Thus silver originated from the tears of the God Śiva, copper from the semen of the God Kārtikeya, lead from the semen of the snake God Vāsuki iron from different part of the semen of the God of fire. Some Vague and remote connection only can be established between this sort of nomenclalture and that of the old Greco-Egyptian alchemists. According to the latter mercury is sometimes designated as the semen of Hermes ; whilst in the Tantras it is the semen of Śiva (हरवीज). Vide "Intro. a l' elude de la chimie des anciens," p. 11.
3. रवि अर्क and other synonyms are as a rule applied to copper but rerely to gold.
4. *Vide* Vol. i, pp. 48, 72 and 127.

that we find seven metals enumerated ;[1] but this author is guilty of a curious inconsistency, since in a subsequent part of his book he adds two more to the list and quietly lays down that there are nine metals, *which derive their names from the nine planets.*[2] This may be read as a sort of vague and indistinct connection between the metals and the astral bodies.

The Arabs who adopted many of the Greek doctrines of alchemy also represented the seven metals by the seven planets. In fact it was through the medium of the Arabs and Jews that a knowledge of the Greek sciences including alchemy filtered on the European nations during the middle ages.[3] We have already seen that Rājputāna had been overrun and even occupied for a brief period by the Mussulman conquerors shortly before the time of Sārṅgadhara and it is reasonable to suppose that he had gleaned some astrological notions from that quarter.

---

1. (आर) brass being taken as the additional metal ; zinc was evidently unknown to Sārṅgadhara.
   स्वर्णताराारताम्राणि नागवङ्गौ च तीक्ष्णकम् ।
   धातवः सप्त विज्ञेयाः ....।
2. Kāṁsya (beel-metal) and "Vṛttaloha" or "Vartaloha" (*cf.* Vol. i, p. 114) make up the number. Rāhu and Ketu are the two mythical planets ; they are in reality the ascending and descending nodes in the ecliptic. Rāhu is supposed to cause the eclipse of the sun and the moon by swallowing it up.
   Sārṅgadhara stops here, but his prolix commentator, Āḍhamalla supplies the deficiency by making copper stand for the Sun, silver for the Moon, brass for Mars, iron for Saturn, gold for Jupiter, tin for Venus, lead for Mercury and bell-metal and vartaloha for Rāhu and Ketu respetively.
   It will also be noticed that lead is here represented by Mercury ; but in the Hindu alchemy mercury always stands as the visible manifestation of the God Śiva, it being generated from his semen, hence the name हरवीज given to it.
3. C'est par l'intermediaire des éscrits arabes et hérbreux que la plupart des conaissances scientifiques des Grecs en mathematiques, en astronomie, en physique et en médecine, ont été transmises au moyen age occidental ; les écrits Grecs proprement dits n'ayaht guere ete connus directement avant la Renaissance."—"La Chimie au Moyen Age," T. i, p. 229.

We now advert for a moment to the main question of the connection of the comparatively modern and scientific phase of Hindu astronomy with the science of Alexandria. Althought it is admitted by all those who are entitled to speak with authority on the subject that the Hindus borrowed from the Greeks and thus improved upon their older system many points have yet to be cleared up. Professor Whitney who has carefully studied the subject is struck with the circumstances that the Sūrya and other Siddhāntas bear resemblance to the system of Hipparchus rather than that in the Syntaxis.[1] In other words the transmittance of astronomical knowledge must have taken place before the time of Ptolemy. Now, the period of lively intercourse is precisely that covered by the age of Ptolemy and that of its followers, and yet the Indian astronomers seem to be unaware of his teachings.[2]

This mystery has never been elucidated ; Dr. Thibaut's arguments are ingenious, but, we are afraid, by no means convincing.[3] Indeed in the purely mathematical sciences the Hindus far excelled the Greeks of the Alexndrian school. Āryabhaṭṭa's knowledge of algebra was superior to that of Diophantus and as for arithmetic it is generally acknowledged to be of Indian origin.[4]

---

1. Intro. to "Sūryasiddhānta". We find from Ptolemy that the observations of Hipparchus extended from B.C. 162 to B.C. 127. Ptolemy flourished at Alexandria in the second century A.D. under the reigns of Adrian and Marcus Aurelius.

   This was also pointed out long ago by Colebrooke, *e.g.* "The whole cluster of astronomers mentioned by Brahma Gupta must be placed in the interval between the age of Hipparchus, and possibly that of Ptolemy."—Essays, pp. 411-412.

2. Āryabhaṭṭa composed his treatise "Laghvāryabhaṭṭīyam," in 496 A.D., and Varāhamihira his "Pañcasiddhāntikā" about 550 A.D. But these astronomers borrowed from the much older treatise entilted "Romakasiddhānta" ascribed to Śrīsena, and Pauliśasiddhānta (*vide* Thilbaut's Intro. to Pañcasiddhāntikā, xxvii-xxxvi).

3. L.c., li and liii.

4. *Cf.*—Colebrooke : "Admitting the Hindu and Alexandrian authors

The last word on the subject has yet to be said and it is doubtful if this dark recess will ever be illumined. In the meantime it would be better to hold our judgment in abeyance. At best we can only launch into hypotheses or indulge in dogmatic assertions.[1]

---

to be nearly equally ancient, it must be conceded in favour of the Indian algebraist, that he was more advanced in the same."—Essays, ii, 429. Again, "in the whole science he [Diophantus] is very far behind the Hindu writers, *ibid.*, p. 438.

Kaye has recently questioned the priority of the claims of the Hindus : *Vide* J.A.S.B. iii [N.S.], July, 1907.

1. One who approaches this subject should divest himself of prepossessions of the "bias of partriotism" as Herbert Spencer puts it. A very agreeable feature in this controversy is that those who are most competent to speak with authority are extremely modest, nay diffident in giving expression of their views. Woepcke in his masterly memoir "on the propagation of the Indian numerical symbols" says:— "La solution parfaite de ces problems ne resulterait que d'une connaissance tellement complete des evenements et de leurs phases, qu'il faut a peu pres renoncer a jamais l'obtenir. D'ur autre cote, remplir les vides par des hypotheses arbitraires ou preconcues serait livre l'histoire des sciences eux hasards et au dogmatisme des inspira-tions individuelles." "Journal Asiatique" (1863).

   Lasssen whose modesty is only equalled by his perfound scholarship observes:—"Es wäre sehr wünschenswerth, dafs ein mit der gehorigen kenntnifs der Sanskritsprache ausgerüsteter Kenner der Mathematik und Astronomie es unternāhme, aus diesen vorhandenen Hülfsmitteln die Lücken unserer Kenntnifs auf diesem Gebiet des Indischen Alterthums aūszufüllen ; fur jetzt mūfsen wir uns darauf baschrānken, die hauptsāchilchsten Ergebnifse der bisherigen Untersuchungen zusammenzustellen "Indische Alterthumskunde." Vol. ii, p. 1150 ed. 1874.

Chapter VI

Some Noted Indian Alchemists and their Works

Aufrecht in his notice of the MS. of Rasaratnasamuccaya,[1] evidently from a mistaken notion that the pursuit of chemistry was unknown in ancient India, goes somewhat out of his way in asserting that the names of the 27 alchemists, mentioned in the beginning of that work,[2] are mostly apocryphal. We have, however, positive proofs that these personages far from being evolved out of the imagination of the author existed in real flesh and blood and contributed not a little to the progress of chemical science. Thus, we read in Rasaratnākara of "chemical operations performed by Māṇḍavya" and also find Ratnaghoṣa taking part in a dialogue on certain processes,[3] Yaśodhara, again, is the author of the comparehensive work named Rasaprakāśasudhākara which we have noticed at some length in the present book. Vyāḍi is a prominent name both a grammarian as well as a chemist. The Garuḍapurāṇa quotes him as an authority on gems.[4] Nandī, as we have already seen, is distinctly referred to by Somadeva as the inventor of the Koṣṭhi apparatus and of the process of sublimation.[5]

1. *Catalogus Catalogorum*, ii, p. 116.
2. *Vide* Vol. i, p. 77.
3. Eng. trans. pp. 6-8.
4. व्याड़िर्जगाद जगतां हि महाप्रभावः सिद्धो विदग्धाहिततत्परया दयालुः।
5. Intro., lv.

There are, indeed, a great many names scattered throughout the mass of chemical and medical literature, some of which have been handed down to posterity, sometimes on account of the important processes they invented and sometimes, again, because of the efficacy of the metallic preparations which they introduced. Thus in Rasendracintāmaṇi we come across the name of Siddha Lakṣmīśvara. The names of the sage Brahmajyoti, Manthānabhairava, Svacchandabhairava and Gahanānandanātha have in this manner been rescued from oblivion. Most of these were Buddhist monks. Śivadāsa in his commentary on Cakrapāṇi quotes besides Patañjali, Bhavyadattadeva as an expect on iron preparations. We give below the names of some of the authors and their works over and above those already noticed it is scarcely possible to submit an exhaustive list.

Name of Author	Name of the work
Ānandānubhava	Rasadīpikā.
Balabhadra	(Unknown).
Bhojadeva	Rasarājamṛgāṅka.
Candrasena	Rasacandrodaya.
Carpaṭa	Carpaṭasiddhānta.
Dhanapati	Divyarasendrasāra.
Garuḍadattasiddha	Rasaratnāvalī.
Gorakṣanātha	Gorakṣasaṁhitā.
(Unknown.)	Baudhasarvasva.
(Unknown.)	Raseśvarasiddhānta.
Harihara	Rasaviśvadarpaṇa
Kaṅkālī	Rasakaṅkālī.
Kapālī	Rasarājamahodadhi.
Keśavadeva	Yogaratnākara.
Mallāri	Rasakautuka.
Narahari	Rasayogamuktāvalī.
Rāmarāja	Rasaratnapradīpa.

(Siddha) Bhāskara	Rasendrabhāskara.
(Siddha) Prāṇanātha	Rasadīpa.
Śrīnātha	Rasaratna.
Trimallabhaṭṭa	Rasadarpaṇa.
Vaidyarāja	Rasakaṣāyavaidyaka.
Vandimiśra	Yogasudhānidhi.
Vāsudeva	Rasasarveśvara.

The Tantric Period

नागार्जुन-विरचितात् रसरत्नाकरात् उद्धृत श्लोकाः
Chemistry in Rasaratnākara of Nāgārjuna

[In the rendering of Rasaratnākara of Nāgārjuna and other Sanskrit texts we have often given only free translations. Such passages as are unnecessary repetitions or are not of sufficient interest to the general student of the history of Chemistry have been left out.]

CHAPTER I

अथ महारसशोधनं व्याख्यास्यामः

किमत्र चित्रं यदि राजवर्त्तकं शिरीषपुष्पाग्ररसेन भावितम् ।
सितं सुवर्णतरुणार्कसन्निभं करोति गुञ्जाशतमेकगुञ्जया।। १ ।।

राजावर्त्तशोधनम्।

I shall now speak of the purification of the important rasas[1] (minerals).

What wonder is it that Rājāvarta digested with the juice of *Acacia sirisa* convert silver of the weight of one guñja into one hundred times its weight of gold of the lustre of the rising sun ? 1

किमत्र चित्रं यदि पीतगन्धकः पलाशनिर्य्यासरसेन शोधितः।
आरण्यकैरुत्पलकैस्तु याचितः करोति तारं त्रिपिटेन काञ्चनम्।। २ ।।

गन्धकशुद्धिः।

1. Regarding rasas see Vol. I, p. 79.

What wonder is it that yellow sulphur, purfied with the juice of *Butea frondosa*, converts silver into gold when roasted thrice over the fire of cowdung cakes? 2

किमत्र चित्रं रसको रसेन क्रमेण कृत्वाम्बुधरेण रञ्जितः
करोति शुल्वं त्रिपुटेन काञ्चनम् ।। ३ ।।[1]

रसकशोधनम् ।

What wonder is it that calamine.....roasted thrice with copper convert the latter into gold.[2] 3

किमत्र चित्रं दरदः सुभावितः पयेन[3] मेष्या बहुशोऽम्लवर्गैः ।
सितं सुवर्णं बहुधर्मभावितं करोति साक्षाद्वरकुङ्कुमप्रभम् ।। ४ ।।

दरदशुद्धिः ।

What wonder is it that cinnabar digested several times

---

1. Agrees nearly with Rasārṇava ; *cf.* Vol. I. Sans. Texts, p. 13.
2. These laconic recipes all refer to the floating mass of traditions current at the time of the author on the transmutation of the base metals into gold. As they are of a fragmentary and disjointed character, the meaning is not always clear. Recipe 3 also occurs in Rasārṇava; *vide* Vol. i, pp. 70-71.Various ingenious methods were resorted to for debasing gold or making an alloy, which would mimic the appearance of this metal. Silver, copper, lead, zinc and mercury were often made into an amalgam, and the latter rendered compact and coloured yellow with the aid of orpiment, No wonder that the law-books prescribe several penalties on the fraudulent debasers of the precious metals. Thus according to Manu "the king shall cause a goldsmith, who behaves dishonestly, the most nocuous of all the thorns, to be cut to pieces with razors." IX, 202.
   The reader who feels interest in this subject may consult Berthe lot's La Chimie au moyen age," t. 1. pp. 31 *et* seq., where will be found several recipes for augmenting gold *aurum plurimum facere*, or for the "fabrication of gold." One example may be cited here. Take gold, copper and mercury, make an amalgam of the whole, afterwards add sulphur, orpiment, "sandarac" (a kind of resin) and the bile of the valture etc. The author of the recipes thus concludes:—"Thou wilt now discover a secret, which is sacred and worthy of praise" (l.c. p. 35).
3. पयसा is the correct form.

with the milk of the ewe and the acids[1] imparts to silver the lustre of gold glowing as saffron ? 4

कुलत्थकोद्रवक्काथे नरमूत्रेण पाचयेत्।
वेतसाद्यम्लवर्गेण दत्त्वा क्षारं पुटत्रयम् ।।५।।

Digestion [of minerals] is to be effected in the decoction of *Dolichos uniflorus, Paspalum scrobiculatum,* the urine of man and the acid juices of [the fruit of] ratan (*Calamus rotang*) and afterwards with the addition to the alkalies (soda, borax etc.) The operation of roasting is to be performed thrice.—*Cf.* the process of Rasaratnasamuccaya (Rasaratnāsamuccaya), Bk ii, 67-68 in Vol. i, p. 83. 5

किमत्र चित्रं कदलीरसेन सुपाचितं सूरणकन्दसंस्थम्।
वातारितैलेन घृतेन ताप्यं पुटने दग्धं वरशुद्धिमेति ।।6।।

माक्षिकशोधनम्।

What wonder is it that the pyrites macerated in the juice of *Musa sapientum* and in castor oil and clarified butter, and placed inside the bulb of *Arum companulatum* and roasted [in a closed crucible] undergoes perfect purificatoin ?[2] 6

द्विगुणा विमला पथ्या रम्भातोयेन संयुता।
लवणैर्वर्क[3] दुग्धेन ताम्रपत्राणि लेपयेत् ।।7।।
अग्नौ सन्तप्य निर्गुण्डीरससिक्तानि सप्तधा।
मासान् वसुरसेनैव शुल्वशुद्धिर्भविष्यति।।10।।

परतः सर्वलोहशोधनम्।

अम्लवेतसधान्याम्लमेषीतोयेन शुध्यति ।।11।।

विमलशुद्धिः।

चपलाद्या धातवः सर्वे जम्बीररसभाविताः।
शोधितास्त्रिदिनं पञ्चमृत्तिकाभस्मलावणैः।
संयुताः संशोधयन्ति पुटपाकेन काञ्चनम् ।।12।।

हेमशोधनम्।

1. Vegetable acids are means here. *Vide* Vol. i, p. 128.
2. Here extraction of copper from the pyrites is no doubt meant. *Cf.* R.R.S. Bk. ii, 89-90 ; *vide* Vol. i, p. 84.
3. The text reads वक्त्र, which is incorrect.

Chapala and other minerals, being macerated in the juice of the lemon for three days, become purified. Gold being smeared with the five earths, the ashes and salts[1] and roasted undergoes purification.12

नागेन क्षारराजेन ध्मापितं शुद्धिमृच्छति[2]।
तारं त्रिवारनिक्षिप्तं पिशाचीतैलमध्यगम्।। 13।।

तारशुद्धिः।

Silver alloyed with lead and fused with ashes becomes purified.[3] 13

अहो नु चित्रं पृथिवीभवेन क्षारेण मेषीपयसा घृतेन।
तैलेन शुद्धं द्रुतषोड़शांशं भवेच्च शुल्वं शशिशृङ्गसन्निभम् ।। 14।।

Lo ! it is not to be wondered at that copper, melted with the alkali derived from the earth[4] and the milk of the ewe, clarified butter and one-sixteenth of its weight of oil, will become pure like the crescent of the moon. 14

मोक्षमारोटपालाश-क्षारगोमूत्रभावितम् ।
वज्रकन्दशिखाकल्क-फलमूलसमन्वितम् ।। 23।।
तत्कल्कं कण्टकं साक्षाच्चूर्णं वैक्रान्तसम्भवम्।
सारघेन[3] समायुक्तं मेषशृङ्गीद्रवान्वितम् ।। 24।।
पिण्डितं मूकमूषास्थं धामितञ्च हठाग्निना।
तत्रैवं पतते सत्त्वं वैक्रान्तस्य न संशयः ।। 25।।

वैक्रान्तसत्त्वम्।

Verses 23-25 : Extraction of the essence of vaikrānta: exactly the same as in Rasaratnāsamuccaya Bk. ii, 71-72. *Vide* Vol. i, pp. 83-84.

---

1. *Cf.* R.R.S. Bk. x, 71, 85 ; *vide* Vol. i, pp. 128-129.
2. This verse also occurs in Rasārṇava.
3. Refers to cupellation. *Cf.* Rasārṇava in Vol. i, p. 68, footnote (1).
4. पृथिवीभवेन क्षारेण may refers to natron or to saltpetre.
5. These couplets also occurs in R.R.S., with this difference that in the latter नवसार has been substituted for सारघ.

क्षौद्रं गन्धर्वतैलं सघृतमभिनवं गो रसं मूत्रकञ्च
भूयो वातारितैलं कदलिरसयुतं भावितं कान्तितप्तम्।
मूषां कृत्वाग्निवर्णामरुणकरनिभां प्रक्षिपेन्माक्षिकेन्द्रं
सत्त्वं नागेन्द्रतुल्यं पतति च सहसा सूर्यवैश्वानराभम्।। 25 ।।[1]

महावृक्षार्कक्षीराभ्यां स्त्रीस्तन्येन सुभावितम्।
मूषायामग्निवर्णायां द्रवेत्ताप्यं न संशयः।। 26 ।।

कङ्कुष्ठटङ्कणाभ्याञ्च ताप्यं स्त्रीस्तन्यमर्द्दितम्।
पश्चात्सत्त्वं निपतति सत्यं मूषा तु अग्निवत् ।। 27 ।।

काञ्जिकं बहुशस्स्विन्नं ताप्यचूर्णं कटुत्रिकम्।
कृत्वाम्बुमधुभ्यां पक्कं वज्रपायसभावितम् ।। 28 ।।

गृहधूमं घृतं क्षौद्रं संयुतं पुनरेव च।
धामितं मूकमूषायां शशिशुल्वनिभं भवेत् ।। 28 ।।

कदलीरसशतभावितं घृतमध्वेरण्डतैलपरिपक्कम्।
ताप्यं मुञ्चति सत्त्वं रसकञ्चैव त्रिसङ्गाते।। 30 ।।

माक्षिकसत्त्वपातनविधिः।

Verses 25-30 : Have reference to the extraction of copper from mākṣika and tāpya (verieties of the pyrites). They agree in all essentials with Rasārṇava, Ch. VII, 12-13. *Vide* Vol. i. p. 70.

क्षारस्नेहैश्च धान्याम्लै रसकं भावितं बहु।
ऊर्णा लाक्षा तथा पथ्या भूलता धूमसंयुतम्।। 31 ।।

मूकमूषागतं ध्मातं टङ्कणेन समन्वितम्।
सत्त्वं कुटिलसङ्काशं पतते नात्र संशयः।। 32 ।।[2]

रसकसत्त्वम्।

Verses 31-32 : extraction of zinc from calamine. *Cf.* Rasaratnāsamuccaya Bk. ii, 163-164. *Vide* Vol. i, p. 88.

विमलं शिग्रुतोयेन काक्षीकासीसटङ्कणैः।
वज्रकन्दसमायुक्तं भावितं कदलीरसैः।। 35 ।।

---

1. *Cf.* Rasārṇava. *Vide* Sans. Texts, Vol. I, p. 12.
2. *Cf.* Rasārṇava, Benares MS. V. 37-38.

माक्षीकक्षारसंयुक्तं धामितं मूकमूषके।
सत्त्वं चन्द्रार्कसङ्काशं पतते नात्र संशयः ।। 36 ।।[1]

विमलसत्त्वम्।

Verses 35-36 : Exactly the same as in Rasaratnāsamuccaya Bk. ii, 103-104. *Vide* Vol. i, p. 85.

दरदं पातनायन्त्रे पातितञ्च जलाशये।
सत्त्वं सूतकसङ्काशं जायते नात्र संशयः ।। 37 ।।

दरदसत्त्वम्।

गन्धकञ्च प्रभावेण सत्त्वभूयं स्वभावतः।
ततः ख्यातं महासत्त्वं रसेन्द्रस्य समं ततः ।। 38 ।।

अभ्रकादिसत्त्वपातनविधिः।

Verses 37 : Essence of darada, *i.e.* extraction of mercury from cinnabar by distillation. *Cf.* Rasaratnāsamuccaya Bk. i, 89-90. *Vide* Vol. i, p. 78.

एक एव महाद्रावी पार्वतीनाथसम्भवः।
किं पुनस्त्रिभिः संयुक्तो वेतसाम्लाम्लकाञ्जिकैः ।। 50 ।।
मुष्काफलानि सप्ताहं वेतसाम्लेन भावयेत्।
पुटपाके ततश्चूर्णे द्रवते सलिलं यथा।
कुरुते योगराजोयं रत्नानां द्रावणे परम् ।। 51 ।।

अभ्रकादिद्रुतपातनविधिः।

तालेन वङ्गं दरदेन तीक्ष्णं नागेन हेमं शिलया च नागम्।
गन्धाश्मना चैव निहन्ति शुल्वं तारञ्च माक्षीकरसेन हन्यात् ।। 52 ।।

× × ×

शुल्वम् अजाक्षीरसुगन्धकेन तारं स्नुहीक्षीरसुमाक्षिकेण।
यद्यस्य धातोर्विहितञ्च युक्तं निरुत्थघातं कथितञ्च तीक्ष्णैः ।। 54 ।।

मृतानि लोहानि रसीभवन्ति रसेन युक्त्यामयनाशनानि।
अभ्यासयुक्त्या पलितादिनाशं कुर्वन्ति तेषाञ्च जराविनाशम् ।। 55 ।।

इति नागार्जुनविरचिते रसरत्नाकरे वज्रमारण-सत्त्वपातन-
अभ्रकादिद्रुतिद्रावणवज्रलोह-मारणाधिकारो नाम द्वितीयः।

---

1. *Cf.* Sans. Texts, Vol. I. p. 12. R.R.S. Bk. ii, 103-104.

Verses 50-51 : Dissolution of gems (pearls etc.) by digestion in vegetable acids, *e.g.* sour gruel (impure vinegar) and the juice of certain acid plants.

Here ends Chapter Second of Rasaratnākara by Nāgārjuna on the killing of diamond and the metals, extraction of the essence of minerals and liquefaction of mica.

तृतीयोऽध्यायः

CHAPTER III

अथातो रसबन्धाधिकारं व्याख्यास्यामः

जम्बीरजेन नवसारघनाम्लवर्गैः क्षाराणि पञ्च लवणानि कटुत्रयञ्च।
शिग्रूदकं सुरभिसूरणकन्द एभिस्संमर्दितो रसनृपश्चरतेष्टलोहान्।। 1।।

चारणजारणविधिः।

I shall now explain [the process of] fixation of mercury. The king of rasas (mercury), rubbed with the juice of lime, sal-ammoniac, the acids, the alkalies, the five salts, *Piper nigrum*, P. *longum*, the dried root of ginger, the juice of *Moringa pter.*, the tuber of *Amorphophallus companulatus*, can readily amalgamate itself with the eight metals. 1

प्रज्ञापारमिता निशीथसमये स्वप्ने प्रसादीकृतं।
नाम्ना तीक्ष्णमुखं रसेन्द्रममलं नागार्जुनप्रोदितम्।। 4।।

Prajñāpāramitā (perfection of wisdom) appeared before Nāgārjuna at midnight in a dream and revealed to him the ingredients of a recipe (consisting of steel, copper, mica, pyrites, etc.) 4

रसं हेमसमं मर्द्यं पीठिकागिरिगन्धकम्।
द्विपदीरजनीरम्भां मर्दयेत् टङ्कणान्विताम्।। 30।।

नष्टपिष्टञ्च मुष्कञ्च अन्धमूष्यां निधापयेत्।
तुषाल्लघुपुटं दत्त्वा यावद्भस्मत्वमागतः।। 31।।

भक्षणात्साधकेन्द्रस्तु दिव्यदेहमवाप्नुयात्।। 32।।

Mercury is to be rubbed with its equal weight of gold and then [the amalgam] further admixed with sulphar, borax etc. The mixture is then to be transferred to a crucible and its lid put on and then submitted to gentle roasting. By partaking of this elixir (*i.e.* sublimate) the devotee acquires a body not liable to decay. 30-32.

गर्भयन्त्रं प्रवक्ष्यामि पीठिकाभस्मकारकम्।
चतुरंगुलदीर्घेण विस्तारेण च त्र्यङ्गुलाम्।। 62।।

मूषां तु मृन्मयीं कृत्वा सुदृढ़ां वर्त्तुलां बुधः।
विंशभागन्तु लोहस्य भागमेकं तु गुग्गुलोः।। 63।।

सुश्लक्ष्णं पेषयित्वा तु तोयं दत्त्वा पुनः पुनः।
मूषालेपं दृढ़ं बद्ध्वा लोणार्द्धमृत्तिका बुधः।। 64।।

कर्षं तुषाग्निना भूमौ मृदुस्वेदेन स्वेदयेत्।। 65।।[1]

I shall now describe the Garbha yantram.[2] 62-65

सूतकस्य पलं गृह्यं तुर्यांशं साक्तुकं विषम्।
तत्समं गन्धकं शुल्वं चूर्णं कृत्वा विनिक्षिपेत्।। 84।।

कृत्वा कज्जलिकामादौ पलं दत्त्वा च गन्धकम्।
घृतपक्वञ्च तच्चूर्णं पचेदायसभाजने।। 85।।

यावद्द्रवत्वमायाति तत्क्षणात् तं विनिक्षिपेत्।
पुटे वा कदलीपत्रे सिद्धं पर्पटिकारसम्।। 86।।

द्वितीयपर्पटीरसः।

Recipe for Kajjali or aethiops mineral.[3] 84-86

x x x x x

प्रणिपत्य सर्वबुद्धान् (सद्बोधान्) सकलदोषनिर्मुक्तान्।
वक्ष्ये सर्वहितार्थं कक्षापुटं सर्वसिद्धिकरम्।।
श्रीशैलपर्वतस्थायी सिद्धो नागार्जुनो महान्।
सर्वसत्त्वोपकारी च सर्वभाग्यसमन्वितः।।
प्रार्थितो ददते शीघ्रं यश्च पश्यति यादृशम्।

1. *Cf.* Rasārṇava. *Vide* Sans. Texts, Vol. I, p. 4.
2. The same as in Rasārṇava. Vide Vol. i, p. 66
3. The process is practically the same as in Cakrapāṇi ; *vide* Vol. i, p. 58.

दृष्ट्वा त्यागञ्च भोगञ्च सूतकस्य प्रसादतः।
सर्वसत्त्वामयावेधी स्वरसेन तथैव च।।
तेषां मध्ये प्रधानञ्च रत्नघोषः प्रचारकः।
कृताञ्जलिपुटो भूत्वा नागार्जुनपुरः स्थितः।
पृच्छते रसकर्माणि विद्यादानं ददस्व मे।।

Having made salutation to all the Buddhas, free from the taint of sins, I shall now deliver the Kakṣapuṭa tantram[1] for the benefit of suffering humanity.

[The most important part, from an historical poin of view, is the dialogue between Nāgārjuna, King Śālivāhana and Ratnaghoṣa.]

Nāgārjuna, whose end (salvation) had been attained benevolent towards all living creatures, blessed with all favours, residing on mount Śrīśaila, before him stood Ratnaghoṣa with folded arms, saying : "Be pleased to communicate to me knowledge on chemical operation."

श्रीनागार्जुन उवाच–

साधु साधु महाप्राज्ञ तुष्टोहं भक्तवत्सलः।
कथयामि न सन्देहस्तत्त्वया परिपृच्छ्यताम् ।।
बलीपलितनाशञ्च तथा कालस्य ध्वंसनम्।
यथा लोहे तथा देहे क्षमते नात्र संशयः।।
सत्त्वानां भोजनार्थाय साधिता वटयक्षिणी।
द्वादशानि च वर्षाणि महाक्लेशः कृतो मया।।
तत्कालदृष्टद्रव्याणां दिव्या वाणी मया श्रुता।
अदृष्टप्रार्थिता पश्चादघ्ष्टा त्वं भव साम्प्रतम्।।

Nāgārjuna said :—

Well done ! well done! I am pleased with your devotion and shall convey to you all that you want to know, namely remedies for warding off wrinkles, grey hair and other signs

3. There is extant a Tantra, which goes by this name, the reputed author of which is Nāgārjuna.

of old age. Mineral preparation act with equal efficacy on the metals as on the body (human system).[1]

श्रीवटयक्षिण्युवाच–

साधु साधु महासिद्ध
किंचित्प्रार्थय मे सिद्ध तत्सर्वं प्रददाम्यहम्।।

For the benefit of living beings I went through all manner of penances for twelve tedious years and worshipped the goddess Yakṣiṇī presiding over the *Ficus religiosus*. Then I heard an oracle.

श्रीनागार्जुन उवाच–

यदि तुष्टासि मे देवि सर्वदा भक्तिवत्सले।
दुर्लभं त्रिषु लोकेषु रसबन्धं ददस्व मे।।

Nāgārjuna said :—

O Goddess ! if you are propitiated, be pleased to comunicate to me the rare knowledge of the fixation of mercury.

x x x x x x x

शालिवाहन उवाच–

सुवर्णरत्नभाण्डारं कुमारी मदसुन्दरी।
निवेदितो मयात्मा मे आदेशो देवि दीयताम्।।
साधु साधु महाप्राज्ञ ममादेशप्रपालक।
साधयामि न सन्देहो युष्मत्सत्येन साधकम्।।
पुनरन्यं प्रवक्ष्यामि माण्डव्येन यथा कृतम्।
रसोपरसयोगेन सिद्धं सूतं सुसाधितम्।।
विद्धशुल्वायनं नागं यथार्थकाञ्चनं कृतम्।
शास्त्रं वशिष्ठमाण्डव्यं गुरुपार्श्वे यथा श्रुतम्।
तदहं सम्प्रवक्ष्यामि साधनञ्च यथाविधि।।
सहायशोभनाः प्राज्ञा निरालम्भा दृढ़व्रताः।
कुलीनाः पापहीनाश्च साधयन्ति जितेन्द्रियाः।।

1. The was a favourite metaphor with the alchemists. *Cf.* Vol i, Intro. lxxvi.

कोष्ठिका वक्रनालञ्च गोमयं सारमिन्धनम्।
धमनं लोहपत्राणि औषधं काञ्जिकं विड़म्।।
कन्दराणि विचित्राणि
सर्वमेलयनं कृत्वा ततः कर्मसमारभेत्।।

Śālivāhana said :—

I have dedicated to you, O Goddess ! treasures of gold and gems, now I await your instructions.

The Goddess said :—

Well done ! well done! O wise ruler ! I shall speak to you of chemical operations performed by Māṇḍavya.[1] .... A disciple should be intelligent, devoted to work, sineless, and master of his passions.[2]

The apparatus known as koṣṭhi, mouth blowpipe, cow-dung, substantial wood (as fuel), a pair of bellows, iron plates, having collected all these, chemical operations are to be performed.[3]

x x x x x x x x x

रत्नघोष उवाच–

साधयित्वा प्रयत्नेन कोटिवेधी महारसः।
शरीरेण विनैतेन सर्वं भवति निष्फलम्।।

Ratnaghoṣa said :—

Having prepared with great care "the powder of projection," which transforms a ten million times its weights of the base metal into gold,[4]

नागार्जुन उवाच–

कथयामि न सन्देहः साकाण्डेन यथा कृतम् ।।
आर्द्रत्वञ्च घनत्वञ्च चापल्यं गुरुतेजसः।

1. A well-known alchemist; one of the 27 adepts mentioned in the opening lipes of Rasaratnāsamuccaya *Vide* Vol. i, p. 77.
2. *Cf.* Initiation into discipleship in Vol. i, p. 115.
3. *Cf.* Rasārṇava in Vol. i, pp. 64-65.
4. कोटिवेधी महारसः–the term वेधः has technical meaning. *Cf.* Vol. i, p. 120.

यस्यैतानि न दृश्यन्ते तं विद्याद्गुरुसूतकम्।।
नानावर्णं भवेत्सूतं विहाय घनचापलम्।
लक्षणं दृश्यते यस्य मूर्च्छितं तं वदन्ति हि।।
गुरुत्वमरुणत्वं वा तेजो भास्करसन्निभम्।
अग्निमध्ये यदा तिष्ठेत् खोटबन्धस्य लक्षणम्।।[1]

Nāgārjuna said :—

I shall convey to you what has been experimented upon by Sākāṇḍa.[2]

अथातो रसेन्द्रमङ्गलानि यन्त्रविधिः

शिलायन्त्रं, पाषाणयन्त्रं, भूधरयन्त्रं, वंशयन्त्रं, नलिकायन्त्रं, गजदन्तयन्त्रं, दोलायन्त्रं, अधःपातनयन्त्रं, भुवः पातनयन्त्रं, पातनयन्त्रं, नियामकयन्त्रं, गमन (?)यन्त्रं, तुलायन्त्रं, कच्छपयन्त्रं, चाकीयन्त्रं, बालुकायन्त्रं, अग्निसोमयन्त्रं, गन्धकत्राहिकयन्त्रं, मूषायन्त्रं, हण्डिकायन्त्रं, कम(?)भाजनयन्त्रं, घोणायन्त्रं, गुड़ाभ्रकयन्त्रं, नारायणयन्त्रं, जालिकायन्त्रं, चारणयन्त्रम्।

Testor for *killed* mercury.

The first two couples are almost the same as in Rasārṇava. *Cf.* Vol. i, p. 74.

When the quicksilver, which has acquired the colour and the lustre of the rising sun, stands the test of fire (*i.e.* is not readily volatilised), then it is to be regarded as fixed.

अथ प्रवक्ष्ये सुगुरूपदेशान् यः पाटलाख्यस्य रसस्य वृद्धः।
यस्य प्रभावात् क्षयमेहकृच्छ्रज्वरादिकुष्ठामयनिग्रहः स्यात्।।
निक्षिप्य खल्ले त्रिपुरान्तकस्य वीजं द्विशुद्धार्द्धपलप्रमाणम्।
क्वाथेन तस्य त्रिपुरोद्भवेन सम्मर्दयेत्तु त्रिदिनानि यावत्।।
कन्यारसेन त्रिदिनं ततश्च सप्तार्चिषा चाथ दिनत्रयञ्च।
चूर्णेस्मिकाया रजनीरजेन[3] सम्मर्दितं तम् उषतोयधौतम्।।
त्रीणि प्रमाणान्यथ सप्तमञ्च सम्पातयेत् पातनयन्त्रयोगात्।
सम्पातितो निर्मलतामुपैति सर्वप्रयोज्योपि हितं प्रशस्तम्।।

1. *Cf.* Rasārṇava, *Vide* Sans. Texts, Vol. I. pp. 17-18.
2. We are unfamiliar with the same, probably the reading is corrupt.
3. The texts seems to be incorrect.

पलं समादाय रसस्य तस्य शुल्वस्य शुद्धस्य च कर्षमेकम्।
कृत्वाष्टपिष्टी विधिवर्त्तयोश्च गन्धाश्मचूर्णं द्विगुणञ्च देयम्।।
घृतप्रयोगेण विपाच्य पश्चान्निक्षिप्य खल्ले परिमर्दितञ्च।
निष्कञ्च मात्रां त्वमृतस्य दत्त्वा गुञ्जाप्रमाणा गुटिका च कार्या।।
श्रीलोकनाथस्य विभोः प्रसादात् ज्ञातं मया पोटलिकाविधानम्।।

।। इति रसेन्द्रमङ्गलं समाप्तम्।।

Here follows a list of apparatus borrowed from Rasendra-maṅgala.—Detailed account of most of these will be found in Rasaratnāsamuccaya Bk. ix. *Vide* Vol. i, p. 67.

Chemistry in Rasahṛdaya of Bhikṣu Govinda

भगवद्गोविन्दपादविरचितात्

रसहृदयः यादुद्धृताः श्लोकाः

A=MS. from Nepal.

B=MS. from India Office Library, with a commentary, named "Bālānvayabodhikā."

C = M.S. from Benares.

द्वितीयपटलः

[टीका–मूर्च्छितरसस्य उत्थापनसंस्कारमाह–]

अमुना विरेचनेन हि सुविशुद्धो नागवङ्गपरिमुक्तः।
सूतः[1] पातनयन्त्रे समुत्थितः[2] काञ्जिके क्वाथात्[3] ।।

Quicksilver digested with sour gruel and then subjected to distillation is freed from [the impurities of] lead and tin. 1

[टीका–अथ मूर्च्छितरसस्य पञ्चमोद्दिष्टपातनसंस्कारं स्पष्टयन्नाह–]

1. A reads पात्य, which makes the sentence incomplete and the metre defective.
2. A reads शुविशुद्धं, which is a tautology as well as a grammatical erros.
3. काञ्जिकेन क्वाथात् is the variant in B. This is a defective metre.

कृत्वा तु शुल्वपिष्टीं[1] निपात्यते नागवङ्गशङ्कातः[2] ।
तस्मिन् दोषान् मुक्त्वा[3] निपतति शुद्धस्तथा सूतः[4] ।।
अष्टाङ्गुल विस्तार[5] दैर्घेण[6] दशाङ्गुलं त्वधोभाण्डम् ।
कण्ठादधः समुच्छ्रितचतुरङ्गुलकृज्जलाधारम्[7] ।।
अन्तः प्रविष्टतलभाण्ड[8] वदनजलमग्ननिजमुखप्रान्ता[9] ।
उपरिष्टाच्चिपिटपटी[10] देयोदरषोड़शाङ्गुलविशाला ।।
तस्मिन्नधोद्र्ध्वभाण्डे[11] निपातितः सकलदोषनिर्मुक्तः[12] ।
सुतरां भवति रसेन्द्रो जीर्णग्रासोऽपि पात्योऽसौ[13] ।।
कृत्वाथ[14] नष्टपिष्टिं त्रिफलाशिखिशिग्रुराजिकापटुभिः ।
संलेप्य चोद्र्ध्वभाण्डं दीप्तैरुपलैरधः पात्यः[15] ।।
अथवा दीपकयन्त्रे[16] निपातितः सकलदोषनिर्मुक्तः ।।
कच्छपयन्त्रान्तर्गत[17] मृन्मयपीठस्थदीपिकासंस्थः ।
यस्मिन्निपतति सूतः प्रोक्तं[18] तद्दीपिकायन्त्रम् ।।

Much the same as above. 2

---

1. A reads सुल्वपिष्टिम्.
2. निपात्यैते नागवङ्ग सदोष is the reading in A, which is incorrect as well as defective metre.
3. A reads सुल्वे दोषांस्त्यक्त्वा.
4. A reads शुद्धं which is tautology as well as a grammatical error.
5. B reads अष्टाङ्गुलञ्च विस्तीर्णं, which destroys the metre.
6. B reads दैर्घ्येण तु, wherein तु is redundant.
7. B reads समं तच्चतुरङ्गुलं कृतजलधारा, which mars the metre.
8. A reads अन्तः प्रविष्टतद्भाण्डं, which is a defective metre.
9. B rads प्रान्तः, which is grammatically incorrect.
10. A has उपविष्टाक्षिपुटपुटित, which is senseless.
11. A reads तस्मिन् धृद्धभाण्डे, which is incorrect.
12. निपातितसकलनिर्मुक्तः, an incomplete variant in A.
13. A has an incorrect reading—ग्रासजीर्णपपात्यासौ.
14. This verse and the next two are not found in B.
15. A reads रपरैरधःप्राप्ता, which has no sense.
16. A has दीपनयंत्रे.
17. A reads कच्छपयंत्रे अन्तर्गत, which is grammatically incorrect.
18. The word तत् is no found in A, thus making the metre defective.

Description of the apparatus for purification of mercury by sublimation and distillation : it agrees almost literally with that given in Rasaratnāsamuccaya Bk. ix, 68. *Vide* Vol. i, p. 121.

सप्तमः पटलः

[टीका–अथ विड़विधानमारभ्यते–]

ग्रासं न मुञ्चति न वाञ्छति तञ्च भूयः
कांश्चिद्गुणान्[1] भजति[2] नित्यमनुक्तमात्रात्[3]।
यज्जीर्यते प्रचुरकेवलवह्नियोगात्
तस्माद्विडैर्मुनिविड़ैः सह[4] जारणा स्यात्।।

[टीका–विड़विधानमाह–]

सौवर्च्चलकटुकत्रयकाङ्क्षी[5] काशीसगन्धकैश्च विड़ैः।
शिग्रो रसशतभाव्यैस्ताम्रदलान्यपि हि जारयति[6]।।

[टीका–विड़ान्तरमाह–]

सर्वाङ्गदग्धमूलकप्रतिगलितं सुरभिमूत्रेण[7]।
शतभाव्यं बलिवसया[8] तत्क्षणतो जार्यते हेम।।

[टीका–अथ क्षारवृक्षगुल्मीषधीराह–]

कदलीपलाशतिल[9] निचुलकनकसुरदालिवास्तुकैरण्डाः।
वर्षाभूवृषमोक्षक[10] सहिताः क्षारा यथालाभम्।।

---

1. A reads काचीद्गुणान् which is incorrect.
2. B reads व्रजति.
3. मुक्तविमुक्तमात्रात् is the variant in A.
4. A reads सुनियतैरिसु, which seems to be incorrect. मुनिविड़ैः means by seven विड़s, mentioned below.
5. A reads (कांजि) which is incorrect.
6. A has शिग्रुरसशतभावितम्। मस्य दनास्यपि हि जारयति, which is incorrect a well as defective in metre.
7. सर्वाङ्गं दग्धमूलकं रसप्रतिमलितसुरभिमूत्रेण, a variant in A, which has no metre at all.
8. A reads शतभावितरसवशया, which has no sense.
9. तिल is not found in B.
10. A reads माक्षीक, which seems to be incorrect.

[टीका–क्षारविधानमाह–]

आनीय क्षारवृक्षान् कुसुमफलशिफात्वक्पलाशैरुपेतान्[1]
कृत्वातः खण्डशस्तान्[2] विपुलतरशिलापिष्टपात्रातिशुष्कान्[3]।
दग्धं काण्डैस्तिलानां करिसुरभिहयाम्भोभिरास्राव्य वस्त्रैः[4]
भस्म त्यक्त्वा जलं तत्[5] मृदुशिखिनि पचेत् हंसपाकेन भूयः[6]।।

[टीका–क्षारजलपाकलक्षणमाह–]

तच्छुष्यमाणं हि सवास्पबुद्बुदान् यदा विधत्ते[7] क्षणभङ्गुरान् बहून्।
तदा क्षिपेत् त्र्यूषण[8] हिङ्गुगन्धकं क्षारत्रयं सल्लवणानि भूखगम्[9]।।
द्रव्याणि संमिश्र्य निवृत्य भूतले व्यवस्थितान्यस्त्रिकटोरिकापुटे[10]।
संस्थापयेत् सप्तदिनानि धान्यगं ततः प्रयोज्यं रसजारणादिकम्।।

[टीका–जारणायां क्षारविधानमाह–]

जम्बीरवीजपूरक[11] चाङ्गेरीवेतसाम्लसंयोगात्।
क्षारा भवन्ति नितरां गर्भद्रुतिजारणे शस्ताः[12]।।

[टीका–रसे विड्योजनमाह–]

विड्मधरोत्तरमादौ दत्त्वा सूतस्य चाष्टमांशेन[13]।
कुर्याज्जारणमेवं क्रमक्रमाद्वर्द्धयेदग्निम्[14]।।

---

1. कुसुमपलशवापत्रकष्टेरुपेतान् is an incorrect text in A.
2. A reads कृत्या तत्खंडसप्तान् which is incorrect.
3. A has विपुलतरुशिलापृष्ठगानातिशुष्कान्.
4. A reads करिसुरभिहयोष्ट्राजकानां च मूत्रे.
5. A reads त, which is incorrect.
6. A reads यावत्.
7. A reads erroneously यथा वइते.
8. A reads क्षिपेत्र्यूखण. B reads क्षिपेन्मूषण. Both the readings are incorrect.
9. A has क्षारत्रितयं पटुलवणं भूमिखगं, which mars the metre.
10. A reads व्यवस्थितान् सस्त्रिकठारिकामुखे। B reads व्यवस्थितशब्दकठोरिकापुटे।
11. A reads बीजपूरकं, which is incorrect. The word वीजपूर is not found in B.
12. गर्भधृतिजारणा सस्ता is an incorrect variant in A.
13. A has दत्त्वा सूतं चाष्टमासेन, which is not correct as it destroys the metre as well as the sense.
14. A reads वर्द्धयेदग्निः, which is grammatically incorrect.

Paṭala vii

Ingredients for making a viḍa for killing metals, *Cf.* Rasārṇava, IX, 2-3. *Vide* Vol. i, p. 72.

Preparation of mild alkali from the ashes of plants :—

The process must be considered as a poor one, compared to the elaborate description in Suśruta. *Cf.* Vol. i, p. 32.

अष्टमः पटलः।

[टीका–अथ रसरागोऽभिधास्यते–अभ्रकजीर्णस्य छायाविशेषमाह–]

जीर्णाभ्रको[1] रसेन्द्रो दर्शयति[2] घनानुकम्पिनीं छायाम्[3]।
कृष्णां रक्तां पीतां सितां[4] तथा सङ्करं मिश्राम्[5]।।

[टीका–अभ्रकयोगाद्वर्णाविशेषमाह–]

कृष्णाभ्रकेण बलवत्[6] सितरागैर्युज्यते[7] रसेन्द्रस्तु।
श्वेतै रक्तैः पीतैः वह्नेः खलु वर्णतो ज्ञेयः[8]।।

[टीका–अथ रञ्जितरसप्रशंसामाह–]

अथ निजमेव हि वर्णं[9] न जहाति यदा स रज्यते[10] रागैः।
क्रमशो हि भक्ष्यमाणो[11] निर्निक्तो[12] रञ्जनं कुरुते।।

---

1. A reads जीर्णाभ्रक, which mars the metre.
2. A reads दुरुसयति, which is incorrect.
3. घनानुकम्पिनी छाया is the variant in A. घनानुक्तपर्णच्छायां is the variant in B. Both the readings seem to be incorrect.
4. कृष्णं रक्तं पीतं शितं is the reading in A, which is incorrect.
5. A reads ददा संकरोर्म्मिश्रा: which is incorrect.
6. श्वेताभ्रकेन बलवान् is the variant in A, which is incorrect.
7. A reads शितरागैः पूर्य्यते.
8. A has नविप्लुतो ज्ञेयं, which has no clear sense. B has वह्निवर्णतो ज्ञेयः, which mars the metre.
9. B reads अथ निजकर्मवर्णं, which is not correct.
10. A reads यथा रसजीर्य्यते, which is grammatically incorrect.
11. A reads भक्ष्यमाणा, which is incorrect. B reads क्रमशो वक्षमाणो, which is also incorrect.
12. A reads नारक्तो.

[टीका–अभ्रसत्त्वादीनां योगे रसे व्यवस्थामाह–]

बलमास्तेभ्रकसत्त्वे जारणरागाः प्रतिष्ठितास्तीक्ष्णे[1]।
बन्धश्च सारलोहे[2] सारणमथ[3] नागवङ्गाभ्याम्[4]।।

[टीका–सर्वकरणं तीक्ष्णमाह–]

क्रामति तीक्ष्णेन रसः तीक्ष्णेन[5] जीर्यते क्षणाद्ग्रासः[6]।
हेम्नो योनिस्तीक्ष्णं[7] रागान् गृह्णाति तीक्ष्णेन।।

[टीका–तीक्ष्णहिङ्गुलयोगेन गुणाधिक्यमाह–]

तदपि च दरदेन हतं हत्वा माक्षीकेण रविसहितम्[8]।
वासितमपि वासनया घनवच्चार्यञ्च जार्यञ्च।।

[टीका–तीक्ष्णवदेतानाह–]

कान्तं वा तीक्ष्णं वा काङ्क्षीं वा वज्रसस्यकं वापि[9]।
एकतमं सर्वं वा रसरजने सङ्करोऽभीष्टः[10]।।

[टीका–स्वे स्वे विकारे वक्ष्यमाणमाह–]

कुटिले[11] बलमप्यधिकं रागस्तीक्ष्णे तु पन्नगे स्नेहः[12]।
रागस्नेहबलानि तु कमले शंसन्ति धातुविदः[13]।।
सर्वैरेभिर्लोहैर्माक्षिकसहितैर्मृतिस्तथा गर्भे।
बिड़योगेन तु जीर्णो रसराजो विशति लोहेषु।।[14]

---

1. A reads erroneously प्रतिष्ठिता तीक्ष्णा.
2. A reads सारणलाहैः, which is incorrect.
3. A reads क्रामणमथ.
4. B reads नागवङ्गमतः.
5. A reads तीक्ष्णे च.
6. A reads ग्रासे, which is incorrect.
7. A reads ज्योतिस्तीक्ष्णे, which has a defective metre.
8. B has omitted हत्वा माक्षीकेण रवि.
9. काङ्क्षीं वा वज्रशस्यकादीनां, a variant in B.
10. रसेपि हि रसायने विद्यात्, a variant in A.
11. This is not found in B.
12. A reads erroneosly पन्नगस्नेह.
13. A reads धातुविधं, which is incorrect.
14. This śloka is not found in B.

[टीका–रसबन्धानां मारणे विधानमाह–]

तालक[1] दरदशिलाभिः स्नेहक्षाराम्ललवणसहिताभिः।
समकद्विगुणत्रिगुणान् पुटो वहेद्वङ्गशस्त्रादीन्[2]।।

[टीका–पुटितधातुकृत्यमाह–]

रक्तस्नेहनिषेकैः सेकं[3] कुर्याद्रसस्य पिष्टिरियम्[4]।
चारणजारणमात्रात् कुरुते रसमिन्द्रगोपनिभम्।।

[टीका–मुख्यत्वेन ताम्रप्रशंसनमाह–]

अथवा[5] केवलममलं कमलं दरदेन वापितं कुरुते।
त्रिगुणं जीर्णं जीर्णं[6] लाक्षारससन्निभं सूतम्।।

[टीका–विध्यन्तरमाह–]

रक्तगणगलितपशुजलबहु[7] भावितताप्यगन्धकशिलानाम्[8]।
एकेन वापितमृतं[9] कमलं रञ्जयति रसराजम्।।

[टीका–रागाधिकारि–गन्धकादीनाह–]

वाह्यो गन्धकरागो[10] विलुलितरागे[11] मनः शिलाताले।
माक्षीक[12]सत्त्वरसकौ द्वावेतौ[13] रञ्जने शस्तौ।।

[टीका–प्रधानयोः ताम्रखर्परयोः कृत्यमाह–]

---

1. A reads तारक.
2. निवाहयेद्वङ्गः शास्त्रादीन्, a variant in A, which has a defective metre.
3. B reads शेषं.
4. A reads कृष्टिर्य, which is incorrect.
5. B reads erroneously अथ.
6. त्रिगुणं हि चीर्णजीर्णं, a variant in B.
7. The word बहु is omitted in B, which tenders the metre defective.
8. A reads गन्धकशिलाभिः
9. A reads वापतमृतं, which seems to be incorrect.
10. A reads गंधकराग, which is incorrect.
11. A reads शितरागो. B reads विलुलितरागेण. Both the readings are grammatically incorrect.
12. B reads माक्षिक which mars the metre.
13. B reads द्राव्ये हि, which seems to be incorrect.

क्रमवृद्धैः रविरसकैः[1] संशुद्धो मूकमूषिकाध्मातः[2]।
त्रिगुणं जीर्णो जीर्णो[3] हेमाभो जायते[4] सूतः।।

[टीका–अभ्रकयोगमाह–]

अथ[5] कृष्णाभ्रकचूर्णं पुटितं रक्तं भवेत्तथा[6] शकलम्।
त्रिगुणं जीर्णो जीर्णो[7] हेमद्रुतिसंनिभः सूतः[8]।।

[टीका–स्वर्णमारणमाह–]

त्रिगुणेन माक्षिकेण तु कनकं च मृतं रसतालयुतम्[9]।
पटुसहितं तत् पक्वं हण्डिकया यावदिन्द्रगोपनिभम्।।

[टीका–एतद्भस्मजीर्णसूतस्य लक्षणमाह–]

तच्चूर्णं सूतवरे त्रिगुणं जीर्णं हि[10] जीर्णं तु।
द्रुतहेमनिभः सूतो[11] रज्यति लोहानि सर्वाणि[12]।।

[टीका–सर्वेषां धातूनां रसानामुत्तरोत्तरविशेषत्वमाह–]

---

1. A reads क्रमवृद्धैरविसेकैः B reads क्रमवृत्तौ रविरसकौ.
We have adopted the above reading after collating the texts.
2. मूषिकासुनिध्मातः, a variant in A, which mars the metre. मूकमूषिकाध्मातौ, a variant in B.
3. A reads जीर्णं जीर्णं. B reads चीर्णो जीर्णो.
4. A reads धृतिहेमनिभो भवेत्, (धृति=द्रुत). B reads हिमवर्णाभो जायते, which mars the metre.
5. A reads अथवा, which destroys the metre.
6. A omits तथा.
7. A reads जीर्णं जीर्णं, which seems to be incorrect. B reads चीर्णो जीर्णो.
8. हेमधृति(?) संनिभो भवेत् सूतः, a variant in A.
9. कनकञ्च रसताप्युतं, a variant in A, which is a defective metre. The reading of B, which we have adopted, is also a defective metre. The word चेत् or स्यात् should be added after the word युतम्.
10. A reads जीर्णं. B reads चीर्णं हि.
11. धृतहेमनिभसूतो, a variant in A, which seem to be incorrect.
12. A has जरति लोहानि सत्त्वानि. B has रञ्जयति लोहानि सर्वाणि. Both the readings have defective matre.

पत्रादष्टगुणं सत्त्वं सत्वादष्टगुणं द्रुतिः।
द्रुतेरष्टगुणं वीजं तस्माद्वीजं तु जारयेत्।।[1]

Paṭala viii

Alum, blue vitriol :—all these severally or collectively are favourable for imparting colour to quicksilver.[2]

Experts [on the properties] of metals are of opinion that tin gives strength, tīkṣṇa colour, led unctuouness, copper all these qualities [to mercury.]

Couplets (in p. 266, Sans, texts): The purport seems to be that when mercury is rubbed with certain sulphides, *e.g.* orpiment, pyrites etc., and sublimed, it assumes the colour of lac or *Coccinella* insect (a poetical and figurative way of expression for red colour) in the shape of the red crystalline sulphide.

नवमः पटलः।

[टीका–वीजप्रशंसनमाह–]

इति रक्तोऽपि रसेन्द्रो वीजेन विना न कर्मकृद्भवति।
द्विविधं तत् पीतसितं[3] नियुज्यते सिद्धयेच्च रसम्[4]।।

---

1. This verse is found in A after the verse तारकदरदशिलाभिः etc. (*vide* ante pp. 28-29) and runs thus:—
   पचादष्टगुणं दत्त्वा सत्त्वादष्टगुणं धृति।
   धृतेरष्टगुणं वीजं तस्माद्वीजं तु जारयेत्।
2. The text is not very intelligible. The other ingredients mentioned are tīkṣṇa and kānta, these being varieties of iron; but iron will not amalgamate with mercury. Tīkṣṇa may also mean a kind of saline earth. The old alchemists racked their brains in devising ingenious processes for robbing mercury of its fludity and imaparting to it various colours. *Cf.* Rasārṇava. IX, 2-3, XI, 197-198. *Vide* Vol. i, p. 74 and also "The Hindu method of manufactring calomol."—*Ibid.*, p. 250.
3. B reads पीतं सितं, which destroys the metre.
4. नियुधये सिद्धयेच्च रसः, an incorrect variant in A. नियुज्यते सिद्धमेव तत्, a variant in B. Both texts are collated in our adopted reading.

[टीका–रसोपरसधातूनां बहुविधत्वात् शोधनमाह–]

तस्य विशुद्धिर्बहुधा[1] गगनरसोपरलोहचूर्णेश्च।
द्विविधं वीजं तैरपि नाशुद्धैः शुध्यते कदाचिदपि[2]।।

[टीका–अशुद्धवीजप्रभावमाह–]

यः पुनरेतैः कुरुते कर्माशुद्धैर्भवेद्रसस्तस्य।
अव्यापकः पतङ्गो[3] न रसे न रसायने योग्यः[4]।

[टीका–रससंज्ञकानाह–]

वैक्रान्तकान्त[5] सस्यकमाक्षिक[6] विमलाद्रिदरदरसकाश्च[7]।
अष्टौ रसास्तथैषां सत्त्वानि[8] रसायनानि स्युः।।

[टीका–उपरससंज्ञकानाह–]

गन्धकगैरिकसुशिला[9]क्षितिखेचरमञ्जनञ्च कङ्कुष्ठम्।
उपरससंज्ञिमिदं[10] स्यात् शिखिशशिनौ सारलोहाख्यौ।।

[टीका–पूतिसंज्ञके आह–]

ताम्रारतीक्ष्णकान्ताभ्रवज्रलोहानि[11] नागवङ्गौ च[12]।
कथितौ च पूतिसंज्ञौ[13] तेषां संशोधनं कार्यम्।।

[टीका–लवणाक्षारसंज्ञे आह–]

सौवर्चलसैन्धवकं चूलिकसामुद्ररोमकविड़ानि।
षड्लवणान्येतानि तु सर्जीयवटङ्कणाः क्षाराः।।[14]

---

1. रस्यति शुद्धिऽर्वाच्या, a variant in A, wherein the first terms seems to be incorrect.
2. B reads नैतत् instead of कदाचिदपि. This is incorrect ; since it contains two negatives.
3. अव्यापकपतंगि, a variant in A, which is incorrect.
4. A reads योज्ये. B reads रसापनयने योग्यः. Both the readings are incorrect.
5. A omits कान्त.
6. A reads माक्षीक, which mars the metre.
7. A reads दलदरसकैश्च.
8. A reads चत्वारि, which is not accurate.
9. B reads शिलालक, which destroys the metre.
10. B reads संज्ञकमिदं, which mars the metre.
11. B reads सत्त्वलोहानि.
12. B reads वङ्गनागौ च.
13. B reads कथितौ कुभूतसंज्ञौ, which is incorrect.
14. This verse is not found in A.

[टीका–शोधकद्रावकगणमाह–]

सूर्य्यावर्त्तः[1] कदली कन्या[2] कोशातकी च सुरदाली।
शीग्रुश्च[3] तज्रकन्दी नीरकणा काचमाची च।।
आसामेकरसेन तु लवणक्षाराम्लभाविता बहुशः।
शुद्ध्यन्ति रसोपरसा ध्माता मुञ्चन्ति सत्त्वानि।।[4]

[टीका–शोधकद्रावकाणां शोधनद्रावणविधानमाह–]

स्विन्नं सक्षाराम्लैः[5] ध्मातं वैक्रान्तकं हठाद्रवति[6]।
तद्द्रुतमात्रं[7] शुध्यति कान्तं शशरक्तभावनया।।

[टीका–तच्चाह–]

सस्यकमपि रक्तगणैः[8] सुभावितं स्नेहरागसंयुक्तम्[9]।
शुध्यति वारैः[10] सप्तभिरतः परं युज्यते कार्ये[11]।।

[टीका–तच्चाह–]

क्षारैः स्नेहैरादौ पश्चादाम्लेन[12] भावितं विमलम्।
शुध्यति तथाच रसकं दरदं माक्षीकमप्येवम्[13]।।

[टीका–ताम्रशोधनमाह–]

---

1. A reads सूर्य्यावर्त्तक.
2. B reads वन्ध्या.
3. A reads शिग्रुश्च.
4. This verse is found in a A only.
5. A reads सक्षाराम्ले.
6. A reads हताद्रवति, which is incorrect.
7. A reads तद्रुतमात्रे, which is incorrect.
8. B reads रसुगणैः, which seems to be incorrect.
9. भावनतः स्नेहरागसंसिक्त, a variant in A.
10. A reads रागैः.
11. सप्तभिस्ततः परं योजयेत् (कार्य्ये, a variant in A, which mars the metre. B reads सप्तभिरभितः परं, wherein अभितः is incorrect as well as a defective metre.
12. B reads पश्चात् उष्णेन.
13. दरदमाक्षिकमप्येवं तथा शुद्ध्यन्ति, a variant in A, wherein in the last two words are superfluous and incorrect.

तनुरविपत्रं लिप्तं[1] लवणक्षाराम्लरविस्नुक्क्षीरैः[2]।
ध्मातं निर्गुण्डीरससिक्तं बहुशो वहेद्द्रवत्वञ्च[3]।।

[टीका–लोहशोधनमाह–]

शुद्ध्यति नागो वङ्गो घोषो रविणा च वारमुनिसंख्यैः[4]।
निगुण्डीरससेकै[5] स्तन्मूलरजः प्रवापैश्च।।

[टीका–तच्चाह–]

रक्तगणगलितपशुजलभावितपुटितं च रज्यते[6] तीक्ष्णम्।
शुद्ध्यति कदलीशिखिरसभावितपुटितं[7] त्रिभिवरैः।।

[टीका–मारणमाह–]

सर्वः शुद्ध्यति लोहो रज्यति[8] सुरगोपसन्निभो वापात्।
माक्षिकसत्त्वे भृङ्ग[9] शुल्वं वा गन्धकेन हतम्[10]।।

Paṭala ix

Vaikrānta, kānta, sasyaka, mākṣika, vimala, adri, darada, rāsaka :—these are the right rasas[11] and their essence are well suited for chemical operations.

Sulphur, gairika, kṣiti, khecara (mica), añjana, Kaṅkuṣṭha :—these are the uparasas. *Cf.* Vol. i, p. 89.

Gold and silver are the *noble* metals.

Copper, brass, tīkṣṇa and kānta (varieties of iron) are the

---

1. B reads तनुरपि पत्रलिप्तं, which has no sense.
2. A reads स्नुहि. B reads स्नुहा. Both the readings destroy the metre.
3. A reads बहुशो वहेद्रक्तं. B reads बहुशो द्रवत्वञ्च. Both the readings are defective.
4. धोसो रविरौधमपि च मुनिसंखैः, a variant in A. B. reads रविणा च वारमुनिभिः. Both the readings are incorrect.
5. B reads सिक्तै, which is not correct.
6. A reads रज्ञते, which is incorrect.
7. शुद्धति कदलीत्वाखिरसभावितं पुटितं, an incorrect variant in B.
8. A and B read रज्ञति, which is grammatically incorrect.
9. A reads सवृसं. B reads दरदेन भृं. Both are unintelligible.
10. A reads शुल्वेन वा गंधकेन हतं. B reads मृतं.
11. *Cf.* Vol. i, pp. 70-81.

essential or substantial metals ; lead and tin emitfoetid odours. *Cf.* Vol. i, p. 113, footnote (1).

The six salts are :—Sauvarcala, saindhava, culika (sal-ammoniac), sāmudra, romaka and viḍa. The alkalies are :—Sarjikākṣāra, yavakṣāra and borax. (*Vide* Vol. i, Index of subjects.)

एकादशः पटलः।

छगास्थिभस्मनिर्मितमूषां कृत्वैवामलकाकारां।
दलयोगे धनरन्ध्रां टंकणविषगुञ्झकृतलेपां।।

[टीका–अथ ग्रन्थकारयितुर्वंशवर्णनमाह–]

शीतांशुवंशसम्भवहैहयकुलजन्मजनितकुल[1] महिमा।
जयति श्रीमदनरथः[2] किरातनाथो रसाचार्यः।।

[टीका–अथ चास्य कारयितुर्गुणवर्णनमाह–]

यस्य स्वयमवतीर्णा रसविद्या सकलमङ्गलाधारा[3]।
परमश्रेयसहेतुः श्रेयसि परमेष्ठिनः पूर्वम्[4]।।
येन चतुर्वर्णम्लेच्छादिव्याधादिलब्धसत्त्वाभम्[5]।
दक्षिणरसा गृहीता[6] आदिवराहेणेव महाप्रलये।।[7]

नष्टशरीरविवर्णा हीनाङ्गाः कुष्ठिनो गुणाढ्यस्य।
अभिनवसोमेश्वरतामापुरपि पुनर्नवैरङ्गैः।।

[टीका–कर्त्ता स्वनाममहत्वं सूचयन्नाह–]

---

1. कुल is not found in B.
2. स जयति श्रीमदनश्च, a variant in B.
3. C reads सकलमङ्गलाचारा.
4. Of the latter half C has only परं सहेतुः.
5. C reads चतुवर्णम्लेच्छादिव्यदातवध्वसत्त्वाभम्, which seems to be incorrect. *Cf.* Paṭala I., in which the following verse in given:—किरातगोपादिकताप-साद्यवनेचरास्ततकुशलास्तथान्ये। विदंति नानाविधभेषजानां प्रमाणवर्णाकृतिनामजातीः।।
6. दक्षनिम्बुरस गृहीत्वा is the incorrect reading in C. [In the case of किरातनाथ रसाः गृहीताः *i.e.* collected रसs—mercury, minerals etc. and दक्षिण is favourable; and in the case of आदिवराह रसा गृहीता, *i e.* lifted up the earth.]
7. This verse and the next are not found in B.

तस्मात् किरातनृपतेः बहुमानमवाप्य रससुकर्मरतः[1]।
रसहृदयाख्यं तन्त्रं[2] विरचितवान् भिक्षुगोविन्दः[3]।।

नप्त्रां मङ्गलविष्णोः सुमनोविष्णोः सुतेन तन्त्रोऽयम्[4]।
श्रीगोविन्देन कृतः[5] तथागतः श्रेयसे भूयात्।।6।।

अष्टादशसंस्कारं रसेन्द्रदेवस्य दिव्यतनु दृष्ट्वा।
लिखितमिदं पुण्यतमं रसहृदयमवाप्यते सकलम्।।

इति श्रीमद्भगवद्गोविन्दविरचितं
रसहृदयं समाप्तम्।[7]

शुभमस्तु

संवत् 1604 समये ज्यैष्ठ वदि 13 भौमे जगन्नाथभट्ट लेखि।।

श्रीराम हाल संवत् 1880 पौष वदि 3 भौमे।

Paṭala xi

A cupel made of boneash (goat's) and lined internally with borax etc.

Colophon giving an account of the author :—

Madanaratha, King of Kirāta, teacher of mercurial lore, in whom the Goddess of the science of mercury and minerals (Chemistry) the source of all good, has taken her abode for

---

1. C reads रसकर्मोपदेशतो दृष्ट्वा, which destroys the metre.
2. C reads रसहृदयेन्मत्तंत्र, which seems to be incorrect.
3. C reads विरचितायां भगवद्गोविन्दः, which is incorrect.
4. C has रसाटव्यनाम तन्त्रीयम्, which is incorrect.
5. C has श्रीमद्भगवद्गोविन्देन कृतं, which mars the metre.
6. This verse and the next are not found in B.
7. MS. A ends abruptly with the colophon, which runs thus :—इति श्रीगोविन्दभगवत्पादविरचिते रसहृदये जीर्णरसः क्रियाधिकार एकविंशतितमः पटलः।। The author's account of himself is altogether wanting in it. B has "इति सकलरसवैद्यकुलकिरीटभूषायमाणारुणपादपद्मराग सकलविद्यापारीण श्रीमद्भिक्षुगोविन्दकृतो रसहृदयः समाप्तः। इति श्रीमत्करलंशपयोधिसुधारकमिश्रमहेशात्मजश्रीचतुर्भुजविरचितायां... ....खण्डे बालान्वयबोधिख्यायां रसहृदयटीकायां रसप्रकाशंसात्मक एकोनविंशोऽवबोधः। इति श्रीः"।।

the benefit of mankind, who can restore to the disfigured patient, afflicted with leprosy, the healthy come liness and lustre of youth :

Bhikṣu Govinda, well versed in chemical operations and loaded with honours by the king of Kirāta, composed this tantra entitled "Rasahṛdaya."

This tantra is written by Govinda, the grandson of Maṅgalaviṣṇu and son of Sumanoviṣṇu :—May Tathāgata pronounce his blessings.

Chemistry in Kākacaṇḍeśvarīmata Tantra

काकचण्डेश्वरीमततन्त्रात्

समुद्धृता श्लोकाः

ॐ नमः सदाशिवाय।

कैलासशिखरासीनमुमा रुद्रं जगद्गुरुम्।
क्रंदनंदीमहाकालौ भृंगचण्डीविनायकौ।। 1।।

योगिनीनामष्ट तत्र गुह्याद्गुह्यतरं परम्।
कपाली कालरात्री च कालचंद्रकलाम्बिका[1]।। 2।।

कराली कालकर्णी च काकचण्डेश्वरी तथा।
एवमादिस्तथा चान्या योगिनीगणमुह्यकैः[2]।। 3।।

नृत्यवाद्यसुखेशानैः सानन्दैर्हृष्टचेतसैः।
तत्रस्थं भैरवं देवं योगिनीगणवेष्टितम्।। 4।।

केचित् स्तुवन्ति संहृष्टाः केचिद्ध्यायन्ति गुह्यकाः।
केचिन्नृत्यन्ति गुह्या ये केचिद्वाद्यं न कुर्व्वते।। 5।।

हृषितं[3] भैरवं देवं पंचवक्त्रं त्रिलोचनम्।
तं दृष्ट्वा भैरवी हृष्टा जगतां त्रिदशेश्वरी।। 6।।

कृताञ्जलिपुटा भूत्वा कृष्णतारकलोचना।
काकचण्डेश्वरी देवी उवाचेदं तु भैरवी।। 7।।

1. The text reads कालयंद्रकमम्बिका, which is incorrect.
2. The text reads गुह्यकौ, which is incorrecct.
3. The text reads हृष्टनं, which is not correct.

काकचण्डेश्वरी उवाच–

भगवन् देवदेवेश सर्वज्ञ सर्बविच्छिव[1]।
सर्वज्ञानप्रकाशाय अहं पृच्छामि शंकर।।8।।
कथं कायस्थितो जीवो जीवकोऽसौ प्रकीर्तितः।
कायस्थः कर्मणा केन स्थितः संसारपंजरे।।9।।
जराव्याधिदारिद्रेण ग्रस्तः संसारबंधनैः।
एको व्रजति यानेन एकः स्कंधे वहेच्च तम्।। 10।।
जयजयशब्दं मांगल्यमेकस्थाने च जंतवः।
किं कुर्वाणाः प्रकुर्वाणाः कस्मान्मे ब्रूहि शंकर।। 11।।

श्रीसर्वज्ञ उवाच

शृणु त्वं काकचामुंडे साधकानां हितं प्रिये।
गृह्याद्गुह्यतरं वाक्यं पृष्टोहं तु वरानने।। 12।।
कथयामि समासेन शृणु त्वं काकचंडिके[2]।
यस्यार्थश्च परो नित्यः अनादिनिधनेश्वरः।। 13।।
सुरूपो निर्गुणः शांतो विश्वव्यापी परात्परः।
अनादिकर्मसंबंधः कायस्थो भवते तु सः।। 14।।
जीवे व्याधिभयं कर्म अज्ञानात् संप्रकीर्त्तितम्।
तस्मात् कर्मभवात् क्लिष्टः संसारे संसरेत्तु सः।। 15।।
करोति[3] विविधं कर्म संसारे काममोहितः।
द्रव्योपायं न जानाति कामभोगार्थहेतुना।। 16।।
काय[4]क्लेशेन मानव्यो द्रव्योपायं करोति सः।
कृषिवाणिज्यसेवायां मेषाविक्रयविक्रये।। 17।।
वहन्ति काष्ठभारांश्च तृणभारमनेकधा।
यानं यवनच्छन्नं[5] च किं करोति[6] करोति सः।। 18।।

---

1. The text reads सर्वच्छिवः, which is not correct.
2. The text reads काकचंते, which is incorrect.
3. The text reads करोतु, which is not accurate.
4. The text reads कालक्लेशेन, which is not correct.
5. The text reads थवनछत्र, which has no sense.
6. The text reads किं कुर्वंति, which is incorrect.

कामासक्तनरः श्रीमान् कुरुते कर्म नैकधा[1]।
कुर्वाणाश्च नराः केचित् कामासक्तार्थचिन्तकाः ।। 19 ।।

जायंते मर्त्यलोकेस्मिन् दारिद्रोपहतचेतसः।
जायंते नात्र संदेहो कुब्जखंजा नपुंसकाः ।। 20 ।।

महासंसारबंधेन त्रिगुणेन सुयंत्रिताः।
अटव्यां घोरसंघट्टघाटिकां तु नयन्ति तत् ।। 21 ।।

एतत्ते कथितं भद्रे यत्वं मां परिपृच्छसि।
किमन्यत् पृच्छसे भद्रे तथा ते कथयाम्यहम् ।। 22 ।।

।। इति काकचंडेश्वरीमते प्रथमः पटलः ।।

Paṭala I

The first paṭala gives the origin (mythical) of the tantra.

श्रीकाकचण्डी उवाच

कथयस्व महादवे कामभोगप्रसाधनः।
अर्थः संपद्यते येन अक्लेशात् परमेश्वरः ।। 1 ।।

तदहं श्रोतुमिच्छमि कथयस्व प्रसादतः।
आकाशगमनं देव खेचरत्वं यथा भवेत् ।। 2 ।।

पादुकाया दले पंचदिव्यस्त्रीकामसाधनम्।
रोचनं अंजनं चैव धातुवादरसायनम् ।। 3 ।।

जलूकाबंधमित्याहुः रसस्य मारणं कथम्।
जारणं रत्नकर्त्तृत्वं वज्रेभद्रावणं कथम् ।। 4 ।।

एतत् सर्वं समासेन ब्रूहि मे त्रिपुरांतक।

श्रीभैरव उवाच

शृणु त्वं काकचामुंडे साधकानां हितं प्रिये ।। 5 ।।
कथयामि समासेन पृथक् सिद्धार्थसाधनम्।
न द्रव्येण विना सिद्धिर्न भोगाः काम एव च ।। 6 ।।

द्रव्यहीना नरा मर्त्ये प्रेतरूपेण संस्थिताः।
अटंति मर्त्यलोकेस्मिन् पर्य्युपद्रवकारणम् ।। 7 ।।

1. The text reads कुरु कर्ममनेकधा, which is incorrect.

* * * विवं (?) तेषां जायतोपि मृतश्रुतेः।
कुलं सौभाग्यरूपं च पांडित्यं ज्ञानसंस्थितम्।। 8।।

द्रव्यहीना न शोभंते प्राणहीना नरा यथा।
कुत्सितो घ्राणहीनोपि केनापि वधिराः स्थिताः।। 9।।

पद्यन्ते मर्त्यलोकेस्मिन् द्रव्यस्थं मकरध्वजम्।
तस्मात् संसाधयेद्देवि रसेन्द्रं साधकोत्तमः।। 10।।

अथातः संप्रवक्ष्यामि रसेन्द्रसाधनं तव।
मारणं जारणाद्यं च द्रव्यसाधनमुत्तमम्।। 11।।

वज्रदंडः सुदंडश्च लोहदंडस्तथैव च।
अयो विना ओषधये रसस्य मारणे हिताः।। 12।।

तान्निबोध समासेन यथा जानंति साधकाः।
वज्रदंडस्तु वज्री स्यात् लोहदंड पुटं विड़ः।। 13।।

सुदंडं ब्रह्मदंडं च समासात् कीर्त्तितं तव।
ग्राहयेत्तं समासेन साधको हृष्टमानसः।। 14।।

तद्रसं रससंयुक्तं एकीकृत्य तु मर्दयेत्।
अंधमूषागतं ध्मातं रसं म्रियेत तत्क्षणात्।। 15।।

सहस्रवेधी कर्त्ता च ज्ञायते स महारसः।
मूषां संलेपयेत् तेन पुरागृहा महौषधीः।। 16।।

अंधयित्वाग्निमध्ये तु रतिबंधः अयं विधिः।
पश्चात् पलसहस्रैकं शुल्वस्य ग्राहयेद्बुधः।। 17।।

अम्लवर्गहतं शुल्वं क्षारवर्गहतं शुभम्।
वज्रीक्षीरार्कक्षीरेण मर्दयेत् शतधा नरः।। 18।।

पुटं दद्यात् समासेन कालिकारहितं भवेत्।
मूषागर्भोदरे क्षिप्त्वा धामयेत् खदिरानलैः।। 19।।

रसेन्द्रः क्षिप्यते तत्र तारं संजायते क्षणात्।
अथातः संप्रवक्ष्यामि मूषाबंधं यथा भवेत्।। 20।।

मृत्तिकाः ग्राहयेत् प्राज्ञस्तृणपाषाणवर्जिताः।
तुषादग्धास्त्रयो ग्राह्या मद्भागास्तु त्रयस्तथा।। 21।।

एकीकृत्य तु संबंधे मूषां तत् कारयेद्बुधः।
गोस्तनाकारसदृशीमथवामलकं कुरु।। 22।।

मूषाबंधमिति ख्यातं साधकानां हिताय वै।
अथान्यत् संप्रवक्ष्यामि रसस्य मारणं यथा।। 23।।

ग्राहयेत् पूर्ववद्भूयः ओषधीस्तिस्रः साधकः।
मूषां संलपयेत् तेन धाम्य कर्म समाचरेत्।। 24।।

रसेन्द्रो म्रियते क्षिप्रं कुरु कर्म यदिच्छया।
वीतोदकेन कामेन हेमकर्माणि वा शृणु।। 25।।

कारयेल्लोहमयं पात्रं तस्योर्ध्वे मृन्मयं शुभे।
अन्योन्यपृष्ठसंलग्नं कारयेत् साधकोत्तमः।। 26।।

अधोयन्त्रेण तं धाम्य आयसं तत्र जारयेत्।
तप्तखल्वोदरे घृष्टं जरते वात्र संशयः।। 27।।

तद्गुणं आयसं तत्र हेमं चाष्टगुणं ददेत्।
क्षीरवज्रं तथा चारी षट्जारी उदिरग्रही। (?)
ग्राहयित्वा रसं तेषां आयसं जारयेन्नरः।। 28।।

हेमं संजायते तत्र धर्मकामार्थसाधनं।
गृह्य खल्वोदरे तप्ते हेमं षड्गुणतां भजेत्।। 29।।

पुरा महौषधैर्युक्तं जरते नात्र संशयः।
जारिते सारितं दद्यात् पुनस्तं जारयेद्यदि।। 30।।
कुरुते कर्मसंघांश्च कर्म सिद्ध्यर्थसाधनं।। 31।।

जारणं मारणं चैव समासात् कीर्त्तितं मया।
किमन्यत् पृच्छसे भद्रे तथा ते कथयाम्यहम् ।। 32।।

इति श्रीकाकचंडेश्वरीमते द्वितीयः पटलः।

Paṭala II

Kākacaṇḍeśvarī said:—Please, give a brief account of the Chemistry of the metals, the killing of mercury and minerals and the fusion of the gems. 3, 4

Śrī Bhairava said:— Mercury rubbed with the Ingredients of the Viḍa (see Vol. i, p. 72) and roasted in a closed crucible, is killed instantly. That mercury now acquires the power of converting a base metal into 1,000 time its weight of gold. 13-16

Copper, treated with the above preparation of mercury and rubbed with the acids, alkalies and the milky juice of *Calotropis gig.*, and roasted in a crucible, turns into gold.[1] 18-20.

Earth for making crucibles. 21-22 *Cf.* Vol. i, pp. 67-68.

Transmutation of iron into gold :–process not clear. 25-32

अथान्यं संप्रवक्ष्यामि अभ्रकस्य यथाक्रमम्।। 20।।

द्रावणं अभ्रकस्यैव समासेन विधीयते।
क्षीरकंचुकिचूर्णेन तद्रसेनाभिभावितं।। 21।।

निर्गुण्डीवज्रदण्डं तु तद्रसे भावितं कुरु।
एकैकं सप्तवारैस्तु भावयेत् साधकोत्तमः।। 22।।

(भावनान्ते) अभ्रकस्य अम्ले क्षिपेद्दिनत्रयं।
वृहत्या सप्तरात्रं तु क्षिपेदभ्रं नरोत्तमः।। 23।।
तेन कल्केन पत्राणि लेपयेदभ्रकस्य च।
कांस्यपात्रेणस्तारे साधकोत्तमः।। 24।।

सूर्यधाम्नि ततो धाप्यः सप्ताहं साधकोत्तमः।
द्रवते नात्र संदेहः प्रियं दृष्ट्वा यथा स्त्रियः।। 25।।

रसस्य भागमेकं तु दुतभागसमं कुरु।
एकीकृत्य तु तद्धाभ्यं म्रियते नात्र संशयः।। 26।।

लक्षांशवेधकोऽसौ हि जायते नात्र संशयः।
अथान्यं संप्रवक्ष्यामि यथा संजायते द्रुतिः।। 27।।

ग्राहयेदभ्रकं प्राज्ञः पीतं वा कृष्णमेव वा।
पुरासंस्कारयुक्तं तु कर्त्तव्यं साधकेन तु।। 28।।

यवासा चित्रकं धारः कंचुकीवज्रदंडयोः।
वज्रकंदार्कक्षीरं च भल्लातं टंकणं तथा।। 29।।

---

1. Refractory silicates are treated similarly at the present day with the "fusion mixture".

क्षारत्रयसमायुक्तं पटुपंचसमन्वितं।
एकीकृत्य तु कल्केन अभ्रपत्राणि लेपयेत्।। 30।।
मूषागर्भोदरे क्षिप्त्वा धामयेत् खदिरानलैः।
द्रवते नात्र संदेहो * * * * *

।। इति काकचंडेश्वरीमते षष्ठः पटलः।।

Paṭala III

Liquefaction of mica:—Thin plates of mica are to be smeared with the three alkalies and borax and strongly heated in a crucible. The whole mass fuses. 29-30.

Chemistry in Rasendracūḍāmaṇi of Somedeva

सोमदेवविरचितात् रसेन्द्रचूड़ामणेः उद्धृता श्लोकाः

रूप्येण सह संयुक्तं ध्मातं रूप्येण चेल्लगेत्[1]।
तदा निरुत्थमित्युक्तं लोहं तदपुनर्भवम्।।

एवं रूप्यं सनागं चेत् ध्मातं ताम्रे लगेन्नहि ।।
निष्कमात्रे तु नागेऽस्मिन् लोहखार्यां हृते सति।
स्वतो लक्षगुणां हैमीं शलाकां ग्रसति ध्रुवम्।।

कुसुम्भतैलतप्तं तत् स्वर्णमुद्गिरति[2] ध्रुवम्।
गुह्यनागोऽयमुद्दिष्टो वक्ति[3] स्वच्छन्दभैरवः।।

Tests for killed iron (rouge).

As Rasaratnāsamuccaya has borrowed these from the present work, it is unnecessary to repeat them here. *Vide* Vol. i, pp. 18-19.

तीक्ष्णं नीलाञ्जनोपेतं ध्मातं हि बहुशो दृढ़म्।
मृदुकृष्णं द्रुतद्रावं[4] वरनागं तदुच्यते।।

---

1. The text reads लघु. But रसरत्नसमुच्चय reads लगेत् which seems to be correct.
2. The text reads स्वर्णं मुद्गिरिं, which is incorrect.
3. The text reads मुद्दिष्टशक्ति, which has no sense.
4. The text has द्रुतद्रावी, which is incorrect. We have adopted the text of रसरत्नसमुच्चय.

मृतस्य पुनरुद्भूतिः संप्रोक्तोत्थापनाख्यया[1]।
द्रुतद्रव्यस्य निक्षेपो द्रवे तड्ढालनं मतम्।।
त्रिंशत्पलमितं नागं भानुदुग्धेन मर्द्दितम्।
विमर्द्य पुटयेत्तावत् यावत् कर्षावशेषितम्।
न तत् पुटसहस्त्रेण क्षयमायाति सर्वथा[2]।
चपलोऽयं समुद्दिष्टो वार्त्तिकैर्नागसम्भवः[3]।।
इत्थं हि चपलः[4] कार्यो वङ्गस्यापि न संशयः।
तत्स्पृष्टहस्तसंस्पृष्टः केवलो बध्यते रसः।।
स रसो धातुवादेषु शस्यते न रसायने[5]।
अयं हि खर्पणाख्येन[6] लोकनाथेन कीर्त्तितः।।
भ्रामकाख्यरजः सूक्ष्मं पञ्चमांशरसान्वितम्।
कुमारीमूलतोयेन मर्द्दयेदेकवासरम्।।
चाङ्गेरीस्वरसेनापि दिनमेकमनारतम्।[7]
एवं भूनागध्मातेन मर्द्दयेद्दिवसद्वयम्।।

Take lead of the weight of 30 palas and rub it with milky juice of *Calotropis gig.* and calcine it till its weight is reduced to a karṣa : this residue even if it be calcined a thousand time, will not undergo further decay (*i.e.* diminution in weight). Metallurgists call this residue derived from lead Chapala.[8]

Tin also similarly treated will yield a chapala which, on merely being touched with the hand, kills (fixes) mercury. This mercury is recommended for metallurgical and not for chemical purpose.

---

1. The text reads साधोक्तस्थापनाख्या, which seems to be erroneous.
2. The text has क्षयमायान्ति सर्वदा, which is grammatically incorrect.
3. The text reads मार्गसंभवः, which is incorrect.
4. The text reads इत्थ क्षीरपलः, which is senseless.
5. The text has रसायनम्.
6. खवर्णाख्येन is the reading in रसरत्नसमुच्चय
7. The text reads दिनमेकमनारसम्, which has no clear sense.
8. Possibly it refers to lead derived form *argentiferous galena*, The lead, being cupelled off in the shape of litharge, will leave a button of silver, which will not undergo oxidation.

अथैकपलमानेन तावता त्रपुणापि च।
दशनिष्करसेन्द्रेण श्लक्ष्णपिष्टीं समाचरेत्।।
योजयित्वाथ कल्केन यथापूर्वं विमर्द्दयेत्।
ततः साररसेन्द्रेण मत्वेन रसकस्य च।
पिष्टीं कृत्वा तु पूर्वेण पूर्वकल्केन योजयेत्।।
अथ प्रक्षाल्य कोष्णेन[1] काञ्जिकेन प्रशोषयेत्।
पलार्द्धशुद्धसस्येन भृष्टगुञ्जारसेन च।
विमर्द्द्य काञ्जिके कुर्यात् मरिचप्रमितां गुटीम्[2]।।
निरुध्य वज्रमूषायां सन्धिबन्धं विधाय च।
शुषिरैर्नवभिः[3] सम्यग्भस्त्राभ्यां च धमेत् खलु।।
ततो मूषागतं सत्वं समादाय समन्ततः।
धमेत् प्रकटमूषायां वङ्कनालेन शुद्धये।।
दशशाणं हि तत्सत्त्वं भस्मना लवणेन च।
सकाञ्जिकेन संवेष्ट्य पुटयोगेन शोधयेत्।।
द्विनिष्कप्रमिते तस्मिन् पूर्वप्रोक्तेन भस्मना।
अशीतिगुणितं नागं ध्मात्वा निर्वाहयेत् खलु।।
इयता[4] पूर्वसूतोऽसौ जायते न कथञ्चन।
चपलोऽयं समुद्दिष्टो लोकनाथेन शम्भुना।।
अनेनापि रसः शीघ्रं बध्यते पूर्ववत् खलु।
कारवल्लीजटाचूर्णैर्दशधा पुटितो हि सः।
भवेन्नागविनिर्मुक्तो ग्रासं गृह्णात्यशेषतः।।
मुखं प्रकटमूषायां भवेच्च विगुणोत्तरम्[5]।
जीर्णग्रासो रसो ह्येष देहलोहकरो भवेत्।।
सोऽयं श्रीसोमदेवेन कथितोऽतीवनिश्चितम्।।

---

1. The text reads सौष्ठेन, which is unintelligible.
2. The text reads मरिचप्रमिता गुटी, which is incorrect.
3. The text reads शिशुरैर्नवभिः, which has no sense.
4. The text reads इयमा, which is incorrect.
5. The text reads विगुणोत्तरा, which is incorrect.

भूभुजङ्गशकृत्तोयैः[1] प्रक्षाल्यापहृत्[2] रजः।
कृष्णवर्णं हि तत् प्रोक्तं धौताख्यं[3] रसवादिभिः।।

[A somewhat complex process described, which is not very clear. We shall try to give the purport thereof. The quantitative measurements are of particular interest.]

One pala of tin should be taken and rubbed with 10 niṣkas of mercury and made into a cake with the addition of zinc ;[4] it is then to be digested with the decoction of certain herbs and plants and warm vinegar and once more rubbed with ½ pala of blue vitriol[5] and the juice of *Abrus precatorius.* The mass is then to be divided into small pills ; these should then be put into a crucible with the lid on the fire to be urged by means of bellows. The essence is then to be taken out of the crucible. This essence of the weight of 10 Śāṇas is then to be enclosed in a mixture of ashes and salts and roasted. This essence is of the weight of 2 niṣka is then to be with 80 times its weight of leads. This (essence) is also termed capala. *Cf.* "the gold-like alloy" in Vol. i, pp. 74-75.

स्वरूपस्य विनाशेन पिष्टतापादनं हि यत्।
विप्लुद्भिर्वर्जितः[6] सूतो नष्टपिष्टः स उच्यते।।
अथ यन्त्राणि वक्ष्यन्ते रसतन्त्राण्यनेकशः[7]।
समालोक्य समासेन सोमदेवेन सांप्रतम्।।
उद्ध्र्वपातनयन्त्रं हि नन्दिना परिकीर्त्तितम्।।
कोष्ठि(ष्टि) कायन्त्रमेतद्धि नन्दिना परिकीर्त्तितम्।।

1. The text reads तोये, which is not accurate.
2. The text has अपहतं, which is not accurate.
3. The text reads धौतास्वं, which is not correct.
4. रसकस्य सत्त्वेन.
5. पलार्द्धशुद्ध सस्येन.
6. But रसरत्नसमुच्चय reads विद्वद्भिर्निर्जितः.
7. But रसरत्नसमुच्चय reads अशेषतः.

विन्ध्याद्रौ हिमपर्वते च मलये गोमन्तके श्रीगिरौ
सह्याद्रावथ पारियात्रकगिरौ किंकिंधिनामालये।
माहेन्द्रेऽप्यथ चाल्पवत्क्षितिधरे तद्रूपनामाधिकं
गोपाभीरकसिंद्धवैद्यमुखतो वैद्यः समावेद्यताम्।।

।। इति श्रीकरवालभैरवपुरवरपतिश्रीसोमदेवविरचिते।।
रसेन्द्रचूड़ामणौ रससूत्रस्थाने रसमहिम-
निरूपणं नाम प्रथमोऽध्यायः समाप्तः।।

When quicksilver has been deprived of its physical properties (lustre, fludity etc.) it is known as naṣṭapiṣṭa. *Cf.* Vol. i, p. 74.

Somadeva will now give a brief account of the apparatuses, after having consulted numerous chemical tantras. (Here follows the description, which has been reproduced verbatim in Rasaratnāsamuccaya *Vide* Vol. i, pp. 121 *et seq.*)

× × ×

Urdhvapātana apparatus us described by Nandin. *Cf.* Vol. i, pp. 124-25.

Koṣṭhī apparatus as described by Nandin. *Cf.* Vol. i, p. 69.

Chemistry of Rasaprakasasudhākara of Yaśodhara

यशोधरविरचितात् रसप्रकाशसुधाकरात्

उद्धृता श्लोकाः

विमलसूतवरो हि पलाष्टकं
तदनु धातुघटीपटकांक्षिकाः।
पृथगिमाश्च चतुः पलभागिकाः
स्फटिकशुद्धपलाष्टसमन्विताः।
सह जलेन विमर्द्य च यामकं
लवणकाम्लजलेन विमिश्रितम्।
उदितधातुगणस्य च मूषिकां
कुरु रसं विनिवेशय तत्र वै।

डमरुकाभिधयन्त्रवरेण तं
द्विदशयाममजाचयवह्निना।
पवनपित्तकफक्षयकारकं
सकलरोगहरं परमं सदा।
गजपतेर्बलवद्बलदो नृणां
द्विजपतीक्षणवन्नयनप्रदः।
युवतिकामविलासविधायको
भवति सूतवरः सुखदः सदा।
सघनसाररसः किल कान्तिद-
स्त्वखिलकुष्ठकरः कथितो मया।।
।।इति कर्पूररसः।।
उदयभास्करनामरसो ह्ययं
भवति रोगविधान्तकरः स्वयम्।
मगधिकामधुना सह गुञ्जिका-
त्रयमितश्च सदा परिसेवितः।
ललितकामविलासविधायकः
स्थविरकोऽपि रतौ तरुणायते।।
गदहरो बलदोऽपि हि वर्णदो
भवति कर्मविपाकजरोगहा।
सकलसूतकशास्त्रविमन्थजो
द्विजवरेण मया प्रकटीकृतः।।
।।इति उदयभास्कररसः।।

Preparation of karpūrarasa (lit. camphor of mercury, *i.e.* calomel) : practically the same process as given in Vol. i, pp. 250-51.

[Note. Calomel is prescribed here only as an aphrodisiac and a specific for leprosy but not as a remedy for syphilis].

अयेदानीं प्रवक्ष्यामि धातुशोधनमारणम्।
अनुभूतं मया किञ्चित् किञ्चिच्छास्त्रानुसारतः।।
सुवर्णं रजतं चेति लोहं शुद्धमुदीरयेत्।
ताम्रं चैवाश्मसारञ्च नागवङ्गौ तथैव च।।
तीक्ष्णलोहं निगदितं द्वितयं रसवेदिभिः।
समिश्रलोहत्रितयं सौराष्ट्रीरीतिवर्त्तकाः।
एतेऽष्टौ धातवो ज्ञेया लोहान्येवं भवन्ति हि।।

।। अथ रसकगुणाः ।।

द्विविधो रसकः प्रोक्तः कारवेल्लक-दुर्दुरः[1]।
सत्त्वपाते परः प्रोक्तः प्रथमश्चौषधादिषु।।
स तु[2] मेहहरश्चैव पित्तश्लेष्मविनाशनः।
रञ्जनः पारदस्याथ नेत्ररोगक्षयापहः[3]।।
पारदो रसकश्चैव देहलोहकरावुभौ।
नागार्ज्जुनेन कथिती सिद्धौ श्रेष्ठौ रसावुभौ।।
कृतौ येनाग्निसहनौ सूतखर्परकौ शुभौ।
तेन स्वर्णमयी सिद्धिरर्जिता[4] च न संशयः।।
रसको द्रावितः सम्यक् निक्षिप्तो रसपूरके।
निर्म्मलत्वमवाप्नोति सप्तवारं निमज्जितः[5]।।
काञ्चिके वाथ तक्रे वा नृमूत्रे[6] मेषमूत्रके।
द्रावितं क्षालितं[7] सम्यक् खर्परं परिशुध्यति।।
खर्परं रेचितं शुद्धं स्थापितं नरमूत्रके।
रञ्जयेन्मासमेकं हि ताम्रं स्वर्णप्रभं वरम्।।
वचाहरिद्रात्रिफला-गृहधूमैः ससैन्धवैः।
भल्लातकैष्टङ्कणैश्च क्षारैराम्लैश्च मर्द्दितम्[8]।।
पादांशसंयुतैर्मूषां वृन्ताकफलसन्निभाम्।
निरुध्य शोषयित्वा च मूषामुखोपरि न्यसेत्।।
प्रध्माते खर्परे ज्वाला सिता नीला भवेद्यदि।
लोहसंदंशके मूषां धृत्वा कृत्वा ह्यधोमुखीम्।।
भूम्यामाढालयेत् सत्त्वं यथा नालं न भज्यते।
तदा सीसोपमं[9] सत्त्वं पतत्येव न संशयः।।

1. The text reads कानो वल्लवदुर्दुरः, which appears to be incorrect.
2. The text reads सत्त्व, which seems to be incorrect.
3. The second hemistich is wanting in the text. We have adopted the reading of रसरत्नसमुच्चय.
4. The text reads सिद्धी राजिता, which is not accurate.
5. The text reads निमज्जितं, which is incorrect.
6. The text has only मूत्रे, which mars the metre.
7. द्राविते छालिते is the reading of the text which is evidently an error.
8. The text reads वर्जितं, which seems to be incorrect.
9. रसरत्नसमुच्चय says वङ्गाभं See Chap. II. Verse 161.

अनेनैव प्रकारेण वारत्रयकृते सति।
विनिःसरेत् सर्वसत्त्वं सहिंगु (?) गुरुणोदितम्।।

Rasaka (calamine) : the author of Rasaratnāsamuccaya has evidently borrowed his description of this mineral from the present work. *Cf.* Vol. i, p. 87.

Extraction of zinc from calamine :

[The process agrees almost word for word with what is given in Rasaratnāsamuccaya The essence possessing the lustre of lead runs of the crucible.][1]

।। अथ सौराष्ट्रीगुणाः ।।

सौराष्ट्रदेशे सञ्जाता खनिजा[2] तुवरी मता।
या लेपिता श्वेतवस्त्रे रङ्गबन्धकरी हि सा।।

फुल्लिका खटिका[3] तद्वत् द्विप्रकारा प्रशस्यते।
किञ्चित् पीता च सुस्निग्धा गरदोषविनाशिनी।।

श्वेतवर्णा परा साम्ला फुल्लिका लोहमारणी।।
कषाया मधुरा काङ्क्षी कटुका विषनाशिनी।
व्रणघ्नी कफहा चैव नेत्रव्यापत्रिदोषहा।।
कण्ठरोगहरा सा तु पारदे वीजजारणी।
धान्याम्ले तुवरी क्षिप्ता शुध्यति त्रिदिनेन वै।।
क्षारैराम्लैश्च मृदिता ध्माता सत्त्वं विमुञ्चति।
तत् सत्त्वं धातुवादार्थे चौषधे नोपपद्यते।।

Saurāṣṭrī of tuvarī (alum-earth) almost indentical with Rasaratnāsamuccaya *Cf.* Vol. i, p. 91.

* * * The essence extracted from the alums (*i.e.* oil of vitriol) is to be used for operations with metals and not as medicine.

।। अथ पुटानि लिख्यन्ते ।।

भूम्यां वै खन्यते गर्तं द्विहस्तं चतुरस्रकम्।
छगणेन सहस्रेण पूरयेत्तदनन्तरम्।।

1. In Rasaratnāsamuccaya we have "the essence possesing the lusture of tin."
2. The text reads खनिता, which is incorrect.
3. रसरत्नसमुच्चय says फटकी फुल्लिका.

औषधं धारयेन्मध्ये तदाच्छाद्य वनोपलैः।
सहस्रार्द्धेन वै सम्यग्वह्निं प्रज्वालयेत्ततः।।
महापुटमिदं प्रोक्तं ग्रन्थकारेण निर्मितम्।।
।। इति महापुटम्।।

एकहस्तप्रमाणं हि चतुरस्रं च गर्त्तकम्।
वनोपलसहस्रेण गर्त्तमध्यं च पूरितम्।।
मूषिकां चौषधेनाथ पूरितां तां तु मुद्रयेत्।
गर्त्तमध्ये निधायाथ गरिण्डानि[1] च निक्षिपेत्।।
उद्ध्वाग्निं ज्वालयेत् सम्यक् सोऽयं गजपुटो भवेत्।।
।। इति गजपुटः।।

उपलं पिष्टकं छाणमुत्पलञ्च गरिण्डकम्।
छगणोपलसारी च तचरा[2] छगणाभिधाः।।
अरत्निमात्रे कुंडे च वाराहीपुटमुच्यते[3] ।।
।। इति वाराहपुटः।।

वितस्तिद्वयमानेन गर्तं चेच्चतुरस्रकम्[4]।
कुक्कुटाख्यं पुटं विद्यादौषधानाञ्च साधनम्।।
।। इति कुक्कुटपुटः।।

छगणैरष्टभिः सम्यक् कपोतपुटमुच्यते।।
।। इति कपोतपुटः।।

तुषैर्वा गोवरैर्वापि रसभम्मप्रसाधनम्।
माणिकाद्वयमानेन गोवरं पुटमुच्यते।।
।। इति गोवरपुटः।।

मृद्भाण्डे तु प्रपूयैव मध्ये द्रव्यं तु विन्यसेत्।
अधस्ताज्ज्वालयेदग्निं मृद्भाण्डपुटमुच्यते।।
।। इति (मृद्) भाण्डपुटः।।

---

1. रसरत्नसमुच्चय reads गिरिण्डानि.
2. रसरत्नसमुच्चय reads वराटी.
3. The text reads अभिमात्रेण रूंधे च वाराहपुटमध्यत which seems to be incorrect.
4. The text reads गर्तं कुर्याच्चतुरस्रकं, which mars the metre.

गर्त्ते तु बालुकापूर्णे मध्ये द्रव्यं तु विन्यसेत्।
उपरिष्टादधस्ताच्च वह्निं कुर्यात् प्रयत्नतः।
तद्बालुकापुटं सम्यगुच्यते शास्त्रकोविदैः।।
।। इति बालुकापुटः ।।

मूषिकां भूमिमध्ये तु स्थापितं द्व्यङ्गुलादधः।
उपरिष्टात् पुटं दद्यात्[1] तत् पुटं भूधराभिधम्।।
।। इति भूधरापुटः ।।

गोवरैर्वा तुषैर्वापि मात्राकर्षमितैः पुटम्।
यत्र तल्लावकाख्यं स्यान्मृदुद्रव्यस्य साधने।।
।। इति लावकापुटः ।।

।। इति श्रीपद्मनाभसूनुश्रीयशोधरविरचिते रसप्रकाशसुधाकरे दशमोऽध्यायः ।।

[Descriptions of various kinds of pits for calcining and roasting purposes are given ; as also their dimensions with the number of cowdung-cakes to be used etc. *Cf.* the typical roasting pit in Vol. i, p. 127.]

x x x x x x

Here ends Chapter X of Rasaprakāsasudhākara by Yaśodhara, son of Padmanābha.

अथातः संप्रवक्ष्यामि धातूनां कौतुकं परम्।
स्वानुभूतं मया किञ्चित् श्रुतं यच्छस्त्रतः खलु।।
तदहं कथयिष्यामि यत् कृत्वा ना सुखी भवेत्।।
रसकं दरदं ताप्यं गगनं कुनटीसमम्।
रक्तस्नुहीपयोभिश्च मर्द्दयेद्दिनसप्तकम्।।
जलयन्त्रेण वै पाच्यं चतुर्विंशतियामकम्।
तेन वेध्यं द्रुतं ताम्रं तार वा नागमेव वा।।
सह (शत) वेधो तु तत्कल्को जायते नात्र संशयः।।
एकभागस्तथा सूतो वज्रवल्ल्याथ मर्द्दितः।
खल्ले त्रिर्निष्पाच्य रसे पञ्चभागसमन्विते।
वेत्रयष्ट्या च रागिण्या पीतकल्कं प्रजायते।।

1. The text reads दत्वा, which is incorrect.

षोड़शांशेन दातव्यं द्रुते ताम्रे सुशोभने।
जायते प्रवरं हेम शुद्धं वर्णचतुर्द्दशम्।।
।। इति हेमक्रिया।।

स्वर्णमाक्षिकं संस्वेद्य काञ्जिके दिवसत्रयम्।
चर्म्मरंग्या (गा) रसेनैव मर्द्दयेद्दिनसप्तकम्।
जलेन धौतं तावच्च यावद्धेमनिभं भवेत्।।

दरदं रसदेशीय[1] गोमूत्रेनैव स्वेदयेत्।
दोलायन्त्रेऽह्निः[2] चत्वारि पश्चाच्छुद्धतमो भवेत्।।
मनःशिला पद्मनिभा रक्त चैव सुशोभना।
स्वेदिता मुनिपुष्पस्य रसेनैव तु दोलया।
याममर्द्धमिता शुद्धा सर्वकार्येषु योजयेत्।।
नवसारस्तथा सूतः शोधितोऽग्निमति स्थले।
समभागानि सर्वाणि मर्द्दयेन्निम्बुकारसैः।।
मातुलुङ्गरसेनैव कुमारीसुरसेन वा।
सूर्यातपे विमर्द्दोऽसौ पाचितो जलयन्त्रके।
दिनानि त्रीणि तीव्राग्नौ कुर्यात्तदवतारयेत्।।

शतांशं वेधयेत्तारं शुद्धहेम प्रजायते।
जलभेदो यदा न स्यान्नात्र कार्या विचारणा।।

शिलया मारितं नागं कुमार्याः स्वरसेन तु।
पुटद्वादशयोगेन नागभस्म प्रजायते।।
शतसंख्यानि वै कुर्यात् पुटान्येवं शरावके।
कुमार्या स्वरसेनैव भावयेद्दिनसप्तकम्।।

पूर्ववत् पुटनं कार्यं शतसंख्यामितं तथा।
सूतगन्धशिलाददुसमं चेन्नागभस्मकम्।
त्रिशंद्वनोपलैर्दद्यात् पुटं वाराहसंज्ञकम्।।
अनेन विधिना सम्यक् शतसंख्यानि दापयेत्।
पुटान्येवं कृते त्रीणि शतानि द्वादशाधिकम्।

1. The text रसदेशीयो, which is incorrect.
2. अह्नि [अहानि] is grammatically incorrect.

पश्चाद्दृढ़े काचमये कूपे द्वात्रिंशयामकम्।
बालुकाग्निं प्रदद्याच्च स्वाङ्गशीतं समुद्धरेत्।।
तलभस्म ग्रहीतव्यं वेधयेत्तारशुद्धके।
शुद्धहेम भवेत्तेन नात्र कार्य्या विचारणा।।
दृष्टप्रत्यययोगोऽयं कथितो नात्र संशयः।।

Hemakriyā or process for the fabrication of gold ; "I shall now speak of some curiosities of metals, partly from my own experience and partly from the classics on the subject. Calamine, cinnabar copper pyrites and realgar are to be rubbed with the milky juice of *Euphorbia neviifolia* for seven days. together and then to be digested another three days. Melted copper or silver or lead, being alloyed with the aforesaid mixture, acquires the power of converting 100 times its weight of base metal into gold."[1] Cf. *ante. p.2.*

Chemistry in Rasacinātamaṇi of Madanantadeva

मदनान्तदेवसूरिविरचितात् रसचिन्तामणिः

उद्धृता श्लोकाः

सैन्धवं तोरिकां सूतं कासीसं लकुचद्रवैः।
विघृष्य खल्वभाण्डस्थं सर्वश्लक्ष्णं दिनत्रयम्।। 76।।
हण्डिकायां तदारोप्य काष्ठवह्निर्विधीयते।
दिनत्रयेऽप्यतिक्रान्ते भस्म श्वेततरं भवेत्।। 77।।

* * * * *

।। इति त्रयोदश भस्मसूतः।।

खर्परो द्वादशांशः स्यात् शिखिग्रीवं च तुत्थकं।
षड्गुणं माक्षिकं नेत्रभागं ग्राह्यं मनःशिला[2]।
भागत्रयमिदं ग्राह्यं शोभनागतककरैः[3] ।। 38।।

---

1. शतवेधी and सहस्रवेधी are technical terms; for the meaning of वेध *vide* Vol. i, p. 120.
2. चापि नेत्रभागा मनःशिला is the correct reading as stated in the margin of the text.
3. त्रयो भागा अपि ग्राह्याः शोभना गतकर्कराः is the correct reading as stated in the margin of the text.

Śvetabhasma : *Cf. ante* p. 16 and also Vol. i, pp. 250-51.

गंधकत्रयमेवास्ते काकमाचीरसैर्भृशं।
दिनसप्तात्र कर्त्तव्या भावना मर्द्दनं क्रमात्।। 39।।
अधोमुखेन यन्त्रेण तेन निःसारयेद्बुधः।
सत्त्वं तद्रसकस्याथ दृढ़ कार्यकरं परं ।। 40।।
अवश्यं तद्दृढ़ सत्त्वं निर्धूमं कर्मसाधनं।।

।। इति रसकसत्त्वपातनं।।

त्रिवारं गन्धकहतं ताम्रहेमद्वयत्रयं।
सत्त्वं तालस्य मासैकं योजयेत् सर्वकामदं।। 43।।
अभ्रवर्णं भवेद्धेमं ताम्रगंधकमारितं।
एतयोर्दीयते रम्यं जायते कांचनं शुभं।। 44।।

।। इति श्रीमदनांतदेवविरचिते रसचिंतामणौ हेमीकरणप्रयोगः।।

पारदं सीसकं गंधं कुनटी तच्चतुष्टयं।
वीजपूरांभसा पिष्ट्वा बाढ़ं दिनचतुष्टयं।। 108।।
अथ सूक्ष्माणि पत्राणि तेन तारस्य लेपयेत्।
वीजपूररसेनैतत् त्र्यतिमात्राग्नितापितं।। 109।।
एकाधिका भवंत्यत्र भावनाश्चास्य विंशतिः।
विशोष्यावर्त्तितं तारं भवेत्तारस्य कांचनं।। 110।।

।। इति चिंतामणौ हेमीकरणप्रयोगः।।

हेमभागद्वयं तारं तथा ताम्रं चतुष्टयं।
एकतः क्रियते पत्रमतिसूक्ष्मं निरामयं।। 111।।
जंबीरनीरसंपिष्टं खर्परस्याष्टटंकणं।
तेन तान्यथ पत्राणि लेपनीयानि वै बहु।। 112।।
आवर्त्तंते पुनर्दत्त्वा मूषके गलितानि च।
तदा तानि भवंत्यत्र हेमरूपाणि नान्यथा।। 113।।

।। इति हेमीकरणप्रयोगः।।

एकभार्गो भवेदारः तारं भागद्वयं भवेत्।
वेदभागं भवेत्तीक्ष्णं सत्त्वं सत्त्वाच्च वै दश।
वंगभागं भवेत्पत्रं सर्वमेकत्र कारयेत्।। 109।।

टंकणेन पुनः ध्मातं पत्रं कृत्वास्य वक्त्रगं।
वेलात्रयमिदं ध्मातं तक्रमध्ये निवेशयेत्।। 110।।
एवं तारं भवेन्नूनं परीक्षाक्षममापणे।।
।। इति तारोत्पादकं।।

Two processes for the fabrication of gold:—much the same as already quoted from Yaśodhara.

Process for the fabrication of silver :—Take one part of bell-metal, two parts of silver, four parts of steel (?) * * * these to be fused with tin and borax ;—by this process will be made silver which will pass current for commercial purposes.

तुत्थकं पिष्यते खल्वे त्रिफलाहिंगुटंकणैः।
मध्वाज्यमिश्रितं ध्मातं सत्त्वं पातयते ध्रुवं।। 69।।
।। इति तुत्थकसत्त्वपातनं।।

खर्परः स्वेद्यते पूर्वं कौलत्थेन जलेन च।
वटारोहजलेनापि पूर्णचूर्णेन शोभनः।। 75।।
गुड़टंकणसंमिश्रस्त्रिफलाक्काथमर्दितः।
मृन्मये कूपके कृत्वा धाम्यमानो भृशं च सः।। 76।।
श्वेतधूमोद्‌गमे जाते तत उत्थाप्य कूपकं।
सावधानं करेणैव भूमौ तं चाध आनयेत्।। 77।।
पुनश्च धाम्यते कूपः तथाजातं च तं पुनः।
कूपकं च पुनर्भूर्मो तथा तं नामयेत् पुनः।। 78।।
सत्त्वं खर्परकस्यैतत् नागरूपं पतत्यधः।।79।।
।। इति खर्परसत्त्वपातनं।।

Extraction of zinc from calamine :—The process is much the same as given above (p. 17), with the difference that, whereas Yaśodhara hits off the completion of reduction when the flame issuing from the pot ceases to be blue, the present author directs that the pot is to be taken out of the furnace as soon as white fumes begin to appear. The crucible to then to be held in an inclined position so that the lead-like essence of calamine drops on the ground.

Chemistry in Rasakalpa (Rudrayāmala Tantra)

रुद्रयामलान्तर्गतात् रसकल्पादुद्धृताः श्लोकाः ।

शिवं नत्वा रसाधीशं चण्डिकाचरणं तथा ।
क्रियते[1] रसकल्पोऽयं रसज्ञानविशारदः[2] ।। 1 ।।

Bowing to Śiva, the lord of mercury, and also the feet of Caṇḍikā, I write this Rasakalpa, the depository of mercurial lore.

अथ परीक्षा

अतेजा अगुरुः शुभ्रो लोहहा(ऽ) चञ्चलो रसः[3] ।
यदा[4] नावर्त्तयेद्वह्नौ नोर्द्ध्वं गच्छेत्तदा मृतः ।। 40 ।।
दन्ते शृङ्गे मणौ वेणौ रक्षयेत् साधितं रसम् ।। 42 ।।

इति श्रीरुद्रयामले रससंकेतकं नाम प्रथमोल्लासः ।

CHAPTER I

Tests of killed mercury:—Killed mercury is that which is devoid of metallic) lustre, not so ponderous, white,[5] destroyer of metals, divested to fludity and non-volatile when stirred over a fire. 40 *Cf.* tests in Rasārṇava in Vol. i, p. 74 ; as also in Rasaratnākara, *ante*, p. 9.

Purified mercury should be preserved in the hollow of a tooth gem or bamboo. 42

1. The text reads करोमि, which is grammatically incorrect.
2. The reading in the text is विशारदम्, which is incorrect.
3. The text reads रस, which is incorrect.
4. The text reads यदामावर्त्तयेत्, which is not correct.
5. The author previously refers to the red, black white and even yellow ash (calx) of mercury ; it is not clear why we should here restrict himself to the white variety only. Chloride or sulphide of mercury, when treated with a metal and heated, *kills* (*i.e.* parts with its chlorine or sulphur to) the latter and in that sense it is a destroyer of metals लोहहा.

हेमरूप्यार्कवङ्गाहि[1] लौहा लोहाः[2] षड़ीरिताः।
अकृत्रिमा इमे येषां वर्त्ताद्याः स्युस्तु[3] कृत्रिमाः।। 1।।

सन्तो बद्धाञ्जलिपुटो युष्मान् विज्ञापयाम्यहम्।
कृपासुधारसयुता दृष्टिर्ग्रन्थे प्रदीयताम्।।
महारसाः स्युस्तत्रादावष्टौ पारदहिङ्गुले।
वैष्णवं शस्यकं शैलं चपलं रसकोऽमलम्।।

–(महारसाः)

अभ्रकं तुत्थकं कान्तं राजावर्त्तमथाञ्जनम्।
वज्रं वैक्रान्तकं चैव टङ्कणं च रसा बहु।।

–(रसाः)

गन्धकं तालकशिले क्षितिखेचरगैरिकाः।
इत्यादयश्चोपरसाः सर्वाचार्यैरुदीरिताः।।

–(उपरसाः)

सितासितारुणं पीतं गन्धकं तच्चतुर्विधम्।।
तालकं द्विप्रकारं स्यात् गोदन्तः पाटलच्छविः[4]।।
रक्ता पीता शिला द्वेधा पूर्वा श्रेष्ठोत्तराधमा[5]।।
बहुप्रकारा सौराष्ट्री कासीसं त्रिविधं मतम्[6]।।
कासीसं पुष्पकासीसं हीरकासीसमित्यथ।।
गैरिको द्विविधः प्रोक्तः सौवर्णो लोहितस्तथा।।
कङ्कुष्ठादीन्[7] रसान् केचिदाचार्या वर्णयन्ति वै।
अस्माभिरिह तन्त्रोक्तं मुनिमार्गानुसारिभिः।।

---

1. The text reads वङ्गाहिः, which is grammatically incorrect.
2. लोहिर्लोहा is the reading of the text, which seems to be incorrect.
3. The text reads रेवार्त्ताद्यासुकृतिमा, which is palpably an error.
4. The text read पटलाछवि, which seems to be incorrect.
5. पूर्वाश्रेष्ठोत्तमाधमा is the reading of the text, which is incorrect.
6. The text reads कासीसस्त्रिविधो मतः, which is incorrect, as it is mentioned neuter in the next line.
7. The text reads कंकुष्टादीन्, which is an error of the scribe.

CHAPTER II

Gold, silver copper, tin, lead and iron—these are the 6 metals : vartaloha, etc., (Vol. i, p. 114) are simply alloys. 1

Honest readers ! I beseech your with folded arms to favour my book with your merciful glance.

The mahārasas are 8 in number; *e.g.* mercury, cinnabar sayyaka, rasaka, etc. ... . the rasas are mica, green vitriol and rājāvarta, etc. ... ... sulphur, orpiment and realgar, etc. are the uparasas as regarded by learned teachers. *Cf.* Vol. i, pp. 79 and 89.

There are four kinds of sulphur :—white, black, yellow and red. *Cf.* Vol. i, p. 90.

The saurāṣṭris (alums) are of various kinds ; the kāsīsas (vitriols) are of three kinds, namely:—kāsīsa proper, puṣpa kāsīsa and hīrakāsīsa.[1] *Cf.* Vol. I, po. 91.

Gairika is stated to be of two varieties,—the one of golden hue, the other red: Kaṅkuṣṭha and other rasas are also described by certain teachers ; but these, however, are not referred to here, as we are treading in the footsteps of sages of old.[2]

इत्येष प्रोदितो मार्गो रसशोधनकर्मणि।
स्वच्छन्दभैरवाद्युक्तो गोविन्दादिसमादृतः[3] ।।
लवणं चूलिकोद्भूतं गन्धकेन समन्वितम्।
सर्वाङ्गदश्च[4] चित्रार्द्रं मूलभस्मप्रगालितम्।।

---

1. Hīrākasa in the name by which green vitriol is known through out India; Royle supposes that this word is derived from the Pesian *hura* green, and Sans. kāsīsa. (Antiq. Hindu Med). There is no necessity for looking upon hīrākasa as a hybrid term, seeing that it occurs in the present work in which Mahommedan influence is nowhere traceable.
2. Evidently Caraka and Suśruta, who mention only a few commonly occuring minerals. *Vide* Vol, i, pp. 29 and 44.
3. The text reads समादितः, which is incorrect.
4. The text reads सर्वांगदश्च, which seems to be incorrect.

गोमूत्रेण शतं भाव्यं तद्रसे जारयेत् शनौ[1]।
तस्य संपर्कतः सूतो राक्षसो भवति ध्रुवम्।।
एतदेव विड़ं दद्यात् सर्वदा हेमजारणे[2]।
मुखं संजायते तेन जीर्यते च[3] विनिश्चितम्।।

।।इति श्रीरुद्रयामले रसकल्पे उमामहेश्वरसंवादे रसशोधनप्रकारः।।

इति संधानयोगोऽयं जारणोऽतिगुणावहः।
प्रकाशितं संप्रदायक्रमप्राप्तः शिवोदितः।।

।।इति श्रीरुद्रयामले रसकल्पे उमामहेश्वरसंवादे रसाधिकरणो[4] नाम द्वितीयोल्लासः।।

महारसानां सर्वेषां रसानां शुद्धिरुच्यते।
तथैवोपरसानाञ्च शास्त्रदृष्टेन वर्त्मना।। 1 ।।
कदलीपत्रजैर्नीरैः माक्षीकं भावयेत् शतम्[5]।
गन्धर्वतैलमध्वाज्य पक्कमेकं दिनत्रयम्।।
तत्ताप्यं वज्रमूषायां[6] पक्कायां निक्षिपेत्ततः।
लोहसन्धानकरणं तत्समं तत्र निष्पचेत्।।
दृढ़ा प्रकल्पयेन्मूषा[7] कोष्ठिके[8] विनिवेशयेत्।
अङ्गारैः खदिरोद्भूतैः धमेद्भस्त्राद्वयेन वै।।
वक्रनालयुजा सत्त्वं ताप्यस्य पतति ध्रुवम्।।
शुक्लदीप्तिः सशब्दश्च[9] यदा वैश्वानरो भवेत्।
तदा सत्त्वं तु पतितं जानीयान्नान्यथा क्वचित्।।

---

1. शनिः, as given in the text, is grammatically incorrect.
2. The text reads हेमजारणौ. which is not correct.
3. The text reads न, which appears to be an error.
4. The text reads रसाधिकारेणा, which is incorrect.
5. The text reads शलं, which is senseless.
6. ततत्ताप्यं वज्रमूषा is the reading of the text, which appears to be incorrect.
7. The text प्रम्लकपेन्मूषां is evidently incorrect.
8. कोष्टिकानि निवेशयेत् is the reading of the text, which has no meaning at all.
9. The text reads सशब्दस्य, which is grammatically incorrect.

रसकं स्वेदयेदादौ[1] पटुचूर्णकृतं बुधः।
चतुर्गुणेन वस्त्रेण दृढ़ बध्वा च डोलिकाम्।।
कृत्वा भाण्डे च सजले[2] स्वेदयेन्मासपञ्चकम्।
उद्धृत्य पश्चाद्रसकं खल्लमध्ये[3] विचूर्णयेत्।।
पादीशान् मालतीजातं सगुड़ं जीर्णगुञ्जकम्।
गृहधूमं रेवकीं च निशामं कुल्लजीरकान्[4]।।
तत् सर्वं चूर्णितं कृत्वा गोपंचकविभावितम्।
कृत्वा तद्वटिकाः पश्चात् छायायां शोषयेत्ततः।।
कोष्ठगेनाग्निना[5] पश्चाद्धमेद्भस्त्राद्वयानिलैः।
सत्त्वं पतत्यसंदेहं स्थिररङ्गं दृढं बहु।।
एकमासं द्विमासं वा रसकं स्वेदितं यदि।
न ध्मातव्यं तच्च कोष्ठे धमेत्तत् नालमूषया।।
इति सम्पादितो मार्गो द्रुतीनां पातने स्फुटः।
साक्षादनुभवैर्दृष्टो न श्रुतो गुरुदर्शितः।।
लोकानामुपकाराय एतत् सर्वं निवेदितम्।
सर्वेषां चैव लोहानां द्रावणं परिकीर्त्तितम्।।
।। इति श्रीरुद्रयामले रसकल्पे उमामहेश्वरसंवादे
सर्वलोहद्रुतिपातनाधिकारः समाप्तः[6] ।।

For the purification of mercury this process is recommended by Svacchandabhairava and Govinda.

Cūlikalavana (Sal-ammoniac), sulphur and the ash of ginger, etc., digested one hudred times with cow's urine ... ... make a viḍa for killing gold. *Cf.* Vol. i, p. 72.

This process for killing [Metals] was revealed by Śiva and has been transmitted by successive schools of adepts.

---

1. रसकांश्चदयेदादौ is the reading of the text, which seems to be incorrect.
2. कृत्वा भाण्डेन सजलै is the reading of the text, which is incorrect.
3. The text reads खल्वमध्ये.
4. The text has जीरकां, which is grammatically incorrect.
5. The text reads कोष्टकेनाग्निना, which is not correct.
6. समाप्तं शुभम् is the reading of the text, which is not correct.

Extraction of the essence of copper pyrites. *Cf.* Vol. i, pp. 70 and 84.

Extraction of zinc from calamine:—

The process is practically the same as given in Rasaratnāsamuccaya *Cf.* Vol. i, pp. 87-88.

I have performed the [aforesaid] experiments with my own hands and have seen them with my own eyes. They are note recorded from mere hearsay or from the dictation of a teacher. These are being promulgated for the benefit of mankind.

Chemistry in Rasarājalakṣmī of Viṣṇudeva

विष्णुदेवविरचितायाः रसराजलक्ष्म्याः प्रारम्भः–

यस्येच्छाङ्कुरवल्लरी मुकुलिता नानाकलाकौतुकैः
चातुर्य्यैकविलोलपल्लववती कीर्त्तिप्रसूनप्रसूः।
संनिर्वापयति ज्वरत्रयमहो च्छायाकलापैर्नृणां
सोऽयं पातु विचारमूर्त्तिरखिलं कालापलीढ़ं जगत्।। 1।।
जयति दुरितसर्पस्तंडमंत्रोद्यदर्पः
प्रबलगदविनाशः सेविविध्वस्तनाशः।
शिव इव विज (वीज ?) मंत्रः संहितायोगमंत्रः
त्रिपुरहरतनूजस्तेजसः कायपुंजः।। 2।।

।। इति श्रीपण्डितमहादेवतनयश्रीविष्णुदेवविरचितायां
रसराजलक्ष्म्यामुल्लासः प्रथमः।।

दृष्ट्वेमं रससागरं शिवकृतं श्रीकाकचण्डेश्वरी-
तन्त्रं सूतमहोदधिं रससुधाम्भोधिं भवानीमतम्।
व्याड़िं सुश्रुतसूत्रमीशहृदयं स्वच्छंदशक्त्यागमं
श्रीदामोदरवासुदेवभगवद्गोविन्दनागार्जुनान्।

।। प्रथमः उल्लासः।।

स्वच्छंदशक्तयागमसारभूतः
समुद्धृतो विष्णुभिषग्वरेण।। 110।।

।। द्वितीयः उल्लासः।।

आलोक्य सुश्रुतं वृन्दहारीतचरकादिकान्।
आत्रेयं वाग्भटं सिद्धसारं दामोदरं गुरुम्।।
।। तृतीयः उल्लासः ।।[1]

Colophon at the end of the Rasarājalakṣmī—

राजन् (सश) शार्वरिवत्सराद्यदिवसे वारे हिमांशोरिदं
चंचद्भूतलपत्तने विजयिनि श्रीबुक्कपृथ्वीपतेः।
शास्त्रं वैद्यकसाररूपमकरोत् श्रीविष्णुदेवः कविः
वाग्देवीचरणारविंदमकरंदामोदसौरस्यवाक्।।

Give an account of common rasas and uparasas. *Cf.* Vol. i, p. 79.

1. Foll, 1-43. Linn. 10. Rasarājalakṣmīs, de medicamentis metallorum et foossilium ope coficiendis liber. a Rāmeśvara (bhaṭṭa), Viṣṇus filio, compositus. Incipit : आनंदैकरसे विकल्पलनानिर्मुक्तबोधात्मकिं प्रागंत्यावधिसंगतिव्यपगते etc. ।। 1।। स्वशिमुखीमयभंयन (l. स्वशेमुषी°) sphurad Rasāmbhodhigabhiragarbhāt । Śrī-Viṣṇudevī (o devī) labhate prayantnā śrī Viṣṇuvatsād Rasarājalakṣmīṁ ।।2।। Dṛṣṭvemaṁ Rasasāgaraṁ, Śivakṛitaṁ śrī-Karkachaṇḍeśvarītantraṁ, Sūtamahodadhiṁ, Rasasudhāmbhodhiṁ, Bhavānimataṁ Vyādi[m], Suśrutasūtraṁ, Īśahṛdayaṁ, Svaccandaṣaktyāgamaṁ, śrī-Dāmodara-Vāsudeva-bhagavadgovinda-Nagārjunāna ।। 3।। आनन्दमानंदकरं प्रसन्नं ज्ञानस्वरूपं निजभावयुक्तम्। योगींद्रमीड्यं भवरोगवैद्यं श्रीमद्गुरुं नित्यमहं नमामि।।4।।
महारसाः स्युस्तत्रादावष्टौ पारदहिंगुले।
वैष्णवं सम्यकं शैलं-चपलं रसकोमलम्।।5।।
अभ्रकं तुछ्कं (tutthakaṁ) कांतं राजावर्त्तमथांजनं।
वट्यकैकांतकं (o vaikrāntakaṁ) चेति टंकणं च रसान्विदुः।।6।।
तालकं गंधकशिले क्षितिखेचरगैरिकाः।
इत्यादयश्चोपरसाः पूर्वाचार्यैरुदीरिताः।।7।।
Aufrecht's "Catalogus," p, 321, No. 761.

Chemistry in Rasanakṣatramalikā of Mathanasiṁha, physician to the king of Malwa

मथनसिंहविरचितायाः रसनक्षत्रमालिकायाः उद्धृता श्लोकाः।

चतुश्चतुः शंखकपर्द्दिकानां सतक्रजम्बीरविमर्द्दितानाम्।
आफेनमाक्षीकविषद्वयानां पलं पलं दन्तिफलान्वितानाम्।। 25।।

यतभस्मगन्धसूता द्वितयं वङ्गस्य शुद्धतीक्ष्णस्य।
दग्ध्वा मृगस्य शृङ्गं त्रिगुणं दत्त्वा रसार्द्धविषम्।।123।।

चित्रकशीतशिवावृषदशमूलदिवनदविल्वगिरैः।
कलहत्तारविकृतकैः प्रत्येकं भावनत्रितयम्।।124।।

दत्त्वा सुसिद्धमात्रो मारिचधूपेन धूपितः सम्यक्।
स्वच्छन्दभैरवाख्यो रसः समस्तामयध्वंसी।।125।।

त्रिकटुकरसेन युक्तो विशेषतः सन्निपातहरः।
ग्रहणिगदामयशूलव्रणविद्रधिवातगुल्मजयी।। 126।।

।। इति स्वच्छन्दभैरवो रसः।।[1]

इयं मालविभूपालभिषजा भिषजां मता।
कृता मथनसिंहेन रसनक्षत्रमालिका।।142।।

।। इति रसनक्षत्रमालिका समाप्ता।। स्वस्ति
संवत् 1557 आश्विन कृष्ण 5 सोमे।

These are several prescriptions given in which mineral preparation play a conspicuous part. We have quoted only two ; in the first, among other ingredients opium occurs ; in the second, the Svacchandabhairavarasa, probably named after the celebrated alchemist, we have the *calces* of tin, iron and mercury along with other drugs.

---

1. The स्वच्छन्दभैरव रस mentioned in the रसनक्षत्रमालिका is quite distinct from that, which is stated in the रसेन्द्रचिन्तामणि. Neither has it any analogy with what is described in the रसरत्नसमुच्चय.

Chemistry in Rasaratnākara of Nityanātha

नित्यनाथविरचितात् रसरत्नाकरात् उद्धृताः श्लोकाः

यदुक्तं शम्भुना पूर्वं रसखण्डे रसार्णवे।
रसस्य वन्दनार्थे च दीपिकारसमङ्गले।।
व्याधितानां हितार्थाय प्रोक्तं नागार्जुनेन यत्।
उक्तं चर्पटिसिद्धेन[1] स्वर्गवैद्यकपालिके।
अनेकरसशास्त्रेषु संहितास्वागमेषु च।
यदुक्तं वाभटे तन्त्रे सुश्रुते वैद्यसागरे।।
अन्यैश्च बहुभिः सिद्धैः यदुक्तञ्च विलोक्य तत्।
तत्र यद्यदसाध्यं स्याद्यद्यद्दुर्ल्लभमौषधम्।
तत्तत् सर्वः परित्यज्य सारभूतं समुद्धृतम्।।
क्वचिच्छस्त्रे क्रिया नास्ति क्रमश्चापि न च क्वचित्।
मात्रायुक्तिः क्वचिन्नास्ति सम्प्रदायो न च क्वचित्।
तेन सिर्द्धिन तत्रास्ति रसे वाथ रसायने।।
वैद्ये वादे प्रयोगे च तस्माद्यत्नो मया कृतः।।
यद्यद्गुरुमुखाज्ज्ञातं स्वानुभूतञ्च यन्मया।
तत्तल्लोकहितार्थाय प्रकटीक्रियतेऽधुना।।

।। प्रथमोपदेशे।।

।। इति श्रीपार्वतीपुत्रनित्यनाथसिद्धविरचिते रसरत्नाकरे
रसखण्डे रसपीठिकानाम प्रथमोपदेशः।।

परीक्षा मारिते सूते कर्त्तव्या च यथोदिता।
अधस्तुषाग्निना तप्तो ह्यक्षीणस्तिष्ठते यदा।
तदा भस्म विजानीयाच्चुल्यां यामं निरीक्षयेत्।।

।। तृतीयोपदेशे।।

दन्त शृङ्गेऽथवा वंशे रक्षयेत् साधितं रसम्।।

।। चतुर्थोपदेशे।।

---

1. धूर्जटिसिद्धेन is the reading in the Sanskrit College MS.

अथातः शुद्धसूतस्य मूर्च्छनाविधिरुच्यते।
मेघनादावचाहिङ्गुशूरणैर्मर्द्दयेद्रसम् ।।
नष्टपिष्टन्तु तद्गोलं हिङ्गुना वेष्टयेद्बहिः।
पचेल्लवणयन्त्रस्थं दिनैकं चण्डवह्निना।।
ऊद्ध्र्वलग्नं समादाय दृढ़ं वस्त्रेण बन्धयेत्।
ऊद्ध्र्वाधो गन्धकं तुल्यं दत्त्वा सोमानले पचेत्।।
जीर्णे गन्धे पुनर्देयं षड्भिवरैः समं समम्।
षड़गुणे गन्धके जीर्णे मूर्च्छितो रोगहा भवेत्।।

।। चतुर्थोपदेशे ।।

Test for killed mercury :—When the substance being heated over a fire of paddy husks, does not further decrease in weight, is should be considered as reduced to a calx.

Purified mercury should be stored in the hollow of a tooth or born or bamboo.

Directions for making factitious cinnabar (Makaradhvaja). *Cf.* Vol. i, p. 132 (footnote).

Chemistry in Dhāturatnamālā

धातुरत्नामलायाः

प्रारम्भः–

प्रणम्य विततीं शक्तिं त्रिसृष्ट्युत्पत्तिकारिणीं।
धातूनां रत्नमालायामभिधायं करोम्यहं।। 1 ।।
ब्रह्मविष्णुहराद्यान् ये मर्त्ता ध्यायंति नित्यशः।
ज्ञानदानप्रदानाय सा मे विश्वेश्वरी मता।। 2 ।।
अथ धातूनां रत्नमालां वक्ष्यामि।
रौप्यं हेमं तथा ताम्रं नागं वंगं तथायसं।
खर्पराभ्रकमौक्तं च प्रबालं तालकं शिला।। 3 ।।
सुवर्णमाक्षिकं सूतं हीरकं च ब्रवीम्यहं।
सर्वधातूपधातूनां लक्षणं मारणं गुणं।। 4 ।।

।। अथ रौप्यमारणं ।।

रौप्यं शुद्धं समादाय नागेन गुरु शोधयेत्।
शुद्धे तारे पनुः पश्चात् सूक्ष्मपत्राणि कारयेत्।। 5।।
निंबचिंचिणिद्राक्षाभिः शोधयेच्च पृथक् पृथक्।
क्षालयेदुदकैः सार्द्धं तथा दुग्धेन शोधयेत्।। 6।।

गंधपारदयोरैक्यं किंचिद्वंगं च घर्षयेत्।
द्राक्षाया द्रवसंयुक्तं तारपत्राणि शोधयेत्।। 7।।
चक्रयन्त्रे विनिक्षिप्य लेपयेद्वस्त्रमृत्तिकां।
क्षिपेद्गजपुटे गर्त्ते ज्वालयेद्बहुछाणकान्।। 8।।

समाप्तिः–

ग्रंथो वैद्यकनामायं रससिद्धांतसागरात्।
धातूनां रत्नमाला च ततो वैद्यस्य हेतवे।। 176।।
मरणेभ्यो भयत्रस्ता रोगग्रस्ताश्च ये नराः।
रत्नमाला कृता तेषां वैद्यानां च हिताय वै।। 177।।

।। इति श्रीवैद्यकशास्त्रे अश्विनीकुमारसंहितायां धातुरत्नमालायां समाप्तोऽयं ग्रंथः।।

The following is taken from Aufrecht's Catalogus"—No. 760.

Incipit :—प्रणम्य सारदां शक्तिं सृष्टेरुत्पत्तिकारकां। धातूनां रत्नमालां च विबोधाय करोम्यहं।। 1।। ब्रह्माविष्णुहरान् ध्यायेद्भक्ता ध्यायंति नित्यशः। तेषां वरप्रदानाच्च सा मयैवमुदीर्यते।। 2।। रूप्यं हेम तथा ताम्रं नागं वंगं तथायसं। खर्परं गगनं प्रोक्तं प्रबालं तालकं शिला।। 3।। माक्षिकं गंधकं सूतं हीरकं च ब्रवीम्यहं। सर्वधातूपधातूनां लक्षणं मारणं गणान्।। 4।। रूप्यं शुद्धं समानीय नागमूषा नु शोधयेत्। शुद्धे तारे पुनः पश्चात् सूक्ष्मपत्राणि कारयेत्।। 5।। निंबुचिंचिणीद्राक्षाभिः शोधनीयं पृथक् पृथक्। क्षालयेदुदकैः सार्द्धं तथा दुग्धेन शोधयेत्।। 6। गंधकं पारदं रूप्यं किंचिद्वंगं च घर्षयेत्। द्राक्षारसेन संयुक्तं तारपत्राणि लेपयेत्।। 7।। नक्तं यन्त्रे विनिक्षिप्य लेपयेद्धवस्त्रद्रमृत्तिकां। क्षेप्यं गजपुटे गर्त्ते ज्वालयेच्च ग्रहोपलैः।। 8।।

In fine libelli disticha haec leguntur:

ग्रंथी वैद्य [क] नामायं Rasasiddhāntasāgarāt।

धातूनां रत्नमाला च कृता वैद्यसुहेतवे ।। 1 ।।
मरणेभ्यो भयत्रस्ता रोगग्रस्ताश्च ये नराः ।
रत्नमाला हि धातूनां कृता तेषां हिताय वै ।। 2 ।।

Jātyā Gurjarakhaṇḍas ca, Devadatto hi dharmavit ।

Harer nāmābhidhānasya सुतस्तस्य भिषग्वरः ।।3 ।।

संहितारसकर्माणि यस्य बुद्धिर्गरीयसी ।
तेन शास्त्रविधिज्ञेन कृता रत्नस्य मालिका ।। 4 ।।

।। इति देवदत्तकृतवैद्यकशास्त्रे धातुरत्नमाला ।।

In the beginning the author after salutation says : "I shall speak to silver, gold copper, lead, tin and iron as also of calamine, mica, pearls, caral, orpiment, relgar, the pyrites, mercury and diamond—in fact, the properties of all the metals and minerals and the mode of their incineration."

Killing of silver. *Cf.* Vol. i, p. 107.

Chemistry in Rasapradīpa

रसप्रदीपादुद्धृताः श्लोकाः

A—MS. from Alahabad.

B—MS. from Benares.

अथ शङ्खद्रावरसः ।

स्फटिका[1] नवसारश्च सुश्वेता[2] च सुवर्च्चिका ।
पृथक् दशपलोन्मानं गन्धकः पिचुसंमितः[3] ।।
चूर्णयित्वा क्षिपेद्भाण्डे मृन्मये मृद्विलेपिते[4] ।
तन्मुखं मुद्रयेत् सम्यक् मृद्भाण्डेनापरेण च ।।
सरन्ध्रोदरकेणैव चुल्ल्यां तिर्यक् च धारयेत् ।
अधः प्रज्वालयेद्वह्निं हठाद्यावद्रसः स्रवेत् ।।

1. A reads स्फाटिका. B reads स्फुटिका.
2. B reads सुसेता, which is not correct.
3. B reads पिसंमिचुतः, which is not correct.
4. मूर्ध्नि लेपिते is the reading in B.

शाणैकं सेवयेद्यत्नात्[1] दन्तस्पर्शविवर्जितम्।
गुल्मोदरयकृतप्लीहग्रन्थियक्ष्मादिशूलनुत्।।
बलपुष्टिप्रदो ह्येष भुक्तं च जारयेत् क्षणात्।
विलोक्यतां महालोका रसमाहात्म्यमद्भुतम्।
कपर्दकाश्च लोहानां यस्मिन् क्षिप्ता[2] गलन्ति हि।।

फिरङ्गव्याधिः-

गैरिकं रसकर्पूरम् उपला च पृथक् पृथक्।
टङ्कमात्रं विनिष्पिष्य[3] ताम्बूलीदलजैः रसैः।।
वट्यश्चतुर्दशास्तेषां[4] कर्त्तव्या भिषगुत्तमैः[5]।
सायं प्रातः समश्नीयात्[6] एकैकां दिनसप्तकम्[7]।।
सघृता योलिका[8] देया भोजनार्थं निरन्तरम्।
फिरङ्गव्याधिनाशाय वटिकेयमनुत्तमा।।[9]

---

1. B reads यस्तु.
2. A reads क्षिप्त्वा.
3. A reads विनिःक्षिप्य.
4. A reads चतुर्दशैस्तेषां. B reads चतुदशस्तेषां.
5. B reads भिषगौतमौ, which is incorrect.
6. B reads समस्तीयात्, which is incorrect.
7. दिनसपूकं, an incorrect variant in B.
8. गोधूमरोटिका इत्यर्थः.
9. The following is taken from योगतरङ्गिणी of Tri-mallabhaṭṭa, printed in Mumbai in the Saka year 1810.
   यंत्रे सुसिद्धे डमरूसमाख्ये निधाय सूतस्य पलानि पंच।
   बल्मीकमुत्स्नाखटिकेष्टिकानां सगैरिकाणां तुवरीयुतानाम्।। 60।।
   ससैंधवानां समभागिकानां चूर्णाढ़कं चोपरितो निदध्यात्।
   अम्लेन दध्ना महिषीभवेन पिष्टं रसोनस्य शरावमेकम्।। 61।।
   समक्रमेणात्र निधाय खंडैराच्छादयेत् खर्परजैर्विसंधि।
   चूर्णप्रलिप्तोदरमूर्ध्वभांडं संस्थाप्य संमुद्र्य दृढ़ं सुचुल्याम्।। 62।।
   प्रज्वालयेद्दह्निमधः क्रमेण संस्थाप्य यंत्रोपरि वस्त्रमार्द्रम्।
   वह्निं प्रदद्याद्दिनषट्कमत्र तत् स्वाङ्गशीतं परिगृह्य बुद्ध्या।। 63।।
   तं द्रोणापुष्पीपयसा प्रपिष्टं कूप्यां विदध्यान्नवसादरं च।
   कर्षप्रमाणं प्रहरत्रयं च वह्निं प्रदद्यादथ शीतलांगीम्।। 64।।
   निष्कास्य कूपीं सिकताख्यंत्रादास्फोट्य कंठस्थममुं प्रगृह्यात्।
   कर्पूरनामा रसनायकोऽयं वल्लः पुराणेन गुड़ेन भुक्तः।। 65।।

चन्द्रोदयाख्यरसवर्णने–

काशे श्वासे फिरङ्गाख्ये रोगे च परमो हितः।

अथ उपायान्तरम्–

मुशल्या[1] कुलकृच्चैव[2] पारसीकयमानिका।
भल्लातकफलं चापि पलमानं पृथक् पृथक्।।

पलार्द्धसानः सूतः स्यात् षट्पलोऽत्र गुड़ः स्मृतः।
एकीकृत्याखिलं कुर्यात् वटीः कर्षप्रमाणतः।।

खादेदेकां वटीं प्रातर्यावदारोग्यदर्शनम्।
गोदध्नश्चानुपानेन फिरङ्गामयनाशिनी[3]।।
निम्बुकेन[4] विना नैव वर्जनीयमिहापरम्।।

अथ उपायान्तरम्–

चोपचीनीभवं चूर्णं शाणमानं समाक्षिकम्।।
फिरङ्गव्याधिनाशाय भक्षयेल्लवणं त्यजेत्।।

Descripton for preparing mineral acids by distillation—"Śaṅkhadrāvarasa or liquid for dissolving conchshells, practically the same as given under "mineral acids" in Vol. i, p. 186. The last line says : "Cowrie-shells and metals dropped into the liquid are at once disolved."

Preparation of calomel and treatment of "phiraṅgaroga" (syphilis) with its application. *Cf.* Vol. i, p. 252.

निर्वातभाजा सरुजा च पथ्यशीलेन कुष्ठामयनाशनः स्यात्।।
फिरङ्गकरिकेशरी सकलकुष्ठाकालानलोऽखिलव्रणविनाशकृद्व्रणजगर्त्तपूर्त्तिप्रदः।
सुवर्णसमर्णकृद्बलहुताशतेजस्करः समस्तगदतस्करो रसपतिः स कर्पूरकः।।66।।
इति कर्पूररसः बौद्धसर्वस्वात्।

1. B reads मुशल्या.
2. B reads कल्लकं चैव.
3. B reads फिराङ्गापविनाशिनी.
4. B reads निंबुकेन.

Chemistry in Dhātukriyā or Dhātumañjarī (Rudrayāmala tantra.)

रुद्रयामले धातुक्रिया धातुमञ्जरी वा[1]

श्रीगणेशाय नमः

पार्वत्युवाच

अथानन्तर[2] देवेश ब्रूहि धातुसुसंस्कृतिम्[3] ।
येन विज्ञानमात्रेण साधयेद्धातुसंज्ञितः ।। 1 ।।

महादेव उवाच

धातुस्तुर्यविधः प्रोक्तः तुर्यस्तुर्ये स्वपक्षतः (?)[4] ।
सत्त्वं रसस्तमश्चैव चतुर्थस्तु निरामयः ।। 2 ।।
राजसे राजसी सिद्धिस्तामसे तामसी सदा ।
सत्त्वसाधनं सात्त्विक्ये आनीते[5] सर्वसाधनम् ।। 3 ।।

पार्वत्युवाच

कथ्यतां देवदेवेश सत्त्वं राजसतामसम् ।
अतीव गुणवत्[6] धातोः कार्यं लक्षणलक्षितम्[7] ।। 4 ।।
उपधातुक्रमेणैव तेषां लक्षणसाधनम् ।
विटकी च मुदा युक्ता तेषां कर्मसु[8] साधनम् ।। 5 ।।
उपकारो[9] गुणलक्षं नाना कार्यक्रिया शुभा ।
वियोगयोगचाञ्चल्यं वर्णभेदैस्तु मारणम् ।। 6 ।।

---

1. The Ulwar MS. (A) reads धातुक्रिया, a part of the Rudrayāmala; but the Benares MS. (B) reads धातुमञ्जरी, a part of the same. The two MSS. are exactly the same in subject matter, differing only in name.
2. A reads, अथानन्तर, which is grammatically incorrect.
3. A reads धात्वस्य संस्कृति, which is grossly incorrect.
4. The sense here is not clear.
5. A reads अनीते, which seems to be incorrect.
6. A has गुणवती, which appears to be incorrect.
7. कार्या लक्षणलक्षिताः, a variant in A, which is not correct.
8. A reads erroneously कर्मस्य.
9. A reads उपकारं, which is grammatically incorrect.

समग्रं देवदेवेश विशेषा धातुकृत्क्रियाः।
येन विज्ञानमात्रेण साधयेत् गन्धकीः क्रियाः ।। 7 ।।

महादेव उवाच

पृथिव्या गर्भमध्ये तु अनेका धातुस्तिष्ठिति[1]।
विस्तृता[2] कियत्कालेन धातुर्नानाविधोदिता[3] ।। 8 ।।
गुणलक्षणसंयुक्तां[4] साम्प्रतं न स्मरामि ह[5]।
कथं ते कथयिष्यामि[6] वद ब्रह्माण्डवासिनि ।। 9 ।।

पार्वत्युवाच

ये केचित् स्मर्यन्त[7] ईश व्यक्तं कथय साम्प्रतम्।
भवतो[8]ऽनुग्रहेणैव साधनार्थश्च[9] सिध्यति ।। 10 ।।

महादेव उवाच

शृणु देवि प्रवक्ष्यामि धातुं नानाविधस्थितिम्।
गुणलक्षणसंयुक्तां यथा बुद्धिः[10] कथिष्यति ।। 11 ।।
मुख्यप्राधान्यता[11] एते रङ्गलोहक[12]ताम्रकैः।
रजतेनैव[13] संयुक्ता धातोरुत्तमता सदा ।। 12 ।।
मध्यमा सत्त्वजा धातुः[14] नीचा च त्रपुसीसयोः।
संयोगे धातुनामा हि नीचा नीचतरा स्मृता ।। 13 ।।

---

1. अनेका धातु तिष्ठति, a variant in A, which treats the word धातु in the masculine as well as in the feminine genders.
2. A reads विस्मृता, which is not correct.
3. धातुनानाविधोदिता, an incorrect variant in A.
4. A reads संयुक्तो, which is grammatically incorrect.
5. A reads स्मरति ह, which is incorrect.
6. कथयामास is the incorrect reading in A.
7. A and B read स्मरते, which is grammatically incorrect.
8. भवता is the reading in A and B, which is not correct.
9. साधनामर्थ is the incorrect reading in A and B.
10. A and B read बुद्ध्या.
11. मुख्यप्रधानता is the correct reading. प्राधान्यता is grammatically incorrect.
12. A and B have रंगलोहताम्रकैः, which mars the metre.
13. A reads रजतेनेव, which is not correct.
14. धातुः is used here in the feminine gender both by A and B.

संयोगान्ते त्रिधा प्रोक्ता उत्तमा मध्यमाधमा।
ताम्रजासत्त्वयोर्योगे[1] नारीधातुः प्रजायते।। 14।।

Tin, iron and copper are to be classed among the superior metals.

In general a metal being alloyed with silver is imporved a quality. Sattvaja metal (*i.e.* and alloy of tin and copper, *vide* below couplet 15) is of middle quality. 12-14

एषा[2] मध्योत्तमा प्रोक्ता कार्यकारणयोगके।
त्रपुताम्रसंयोगेन[3] जाता धातुश्च मध्यमा।। 15।।
सीसके[4] भङ्गदा प्रोक्ता कार्यकाले सदोदिता।
अनेनैव प्रकारेण येन येन च कर्मसु।। 16।।
कर्त्तव्या जाघवो विद्या ज्ञात्वा[5] बुद्धिविशारदैः।
अभङ्गं सीसके जातं शुद्ध उत्पत्यकारकम्।। 17।।
हाटके श्वेतता जातेः[6] अक्षया[7] रजते यदि।
लघुद्रावः शुभे शुल्वे लोहे च द्राविणी कला।। 18।।
जासत्वे अङ्गहीना च काठिन्याभ्यधिकौ परौ।
कला एताश्च धातूनां जातेर्भवति शुद्धता[8]।। 19।।
कला एतादृशी दिव्याः प्राप्तः[9] सिद्धस्तु जायते।
शुल्वजातसत्त्वसंयोगे नारीधातुस्तु[10] जायते।। 20।।
कार्यानेककरी प्रोक्ता पात्रामोदसु[11] कोमला।
मारणे जारणे दिव्या चारणे त्रोटने तथा।। 21।।

---

1. *Vide* verse 20.
2. Both A and B read एतत्, which is grammatically incorrect.
3. संयोगे is the reading in A and B, which mars the metre.
4. A and B read शीशके.
5. A reads त्रात्वा, which is incorrect.
6. A and B read जाते, which is incorrect.
7. A reads अक्षयी. B reads अक्षमी. Both the readings seem to be incorrect.
8. जाते भवति शुद्धिताः, a variant in A. जाते भवति रशुद्धिवाः, an incorrect variant in B.
9. A and B read प्राप्ते, which is grammatically incorrect.
10. A reads नारीधन्यत्वसु. B reads नारीधन्यवस्तु. Both seem to be incorrect. (*Vide* verse 14.)
11. A reads पात्रामोदस्रु, which is unintelligible.

निर्गन्धा वर्णरहिता जातिभेदकरी स्मृता[1]।
एतावद्[2] गुणसंयुक्ता कार्यकारणसंयुता।। 22।।
संस्कारैः संस्कृता सर्वैः[3] सर्वसिद्धिकरी स्मृता।
नानासुवर्णकार्येषु रूप्यकार्येषु संयुता।। 23।।
धातुधौता च या[4] धातुः सुधामायः सुसंस्कृता।
कार्यसिद्धिकरी[5] श्रेष्ठा ज्येष्ठकार्यविधायिका[6]।। 24।।
सर्वथाशु सुधामायः[7] ग्रासग्रहणग्राहकः (?)।
ग्राहयेत् ग्रासमात्रेण पाचयेदशनं[8] स्त्रियम्।। 25।।
स्वकीयाशनसंपर्के[9] जाते भवति सिद्धिदा।
अन्यथा नैव सिध्यन्ति जायते मृग-अम्बुवत्।। 26।।
तस्मात्तेनैव यत्नेन सुधामायः[10] सुसंस्कृतिः।
यथोक्तसंस्कृतियुता[11] सर्वतो धातुरुत्तमा[12]।। 27।।
क्रियोचितसुसंस्कारान्[13] कारयेद्यत्नतः सदा।
यथोदितैस्तु संस्कारैर्युक्ता कार्यकरी सदा।। 28।।
सर्वलक्षणसंयुक्ता धातुधौता सदैव हि।
जायते नैव सन्देहः कार्यकारणकारिका[14]।। 29।।

पार्वत्युवाच

कियत्परिमिता धातुः संख्या वदतु मे विभो।
उत्पत्तिर्लक्षणं तेषां स्थानं चैव[15] पृथक् पृथक्।। 30।।

---

1. A reads मता.
2. Both A and B read एतद्, which destroys the metre.
3. A and B want सर्बैः, thus rendering the foot incomplete.
4. A and B reads ये, which is not correct.
5. कार्यसिद्धिकरा is the variant in A and B.
6. A and B reads विधायका, which is grammatically incorrect.
7. A has no incomplete foot सर्वतेशुधा. B also has सर्वो तेसुधा.
8. A reads पाचयेत् शन. B reads पाचयेत् शनं. Both of them seem to be incorrect.
9. B reads संयुक्ते.
10. A reads सुधाम्नाय. B reads शुधाभ्राय. Both are unintelligible.
11. Both A and B read संयुक्ता, which mars the metre.
12. सर्वे ते धातुरुत्तमाः, a variant in A and B, which is grammatically incorrect.
13. A and B read only संस्कारा, which is incorrect.
14. Both A and B read कारकाः, which seems to be incorrect.
15. B reads स्थानं तेषां.

कथ्यतां देवदेवेश लक्ष्यलक्षणसंस्कृतिः।
भेदोऽभेदोऽथ[1] योगश्च मारणं चारणं तथा।। 31।।

पातनं द्रावणं चैव त्रोटनं जारणं तथा।
सारणं जीवनं षण्ढं गोपनं लेपनं पुनः[2]।। 32।।

जातिभेदः सुगन्धश्च मेलनं जरणं परम्।
समग्रं कथ्यतां देव उपधातुकरी क्रिया।। 33।।

महादेव उवाच

शृणु देवि प्रवक्ष्यामि धातुस्थानानि लक्षणम्।
उपधातुरनेका हि[3] तेषां मारणचारणम्।। 34।।

अक्षयं राग(?)[4] कारं च शोधनं बन्धनं तथा।
मारणादि च यत् प्रोक्तं तत् सर्वं श्रूयतां शुभे[5]।। 35।।

पार्वत्युवाच

प्रथमं वद देवेश धात्वनुक्रमलक्षणम्।
प्रत्येकस्थानरूपं हि गुणसम्बन्धकारकम्।। 36।।

महादेव उवाच

शृणु देवि प्रयत्नेन सिद्धान्तवचनं मम।
येन विज्ञानमात्रेण जायते धातुकृत्क्रिया।। 37।।

प्रथमे धातुनामानि संक्षेपात् कार्यकारणम्।
कथयामि च ते भद्रे पश्चात् स्थानस्य संपदः[6]।। 38।।

आदौ सुवर्णनामानि संक्षेपात् शृणुतां प्रिये।
स्वर्णं सुवर्णं सूर्यश्च हाटकं वह्निरोचनम्।। 39।।

---

1. भेदाभेदाथ is the incorrect reading in A and B.
2. A reads युतः, which is senseless.
3. उपधातुरनेकानि, an incorrect variant in A and B.
4. Both A and B want this word, which may be supposed to be राग or बल here.
5. तत सर्वं कथयामि ते, a variant in A.
6. Both A and B reads संपदि, which has no sense.

सु(स) रङ्गं च तथा लोहं देवधातुर्मनोहरम्।
वेशी[1] विश्वासगं चैव धर्मदं कृषिदेवता।। 40।।

जीवनम् अमृतं चैव हेमं[2] हिमवतोद्भवम्[3]।
देववल्लभगं दिव्यं संसारोत्तारणं महत्।। 41।।

जीवनं सर्वलोकानां[4] नारीणां रञ्जनं शुभम्।
अन्यानि विविधनामानि[5] गोप्यगोप्यतराणि च[6]।। 42।।

रजतं च तथा रूप्यं चन्द्रश्चन्द्रस्य दीपकम्।
शुभ्रज्योतिकरं[7] स्वर्णोषद्वीजं तारकं तथा।। 43।।

अनन्ता वर्त्तते तेषां कथा वृद्धिकराश्च ये।
स्थापितं लौकिकं चैव[8] येनाग्रन्थसमूहनम्।। 44।।

शीतवीर्यं लघुवीर्यं विषघ्नं वातनाशनम्।
रसपुञ्जीकरं श्रीदं धातुपुञ्जनकारकम्।। 45।।

रतिदं बहुवीर्यं च शुभधाम च संक्षयम्।
एतानि रजतनामानि शतान्यन्यानि सन्ति वै[9]।। 46।।

ताम्रं च[10] त्र्यम्बकाधिष्ठं शुल्वं नागस्य मर्द्दनम्।
नीलं च नीलविभृमं वमनं[11] चैव पातकम्।। 47।।

---

1. A reads वशी.
2. हेम is generally used.
3. हिमवदुद्भवं is the correct term.
4. The first term in the previous verse is only जीवनं. But here it is सर्वलोकजीवनं, Hence there is no repetition.
5. अन्यानि विविधान् नामान्, an incorrect variant in A and B.
6. गोप्यागोप्यतराणि च is the reading in A and B, which has no clear sense.
7. A and B read शुभ्रयोतिकरं, which is incorrect.
8. A reads लौकिका येना, which leaves out चैव.
9. A has शतान्यानि वर्त्तते, B has शतानानि वर्त्तते. Both the readings are grammatically incorrect.
10. च is dropped in A and B.
11. The sense of नीलविभृमं seems to be नीलविभ्रमं=नीलशोभम्. A reads चमनं, which is incorrect.

उष्णाम् उष्णकरं सिद्धं कृकशं कालमर्द्दनम्।
हेमगर्भं[1] च कामोदं विषदं विषनाशनम्।। 48।।
आरक्तश्च सहायश्च[2] मन्मथो मार एव च।
एतानि शुल्वनामानि अन्यानि विविधानि च।। 49।।

Synonyms of the different metals. 39-49

जासत्वं च जरातीतं राजतं यशदायकम्।
रूप्यभ्राता वरीयश्च त्रोटकं कोमलं लघु।। 50।।
चर्मकं खर्परं चैव रसकं रसवर्द्धकम्[3]।
सदापथ्यं बलोपेतं पीतरागं सुभस्मकम्।। 51।।

Synonyms zinc : Jāsatva, Yasadāyaka,[4] rūpyabhrātā (lit. brother to silver), charmaka, kharpara, rasaka, etc. 50-51

एतत्तु[5] खर्परनाम कार्यकर्मसु सिद्धिदम्[6]।
रसराजार्द्रकं कर्म[7] कृतं सिद्धिषु[8] जायते।। 52।।
त्रपुस्तापहरं वङ्गं रजतारिश्च[9] नीरदम्।
शीतवीर्यकरं श्रीदं रतिदं तापहारकम्।। 53।।
मेहघ्नं माहलाद(?) भूतं सहायं धातुशेषयोः।
एतानि त्रपुनामानि सुसिद्धं कार्यकारकम्[10]।। 54।।
सीसकं[11] धातुभङ्गं च मृतकं रोचकं परम्।
रक्तं च रक्तवीर्यं च[12] रक्तवस्तुविवर्द्धनम्।। 55।।

---

1. A reads हेमगर्भं.
2. A and B read सहायं च.
3. रसचार्द्धकं, the reading in A and B, is not grammatically correct.
4. Yasada is the name commonly given to the metal extracted from calamine. *Vide* Vol. i, p. 156.
5. तत्ते is the reading in A and B, which is incorrect.
6. Both A and B read कार्यकर्मेषु सिद्धिदान्, which is grammatically incorrect.
7. रसराजार्द्धकान् कर्मान्, an incorrect variant in A and B.
8. A reads कृते सिद्धिषु. B reads कृते सिद्धिषु. Both are incorrect.
9. Both A and B read रजतानि च, which is senseless here.
10. सुसिद्धे, कार्यकारकान्, an incorrect variant in A and B.
11. A and B read शीशकं.
12. B has only रक्तं वीर्यं च, which is incomplete.

अन्तरूप्यं च गम्भीरं धातुभञ्जनकारकम्।
ग्राहकं रसराजेन अक्षोभ्याक्षोभणं परम्।। 56।।

विचित्रं विनतं चैव नानारङ्गप्रदायकम्।
नागं नगालयं प्रोक्तं सिक्तकं कण्ठशोधनम्।। 57।।

षण्डं षण्डकर[1] चैव दुरितं षण्डकारकम्।
एतानि[2] नागनामानि कार्यकाले प्रयोजयेत्।। 58।।

लोहं च आयसं शूरं सूर्यकर्मविशारदम्।
हन्तकं खननं स्वर्णमारकं[3] ताटकं मलम्।। 59।।

अजरं च जरायुक्तं धात्वाद्यम् अवनीसुतम्।
कालेयं जीषणं कालं रुधिरं रोगनाशनम्।। 60।

ग्रन्थिदं सर्वधातूनाम् अक्षोभ्यं जलवह्निषु।
वह्निदं वह्निवीजं च सर्वधातुसहायकम्।। 61।।

एतानि लोहनामानि ज्ञात्वा कर्माणि कारयेत्।
जायते[4] सकला सिद्धिर्वियोगसंयुता[5]।। 62।।

नामानि धातुसंयोगे जातानि तानि च शृणु[6]।
शुल्वखर्परसंयोगे जायते पित्तलं शुभम्।। 63।।

Pittala is an alloy of copper and zinc. 63

पित्तलं चैव नारीकं कफदं गन्धदुर्ज्जयम्।
वर्णदं दुर्बलं चैव राजतं वल्लभं लघु।। 64।।

वङ्गताम्रसंयोगेन जायते तेन[7] कांस्यकम्।
कांस्यकं कोमलं चैव नैरसं रसकाठिनम्[8]।। 65।।

---

1. A reads खडं खंडकरं चैव.
2. A and B read एतत्ते, which is incorrect.
3. A has स्वर्णं प्रारकं B has स्वर्णं मारकं.
4. A reads नीयते.
5. वियोगायोग is the variant in A. विषोगायोग is the variant in B. Both the readings seem to be incorrect.
6. जातानि च ते शृणु, an incomplete variant in A and B.
7. A has जातेनैव. B has जायते तेनैव.
8. A and B read erroneously कठिणं.

Kāṁsyaka is an alloy of tin and copper. 65

तीव्रनादं महाशब्दम्[1] उग्रकान्ति रविग्रहम्।
ज्वालापिधानं रौद्रं च ग्राहकं[2] धूम्ररोधनम्।। 66।।

एतानि[3] कांस्यनामानि कार्यकाले प्रयोजयेत्।
बन्धने धारणे चैव मुद्रणे धूमबन्धने।। 67।।

खर्परैः सह पारदं दिव्यं[4] किञ्चित् प्रमेलयेत्।
जायते रसको नाम नानारोगहरो भवेत्।। 68।।

Zinc being amalgamated with mercury gives rise to rasaka. 68

[*N.B.*—Rasaka and kharpara are the names generally applied to the mineral calamine (*vide* Vol. i, p.158), but here they stand for the metal (zinc) or its amalgam.]

नागस्तु रहते[5] हीनो मृतधातुस्तु जायते।
स एव कोमलाग्निस्थः[6] सिन्दूरं जायते ध्रुवम्।। 69।।

Lead being killed with the aid of gentle heat gives rise minium (red lead). 69

अनेकाः[7] साधयेद्विद्या मन्त्रयन्त्रकरीः क्रियाः।
पूजनार्थे मम गुणाः सर्वे कल्पितरोचनाः[8]।। 70।।

ताम्रदाहजलैर्योगे जायते तुत्थकं शुभम्।
नाना रसायनी विद्या साधयेत् सा[9] सदैव हि।। 71।।

Copper in combination with the "burning water" gives rise to tuttha (green vitriol):—Thus manifold operations are

---

1. A reads महच्छब्दं, B read महत्शब्दं, Both the readings are incorrect.
2. A reads व्याहकं, which has no clear sense.
3. A and B read एतत्ते, which is incorrect.
4. पारदा दिव्यो is the reading in A and B, which is incorrect.
5. B reads रहरते, which is not correct.
6. कोमलाग्निस्थे is the reading in A and B, which is not correct.
7. A reads एनेका. B reads एतेका. Both are incorrect.
8. A and B read कल्पितरोचनं.
9. सा is not found in A and B.

performed with the aid of *mantras* and apparatus : all these belong to the province of Chemistry. 70-71

रसरसक(?)[1] योगेषु राजमर्यादा जायते।
राजकं जायते नाम नारी राजकलागतिः[2] ।। 72 ।।
संयोगे साधनं दिव्यं वियोगे साधनं शुभम्।
वियोगं पुनर्नैर्योगं रसताम्रे शुभो विधिः[3] ।। 73 ।।
अनेका साधनी विद्या साधयेद्द्रव्यकाङ्क्षितम्[4] ।
लभते रसायनीं विद्यां स्वस्थोऽपि विकृतोद्यमी ।। 74 ।।
अनेनैव प्रकारेण ज्ञायते नाम कर्मभिः।
नामकर्मज्ञता[5] जाते चित्ते निर्मलता[6] भवेत्।। 75 ।।
एतत्ते कथिता धातोरुपधातोस्तु तां[7] शृणु।
येन विज्ञानमात्रेण कालकाल्या[8] न लिप्यते ।। 76 ।।
उपधातोस्तु मध्येषु श्रेष्ठा माता रसायनी।
तस्या नामानि वक्ष्यामि पश्चात् स्थानगुणानपि[9] ।। 77 ।।
हरितालं विसङ्गा च रङ्गदीप्तिकरी सदा।
तालं च हरिता हंसी श्वेतपीतविधायिनी ।। 78 ।।
नर्त्तकस्य प्रिया देशी निशा चैव निशाचरी।
पारदस्य क्षयकरी पक्षच्छेदकरी स्मृता[10] ।। 79 ।।

Synonyms of haritāla (orpiment) : Orpiment is a consumer of mercury and also a clipper of its wings.[11] 79

---

1. A and B read रसारसयोगेषु, which is incomplete.
2. नाराराजकलागनिः, a variant in A. नाराजकलागतिः, a variant in B. Both the readings seem to be incorrect.
3. A and B read शुभविधि.
4. A reads द्रव्यकोखिनः. B reads द्रव्यकांखितः.
5. A and B read नामकर्मरदाजाते, which is unintelligible.
6. A and B read निर्मला, which is not grammatically correct.
7. A and B read ते, which is incorrect.
8. A reads कालकाप्यू. B read कालकाप्पू. Both the readings are incorrect.
9. Both A and B read गुणानि च, which is incorrect.
10. The 78th and 79th ślokas are not found in A.
11. The sulphide of arsenic or being rubbed with mercury and gently heated converts the latter itself into the sulphide. The mercury is thereby *killed* and *fixed, i.e.* deprived of its volatility (clipped of its wings–the Hindu alchemist's favourite figure).

रोगहा जीविनी जेत्री जरादारिद्रनाशिनी।
एतानि तालनामानि संक्षेपात् कथितानि ते[1] ।। 80 ।।

युक्तो ज्ञानेन यः[2] कर्त्ता अक्षोभ्यो[3] जायते सदा।
अनेनैव प्रकारेण कर्त्ता कर्माणि साधयेत् ।। 81 ।।

मनःशिला शिला शृङ्गी कुनटी च कृतोद्यमा।
दरदा[4] भगिनी देशी सिन्दूरस्य सखी सदा ।। 82 ।।

पुष्पा पुष्पवती[5] पत्री शालिनी रसिनी गुणा।
एतत्तु च शिलानाम[6] कार्यकाले प्रयोजयेत् ।। 86 ।।

अभ्रकं चैव व्योमं[7] च गगनं ग्राहकं परम्।
दुष्टवीर्यञ्च वातञ्च वह्निनादोन्नताकृति[8] ।। 54 ।।

अजरम् अमरं घोरं घोरषण्ढत्व[9] नाशनम्।
रसमूलरसातीतं रसज्ञे राज्यवर्द्धनम् ।। 85 ।।

अमृतम् अमरं[10] चैव महामृत्युविनाशनम्।
एतान्यभ्रकनामानि ज्ञात्वा कर्माणि कारयेत् ।। 86 ।।

सोमलं मल्लराजं च विषं वातविनाशनम्।
श्वेतवीर्यं बलग्रन्थि नीलउज्ज्वलकारकम् ।। 87 ।।

स्वेदनं भेदनं चैव कामवीर्यविवर्द्धनम्।
बन्धनं सर्वधातूनां मारणं पारदे रसे ।। 88 ।।

गगनग्राहके दिव्ये रसयुग्मे प्रमेलनम्।
एतानि मल्लनामानि कार्यकाले प्रयोजयेत् ।। 89 ।।

---

1. संक्षेपे त्रिकथानिजे, a variant in A and B, which has no sense.
2. युक्तजानेन ये, an incorrect variant in A and B.
3. A reads अक्षोन्यो, which is incorrect.
4. B reads देरदा.
5. A reads पुष्पवर्त्ता.
6. एतत्ते च शिला नामान्, an incorrect variant in A and B.
7. Generally used as व्योम.
8. A reads वह्निनादे एतः क्षतिः, which is incorrect.
9. A and B read खंडत्व, which is incorrect.
10. A repetition of the name.

Synonyms of realgar, cinnabar, mica, etc. 81-89

हीनधातोश्च नामानि संक्षेपात् तानि च शृणु[1]।
येन विज्ञानमात्रेण जायते पारदी क्रिया।। 90।।
मुक्ता मुक्ताफलं वारि वारिजं स्वातिसम्भवम्।
शुक्तिगर्भं[2] च सगरं शृङ्गारं मङ्गलोदयम्।। 91।।
तैजसम्[3] अमृतोद्भूतं शीतलं शिलीमं मनः[4]।
स्वयम्भुवं कलं दीप्तं स्त्रियि सौभाग्यदायकम्[5]।। 92।।
रतिराजग्रहः शोभाकामसन्दीपनं तथा ।
मुक्तानामानि रम्याणि[6] ज्ञात्वा कर्माणि कारयेत्।। 93।।
विद्रुमं वरदं[7] वेलावल्लीजातं प्रबालकम्।
नीरसम्[8] अधरेसीमा शिखरं शिखरोद्भवम्।। 94।।
वर्णदं विवरं चैव ग्रैवेयं[9] च विभूषणम्।
विद्रुमस्यैतन्नामानि[10] कार्यकाले प्रयोजयेत्।। 95।।
चर्मकं कूर्मकं चैव कमठपृष्ठं सकालकम्।
कमठं कर्कशं चैव कलिङ्गजात[11]माशुभम्।। 96।।
कचकं कद्रुहुः[12] कर्म लोके क्रचकटं तथा।
एतानि कूर्मपृष्ठस्य नामानि च विशेषतः।। 97।।
शङ्खं जलजं चैव देववादित्रमेव च।
विष्णुवल्लभकं[13] चैव सोमदैवतकं शुभम्।। 98।।

---

1. A reads वते शृणु, which is incorrect.
2. B reads मुक्तिगर्भं, which is not correct.
3. B reads तैनसं, which is incorrect.
4. B reads ननः, which is no correct.
5. स्त्रीणि सौभाग्यदायिनं, an incorrect variant in A and B.
6. A and B read ते रम्यान्, which is incorrect.
7. A reads विद्रुमे वरदं. B reads विद्रमे वरदे.
8. A reads नीरसे.
9. A and B read ग्रावेयं.
10. A has विद्रुमैनावि. B has विद्रुमैनानि. Both are incorrect.
11. कलिंगमाशुभं is an incomplete variant in A and B.
12. A has कन्वककद्रुहुः B has कचकर्हहुः.
13. B reads erroneously विष्णुवल्लभाभकं.

वरायुधं च वरदं शुभ्रं माङ्गल्यदायकम्।
अजेयं सर्वजन्तूनां दानवास्थि वरोदयम्[1] ।। 99 ।।
गम्भीरं घोषशब्दं च पवित्रं पारदं गृहम्।
एतानि[2] शंखनामानि वामदक्षिणकावुभौ[3] ।। 100 ।।
शुक्तिका सीफिका[4] कान्ती मुक्तामाता गरीयसी।
चन्द्रपत्नी पयोग्राही संपुटी पटवादिनी ।। 101 ।।
रजताभासिनी दीप्ता वङ्गवारिविशोषणी।
पद्माकरीगतिः पद्मगर्भ[5]समुद्रसम्भवा ।। 102 ।।
उत्तमा मध्यमा नीचा त्रिविधा सागरोद्भवा।
नद्युद्भवा तु सा नीचा[6] नीचकर्मविशारदा ।। 103 ।।
एतत्तु लक्षणं शुक्तेः[7] कर्मकाले प्रयोजयेत्[8]।
ज्ञात्वा यः कुरुते कर्म सिद्धयेत् सकलं फलम् ।। 104 ।।
गजोद्भवं रद[9] दन्तं द्विजं चैव तु पुष्पकम्।
नारीसौभाग्यदं चैव जगभूषणसम्मुखम् ।। 105 ।।
तैलरक्षाकरं श्रीदं संपुटं पुटिकं तथा।
नामानि गजदन्तस्य अन्यानि मध्यमानि च[10] ।। 106 ।।
पिच्छकं कुर्कुठं चैव शिखिपिच्छं सुचित्रकम्।
हरिमौलिधरं श्रीदं मोहनं चन्द्रकं तथा ।। 107 ।।
नागारिं सर्पदमनं विषदं विषनाशनम्।
एतानि शिखिपिच्छस्य नामानि विविधानि च ।। 108 ।।

---

1. Both A and B read वरं देयं, which destroys both grammer and metre.
2. A and B read एतत्ते, which is grammatically incorrect.
3. A reads वामदक्षिणथाकभौ. B reads वामदक्षिणमावुभौ.
4. B has only सीका.
5. A and B read पद्मेगर्भ.
6. नद्योद्भवा तत्रीचा, an incomplete variant in A. नद्योद्भवा तत्रीवा, an incorrect and incomplete variant in B.
7. A and B एतत्ते लक्षणा शुक्ति, which is an incorrect variant.
8. कर्मकार्ये प्रयोजने, a variant in A, which seems to be incorrect.
9. A reads वरदं, which is incorrect.
10. अन्यस्ते मध्यगानि च, variant in A and B, which is grammatically incorrect.

Synonyms of pearls, coral, tortoise-shell, conch-shell, tusk of elephants, the tail of the peacock. 90-108

नखं सहृदयपुरं[1] धातुकोमलकारणम्[2]।
वज्रं[3] दुर्गन्धिकं चैव तुरीटापस्तु क्षोभकम्[4] ।। 109 ।।
कचं केशं च बालं च मूर्द्धजालकमेव च।
श्मामं चैन धनं प्रोक्तं कर्कशं नर्मदं लघु।। 110 ।।
एतानि धातुनामानि यथाकर्माभिधानकम्।
ज्ञात्वा आरभ्यते कर्म स्वस्थे सिद्धिः प्रजायते।। 111 ।।

पार्वत्युवाच

शम्भो शङ्कर विश्वेश विश्वनाथ जगद्गुरो।
श्रुतानि चैव नामानि स्थानानि कथ्यन्तां[5] प्रभो।। 112 ।।

महादेव उवाच

श्रृणु देवि प्रयत्नेन स्थानानि विविधानि च।
सुवर्णस्योदयो धातोः सर्वव्यापकपर्वते[6] ।। 113 ।।
पार्थिव्यं हि च तत् सर्वं[7] मृदामध्यात् प्रजायते।
तस्मात् स्थानस्य संक्षेपः श्रूयतां सादरं[8] शुभे।। 114 ।।
हेमं च प्रथमं कल्पं मेरुमध्ये सदैव हि।
दुर्लभं मर्त्यलोकेषु कष्टेनैव तु लभ्यते।। 115 ।।
पुनर्हेमाद्रिणा तेन जायते सर्वदा शुभे।
महाशीतकृते[9] तेन दुर्लभं स्यात् सदैव हि[10] ।। 116 ।।

---

1. न हृदयषुरं, a variant in A. न हदयपुरं, a variant in B. Both are incorrect.
2. B reads कारकं.
3. B reads वज्रे, which is not correct.
4. A reads तुरीटापसुक्षोभकं.
5. A and B read कथ्यतां, which is grammatically incorrect.
6. A reads वत्तीत. B reads पर्त्तत. Both the readings are incorrect.
7. पार्थिव्यानि न ते सर्वे, an incorrect variant in A and B.
8. A and B read सादरात् which is incorrect.
9. A reads महत्सीतकृते, which is incorrect.
10. B reads सदेव हि, which is incorrect.

पश्चाज्जम्बूनदोद्भूतं जम्बूद्वीपेषु जायते।
तच्च कष्टतरं मन्ये दुर्लभं तत् सदैव हि।। 117।।
पृथिव्यां धातुमध्येषु सर्वेषामधिकं सदा।
लोहाधिकतरं तत्र ज्ञातव्यं सर्वदा शुभे।। 118।।
कष्टे निःसारणं तेषां मौल्यादधिकं जायते।
तस्मात्तेनैव यन्त्रेन साधनं क्रियते नरैः[1] ।। 119।।
असाध्यसाधनं तस्य जातं सर्वयुगे शुभे।
तस्मात्तु[2] बहुमूल्येन सर्वधातूपरिस्थितिः[3] ।। 120।।
पुनः स्थानं कृतं तेन लङ्कायां च सदैव हि।
उङ्कितं तेन यत्नेन मर्त्ताद्[4]भीतं मया शुभे।। 121।।

Localities for different metals. 113-121

[The information is very commonplace.][5]

पार्व्वत्युवाच

महारुद्र शङ्कर श्रीमान्[6] लोकनाथ जगत्पते।
कथं ते मानुषे भीतिः[7] कथ्यतां परमेश्वर।। 122।।

महादेव उवाच

सुवर्णस्यैषा महती[8] भीतिर्मानुषसम्भवा।
उच्चो नीचपदं सम्यक् जायते बहुले रतिः।। 123।।
मया मानुषभीतेन[9] उङ्कितं सादरेण च।
दुर्लभं च विशेषेण प्राक् कलियुगे इदम्[10] ।। 124।।

---

1. A has कृयते नराः. B has ज्ञयते नराः, Both are incorrect.
2. A and B read तस्मात्ते, which is incorrect.
3. सर्वधातोपरिस्थितिं, an incorrect variant in A and B.
4. A and B read erroneously मृत्याद्.
5. All these have been placed the category of (हीन inferior dhātu. Though the term dhātu generally means a metal, it is often applied to a mineral; here, however, it covers a much wider field.
6. श्रीमान्, a reading in A and B, is not correct.
7. मानुषाद्भीतिः is the correct form.
8. A and B read सुवर्णस्येषां महतान्, which is not correct.
9. मां च मानुषोभीतेन, an incorrect variant in A and B.
10. A and B read इमे. This foot is incomplete.

पार्वत्युवाच

मदीये मानसे नाथ जातोऽसौ संशयो महान्।
भवतां मानुषी भीतिः कथं तु जगदीश्वर[1] ।। 125 ।।

महादेव उवाच

कारणं वर्त्तते देवि अव्याख्येयं सदैव हि।
कथनीयं त्वयि भद्रे यतस्त्वं[2] मम वल्लभा ।। 126 ।।

बुद्धिचतुरतावृद्धिर्जायते च कलौ युगे।
मानुषे च भवतीह[3] मम क्षोभानुवर्त्तिनी ।। 127 ।।

तस्मात्तैर्बहुले[4] द्रव्ये साध्यते गन्धकी क्रिया।
अथवा पारदी चैव मम क्षोभानुवर्त्तिनी ।। 128 ।।

पार्वत्युवाच

श्रुता सा गन्धकी विद्या सर्वसौभाग्यदायिनी।
सङ्गभक्षकृता सा च[5] न सिध्यति गरीयसी ।। 129 ।।

भवता पूर्वमुक्तं हि सङ्गभक्षणयोगतः।
न सिध्यति च ते विद्या सदा गन्धकपारदी ।। 130 ।।

तस्मात्तु कथ्यतां देव मनो[6] म्लानिविवर्जितम्।
सुवर्णसाधिनीं विद्यां[7] तस्माद्वदतु मे विभो ।। 131 ।।

महादेव उवाच

मानसं मदीयं देवि जातं सम्यक् सुनिर्मलम्।
तथापि मानुषे भीतिर्विद्या[8] गन्धकपारदी ।। 132 ।।

---

1. A reads कथं ते जगदीश्वरं. B reads कथं ते जगदीश्वरः Both are incorrect.
2. यत्तेत्वं is an incorrect variant in A and B.
3. A and B read भवतीह, which is not correct.
4. A and B read तस्मात्ते, which is incorrect.
5. A and B read ते च.
6. A reads मत.
7. A and B read सुवर्णसाधिनी विद्या.
8. A and B read विधे, which is senseless.

सङ्गभक्षणजं कष्टम्[1] उभे मेलापनं महत्।
सुष्ठु मन्त्रकृते वह्नौ ज्वलते तत्क्षणात् ततः ।। 133 ।।

मन्त्रवत् साधयेद्विद्यां साधयेद्यत्नतः क्रमात्।
तस्मात्तेनैव महता गोप्या रिद्धिकरी क्रिया ।। 134 ।।

पार्वत्युवाच

पुरा प्रोक्तं त्वया नाथ पटले पञ्चदशके[2]।
कैलासात् परमं सौख्यं भजते मन्त्रविज्जनः ।।135 ।।

क्रिया गन्धकिनी[3] मध्ये अथवा पारदी शुभे।
कैलासादधिकं सौख्यं भुञ्जते विधिवन्नराः[4] ।।136 ।।

महादेव उवाच

निर्मलं मानसं जातं तव वाक्यविमोहितम्।
कथयामास ते भद्रे गुणलक्षणसंयुतम्[5] ।।137 ।।

स्वर्णोत्पत्तिस्थलं[6] नित्यं सर्वत्रापि विधीयते।
उत्पादनगतिस्तस्य स्थाने स्थाने पृथक् पृथक् ।।138 ।।

श्रूयतां वरदे देवि स्थानं चैव यथोदितम्।
तस्य वातोत्थजा रेणुर्धमने स्वर्णं[7] जायते ।।139 ।।

पावकाद्रेर्भुवावाहुः[8] सप्तरत्नाकरे तथा।
समुद्रस्य तटे दिव्ये स्थानं स्यात् स्वर्णकस्य च ।। 140 ।।

स्वल्पं च सिन्धुदेशेन कामरूपैस्तथैव च।
अन्यैस्तु विविधैः स्थानैः साम्प्रतोत्पत्तिर्जायते ।। 141 ।।

---

1. A and B have काष्ठं, which is incorrect.
2. पटले पंचाधिके, a variant in A and B, which is incomplete.
3. A and B read गंधकिना.
4. A and B read नरः, which is incorrect.
5. A and B read संयुतः, which is not accurate.
6. Both A and B read स्वर्णोत्पत्तिः स्थलं, which is not correct.
7. A and B add च after स्वर्णं.
8. A reads पावकाद्रस्य शुभावाहु. B reads पावकाद्रस्य भुवाबाहुः.

तत्र तत्रोत्पत्तिस्तेषां यत्र यत्र हिमद्रवः।
प्रसरन्ति भुवि संघाः[1] संशयो नास्ति ते सदा॥ 142॥
ताम्रोत्पत्तिश्च महता सुखेनैव प्रजायते।
तेषां स्थानानि वक्ष्येऽहं[2] याथातथ्येन न शृणु[3]॥ 143॥
नेपाले कामरूपे च[4] वङ्गले मदनेश्वरे[5]।
गङ्गाद्वारे मलाद्रौ च म्लेच्छदेशे तथैव च॥ 144॥
पावकाद्रौ जीर्णदुर्गे रूमदेशे फिरङ्गके।
एतान्युदितस्थानानि[6] सर्वपर्वतके[7] सदा॥ 145॥

Localities for copper specified : Nepāl, Kāmarūpa (the eastern portion of Assam), Baṅgala,[8] the country of the Mlechhas. Rūma and the country of the Phiraṅgas, etc. 143-145

जासत्वं यत्तु[9] दिव्यं हि स्थानानि तस्य च शृणु।
कुम्भाद्रावथ[10] काम्बोजे रूमदेशे बलक्षति॥ 146॥

Localities for zinc : Kāmboja, Rūma, Balkh, etc. 146

एतान्युभयोः स्थानानि[11] रजः स्थानं च[12] उच्यते।
जासत्वं वङ्गले नागं नेपाले च सदैव हि॥ 147॥
केदारे कर्मकाण्डे च त्र्यम्बके त्रिपुरे तथा।
एतत्स्थानानि दिव्यानि[13] गोप्यान्यन्यानि तानि ह[14]॥ 148॥

---

1. स्मंगे is the variant in A. स्मंगे is the variant in B.
2. स्थानान् प्रवक्ष्येहं is the incorrect variant in A and B.
3. A and B read ते शृणु.
4. नेपालं कामरूपं च, a reading in A and B.
5. वङ्गालं मदनेश्वरं, a variant in A and B.
6. एतानि उदितस्थानात्, an incorrect variant in A.
7. A and B read सपर्वतगे, which is incorrect.
8. This is the name given to Bengal in many Tantras.
9. A and B read यत्ते.
10. A and B read कुंभाद्रेथ, which is incorrect.
11. एति उभयोः स्थानात्, a reading in A. एति उनयोः स्थानात्, a reading in B.
12. A and B read रजस्थानानि.
13. A and B read स्थानकान् दिव्यान्.
14. अन्या गोप्या तितेति ह, a variant in A and B, which is senseless.

लोहाद्रौ[1] लोहकृतकर्म गयाद्रौ[2] गौतमाद्रिके[3]।
विन्ध्यमध्ये[4] हि सर्वत्र नलाद्रौ[5] निष्कलेऽपि वा।। 149।।

त्र्यम्बके विमले[6] चैव लौहाकशालिवाहने।
समुद्रस्य तटे रम्ये आद्ये अन्ते च मध्यगे।। 150।।

लोहस्यैतानि स्थानानि[7] हाटकैश्च[8] सदैव हि।
किञ्चित्[9] गोप्यानि दिव्यानि संस्कारैः रहितानि च[10]।। 151।।

युक्ते तु संस्कृते तेषु स्वर्णी सिद्धिः प्रजायते।
अयुक्ते नैव लभ्या तु[11] युक्ते प्राप्तिः सदैव हि।। 152।।

पारदे गन्धके धात्वोः योग औषधकृत्[12] सदा।
मानसे निर्मले चैव लभ्यते[13] हाटकी क्रिया।। 153।।

अन्यथा नैव लभ्या तु[14] विना मन्त्रेण सिध्यति।
तस्मादेव[15] प्रयत्नेन निर्मलं मानसं शुभम्।। 154।।

अशुभं समलं चैव मानसं सर्वदा भवेत्।
तस्मात्तु चतुराहारे[16] सर्वदा शुभमानसः।। 155।।

जायते च क्रिया सिद्धा हाटकी भवति सर्वदा।
मन्त्रसिद्धिःप्रजायेत् सर्वकर्माणि साधयेत्।। 156।।

---

1. A and B read लोहाद्रे, which is incorrect.
2. Do. Do. गयाद्रे, Do.
3. Do. Do. गौतमाद्रिणे, Do.
4. Do. Do. वंध्यमध्ये, Do.
5. Do. Do. नलद्रे, Do.
6. A reads त्रिमले.
7. लोहस्येतानि च स्थानात्, a variant in A and B, which is incorrect.
8. A reads हाटके च.
9. केचित् is the reading in A and B, which is incorrect.
10. A reads ले. B reads ते.
11. A and B read लभ्यते, which is not correct.
12. Do. Do. योग्यौषधकृत्, Do.
13. Do. Do. लभते, Do.
14. A and B read लभ्यन्ते, which is not correct.
15. A and B read तस्मादेनं, which is not correct.
16. A reads चतुरादारे, which unintelligible.

अनेनैव प्रकारेण उत्तमा हाटकी क्रिया।
सर्वकार्यकरी नृणां चतुर्वर्गफलप्रदा।। 157।।

Metals in combination with mercury and sulphur becomes fire for administration as medicine. 153

x x x x x x x

महादेव उवाच

प्रोक्तसंपुटके चैव स्थूलकाचीयकेन च।
उष्णाङ्गारकृते अग्नौ[1] मुर्मुरं छागरेकजम्।। 9।।
खर्पराख्ये पुटे चैव आरण्योत्पलकेन च।
सार्द्धं विनास्ति(?) गति च ज्वलिते अर्द्धमर्द्धगे[2] ।। 10।।

Process of roasting a substance in a shout galss-vessel using goat-and-cow-dung-cakes and husks of paddy grains as fuel. 9-10

एवानुक्रमतो योगे पुटमेकं च जायते।
चतुरंशेन जासत्वं सुवर्णाद्दापयेत् सुधीः।।11।।
प्रति संपुटपुटे चैव उद्धयेत् कुण्डखर्परे।
स्वाङ्गशीतं[3] समृद्धृत्य लेपमुत्तारयेत्ततः[4]।।12।।
पुनर्लेपः प्रदातव्यः[5] पुनः क्षारान् प्रदापयेत्।
पुनरेव घोटनं देयं सादरं पुनरेव हि।। 113।।
अनेनैव प्रकारेण तुर्यांशसंपुटेन हि।
सुवर्णसंयुतं तच्च[6] गालयेत् क्षारसंयुतम्।।14।।
ज्वलिते तोलयत्येव तिलार्द्धं संपुटं प्रति।
नम्रीभूतं च रक्षेयं संस्कारैः शुध्य साम्प्रतम्।।15।।
एतत्क्रमक्रमेणैव सुवर्णं[7] मनुसंक्षय।
तद्गुणे शोणिमापीते महत्ततः प्रजायते।।16।।

1. A reads कृतेरग्नि. B reads क्षतेरग्नि.
2. A and B read ज्वलितेरर्द्धरर्द्धगा.
3. स्वांगशीतेन is the reading in A and B, which is not accurate.
4. A and B read लेपमुत्तारणां ततः, which is incorrect.
5. A and B read पुनर्लेपं प्रदातव्यं, which is not correct.
6. A and B read ते च, which is incorrect.
7. A and B read सुवर्णान्, which is not correct.

आरक्तं पीतसंयुक्तं जायते हाटकं महत्।
हाटकं मत्तकं[1] चैव न च हीनेन योजयेत्।।17।।

One part of gold and four parts of zinc are to be melted together and the alloy roasted in a closed crucibe...the process repeated with the addition of alkalies...... gold of raddish yellow colour will thus be generated. 11-17

कलागुणगते वृद्धिर्जायते उत्तमे नरे[2] ।
अनेनैव प्रकारेण मात्राद्विगुणं दापयेत्।।18।।
कलात्रय[3] मिते वृद्धिर्दापिता वृद्धिरिष्यते।
निर्द्दोषं जायते दिव्यं द्रव्यसंपत्तिदायकम्।।19।।

पार्व्वत्युवाच

अधिकेन कृते तत्र[4] शोणपाने सुशोभने।
जायते कीदृशं देव संयोगे अधिके सति।।20।।

महादेव उवाच

शोणपाने च संयोगे भङ्गदः अस्ति[5] निश्चयः।
द्विगुण[6] शोणपानेन कलंकं जायतेऽशुभम्[7]।।21।।
वेधसंगं सुधनार्थे त्रिंशांशे दिक्पलोपमम्।
अतिरिक्तकृतं तच्च जायते सगुणं महत्।।22।।
गालयेद्यन्त्रपूर्वेण छायेद्रसमांशके।
समांशे गन्धकं देयं शुद्धं पीतक[8] रागकम्।।23।।
त्रयतुर्यांशकं तालं दापयेद्रसमोदितम्।
तदर्द्धं सादरं देयं मर्द्दयेत् कन्यकारसे।।24।।

---

1. A and B read हाटके मस्तके चैव.
2. उत्तमो नरः, a variant in A and B.
3. A and B have कलातर्क.
4. A and B haver ते च, which is incorrect.
5. A and B read भंगदा अस्ति, which is incorrect.
6. Do. द्विगुणे.
7. Do. शुभं, which appears to be incorrect.
8. Do. पीयत, which is senseless.

यामत्रितयमामर्द्द्य[1] छयाशुष्कमण्डजे रसे।
अतिशुल्वेतरे खल्ले अथवा लोहसंभवे।।25।।
मर्द्दयेद्यत्नपूर्वेण छायाशोषितं कारयेत्।
काचकूप्ये प्रदातव्य[2] वह्निर्बालुकयन्त्रगः[3]।।26।।
नखसंख्यामिते यामे पाचयेद्रसं शोभने।
स्वाङ्गशीते तु संजाते पुनः खल्ले निधापयेत्।।27।।
रसेनानेन आमर्द्द्य छायाशोषितं कारयेत्।
उक्तयामे रसे[4] दत्ते पुनः शोषं[5] प्रदापयेत्।।28।।
उक्ताग्नौ दापयत्येव पुनः संस्कारमाचरेत्।
संस्कारे त्रिविधे जाते अग्निपूर्वे क्रमोदिते।।29।।
जायते रसराजोऽसौ सर्वसामर्थ्यदायकः।
संचयेत्तण्डुलार्द्धस्तु अतिविक्रमवेगवान्।।30।।

A process for making factitious cinnabar by pouring mercury into its own weight of molten sulphur and then adding three-fourths its weight of orpiment....The mass to be rubbed in a mortar and subjected to sublimation. 23-30

जायते प्रबला बुद्धिः त्रिदिने सेवते यदि[6]।
सार्द्धवृद्धिक्रमेणैव दिने मासफलं सदा।।31।।
जायते नात्र सन्देहः सिद्धिगन्धकसम्भवा।
अनेनैव प्रकारेण रससिद्धिः प्रजायते।।32।।

* * * * *

पार्वत्युवाच

श्वेतता कर्मणा केन[7] जायते रजतेषु च।
क्रमेण राजती सिद्धिः पश्चाद्घाटकसम्भवा।।38।

1. यामत्रयमामर्द्द, an incomplete variant in A and B.
2. प्रदातव्या, a reading in A and B.
3. यन्त्यागा, a reading in A and B.
4. रसो is the variant in A and B.
5. शीश्ये is the incorrect in A and B.
6. A and B read सेवनेत्यपि, which is grammatically incorrect.
7. केन कर्मेण, an incorrect variant in A and B.

महादेव उवाच

अथानन्तरता विद्या धातुसंस्कारकारिणी।
जासत्वे[1] कठिने जाते शुद्धे कोमलकारिणी।। 39।।

तथाच पीतशुल्वेषु जायते शुद्धनम्रता।
शृणु यत्ने क्रिया दिव्याः शुद्धजासत्त्वसम्भवाः।। 40।।

पिष्टं हयखुरोत्थं च द्विगुणं धातुतः सदा।
अर्द्धन्तु दापयेत् स्रावे अर्द्धं निम्ने च[2] रक्षयेत्।। 41।।

निर्मलं धातुजं द्रावं स्रावचूर्णेन ढालयेत्।
यदर्द्धं च स्थितं निम्नं तत्कालं दापयेत् सुधीः।। 42।।

उपरि संपुटं देयं समुद्रकुक्कुटे पुटे[3]।
स्वाङ्गशीतं समुद्धृत्य जायते निर्मलं महत्।। 43।।

जासत्वे सोमलं देयम् उक्तभागेन[4] साम्प्रतम्।
चूर्णान्तरगतं कृत्वा उज्ज्वलं जायते महत्त्।। 44।।

तदेव[5] जासत्वं चैव दिव्यम् उज्ज्वलगं सदा।
रूप्यार्द्धे मेलयत्येव जायते शुद्धता शुभा।। 45।।

अनेनैव प्रकारेण धनं किञ्चिच्च जायते।
धने चैव तु संज्ञाते कारयेदूर्द्धजाः क्रिया।। 46।।

* * * * *

By using the *augmented*[1] gold as a means of exchange one can amass wealth. 52

अतीव शोभना विद्या रसगन्धकजा सदा।
अतिसौभाग्यसदा सा[3] च साधके सुखदायिका।। 67।।

---

1. जासत्वं is the reading in A and B, which is not correct.
2. A and B read निम्नेन.
3. समुद्रं कुकुटे पटे, a variant in A and B, which seems to be incorrect.
4. उक्तभोगेन, a variant in A and B.
5. A and B read स एव, which is not correct.
6. *Cf.* footnote, p. 2.
7. A and B read ते, which is not correct.

एतादृशं च जासत्वं शोधितं गन्धकेन च।
धालधौतेन ताम्रेण पुञ्जितं[1] सुभगं सदा।। 68।।

पुञ्जितं जायते दिव्यं शुल्वपीतं शुभोदितम्।
निर्मलं जायते तच्च हीनहाटकसन्निभम्।। 68।।

लघुता जायते तस्य किञ्चिन्नार्या गुरुता सदा।
जायते गुरुता दिव्या पुञ्जिते हाटकेन च।। 70।।

29 शपटलादुद्धृताः श्लोकाः।

पार्व्वत्युवाच

पूर्वमुक्तं त्वया नाथ धने सिद्धिस्तु जायते।
लघुद्रवकृते धातौ[2] वद[3] किं तत्र मुह्यते।। 43।।

महोदव उवाच

क्रिया अग्नेः शुभा तेषां द्रव्यसिद्धिविधायिनी।
शृणु यत्नेन तां भद्रे क्रियां रंगीति नागजाम्।। 44।।

आनयेत् पारदं दिव्यम् अधऊर्द्ध्वविपातितम्।
क्षालयेद्यन्त्रपूर्वेण खल्ले पिष्टं तु कारयेत्।। 45।।

तैलेनानेन आमर्द्य भावना क्षणमात्रतः।
निम्ने डमरुके यन्त्रे अग्निं दद्याद्विचक्षणः।। 46।।

उत्थयेद्यामयुग्मेन स्वाङ्गशीतं समुद्धरेत्।
पुनः समाननागेन पूर्ववत् क्षालयेत् सुधीः।। 47।।

पुनरेव हठादग्नी उक्तयामेन सिद्ध्यति।
रसे द्विवाग्नेः चैव हीनहेमोत्तमोत्तमः[4]।। 48।।

तौलिके रक्तिकायुग्मं दापिते वर्णपञ्चकम्[5]।
अनेनैव प्रकारेण जायन्ते षोड़श कलाः।। 49।।

---

1. A and B read पुंजिते.
2. A and B read धाते, which is incorrect.
3. A reads वच. B reads वचः.
4. हीनहेमोत्तमः, an incomplete variant in A.
5. पंचकः, is the reading in A and B, which is incorrect.

द्रव्यवृद्धिकरं तच्च जायते नात्र संशयः।
पुनरेव क्रिया रम्या पारदे नागचारिणी।। 50।।

द्विगुणो चारिते तत्र[1] जायते चन्द्रिका शुभा।
आभासेन शिखिग्रीवा जायते नात्र संशयः।। 51।।

पुञ्जिते हीनहेमेन जायते क्रयविक्रयः।
अनेनैव प्रकारेण जायते धनसंपदः।। 52।।

अतऊर्द्ध्वं क्रिया या च[2] विद्या नागस्य सम्भवा।
रसे तु नागजे चैव विधिनानेन वर्त्तते।। 53।।

तायते[3] सकला विद्या नागतैलस्य संभवा।
सर्वा सा[4] शोभना विद्या तीव्रदारिद्र्यनाशिनी।। 54।।

अथातः संप्रवक्ष्यामि त्रपूणां विधिमुत्तमम्[5]।
येन विज्ञानमात्रेण जायते रजतं महत्।। 55।।

अनेका राजती सिद्धिः जायते परमोत्तमा।
जायते धनसिद्धिश्च[6] जायते नात्र संशयः।। 56।।

पार्वत्युवाच

वद वैभवदातारं[7] वङ्गं विज्ञानदायकम्[8]।
ज्ञात्वा सिद्धिमवाप्नोति गुणज्ञैः सौख्यकं च यम्।। 57।।

महादेव उवाच

आनयेच्छोभनं वङ्गं विन्ध्याचलसमुद्भवम्।
गालयेद्यत्नपूर्वेण ढालेत् कुष्माण्डजे रसे।। 58।।

---

1. ते च is the incorrect reading in A and B.
2. A and B read ये च, which is incorrect.
3. A and B have ताते, which is incomplete.
4. A and B have ता, which is not correct.
5. A and B have विधिरुत्तमा, which is grammatically incorrect.
6. A and B have स्यात् instead of च, which is a tautology.
7. दाता च, a variant in A and B.
8. दायक, a variant in A and B.

त्रिः सप्तककृते तत्र[1] जायते निर्मलं शुभम्।
तदुत्थान् कारयेत् पत्रान्[2] शुचिविद्वान् यथोदितान्।। 59।।

रक्षयेद्यन्त्रपूर्वेण आनयेद्वर्किकं[3] पयः।
भावयेत् शुक्तिकाचूर्णम् उज्ज्वलं वह्निशोधितम्।। 60।।

मर्द्दयद्यन्त्रपूर्वेण यावल्लेपः सुलक्ष्यते[4]।
यवार्द्धं लेपयत्येव पत्रान् तु यत्नपूर्वकम्।। 61।।

छयाशुष्के च संजाते दापयेच्छुक्तिसंपुटे।
चूड़ाधो लेपयत्येव सन्धी यत्नेन मुद्रयेत्।। 62।।

मेलिती माषगोधूमौ पिष्टलेपं प्रदापयेत्।
छयाशुष्के च संजाते मृन्मयं लेपं दापयेत्।। 63।।

सुशुष्के च[5] गजे पाच्यं स्वाङ्गशीतं[6] समुद्धरेत्।
जायते उज्ज्वलं वङ्गं वातशुष्कं तु कारयेत्।। 64।।

यः कश्चित् मूर्च्छितश्चैव[7] जीवयेज्जीवनक्रियाम्।
स हि वङ्गाश्रयेणैव[8] उज्ज्वलः कठिनो[9] भवेत्।। 65।।

अतिशुद्धतरं तच्च जायते नात्र संशयः।
सर्वकर्मकरं श्रेष्ठं नानाभागेन पुञ्जितम्।। 66।।

भोगभागसमायुक्तं वङ्गं च गुणसंयुतम्।
लक्षणार्थे[10] धनार्थे च[11] संयोगे साधयेत् सदा।। 67।।

---

1. ते च is the incorrect variant in A and B.
2. A and B read तदोत्थान्, which is incorrect. The word पत्र is used here in the masucline gender.
3. B reads वद्रिकं.
4. यावल्लेपस्य लक्षते, a variant, which seems to be incorrect.
5. सुशुष्केन, a variant in A and B.
6. स्वांगशीते, a variant in A and B.
7. ये केचित् मूर्च्छिते ते च, an incorrect variant in A and B.
8. सहैवंगालयेत्येव a variant in A and B, which is unintelligible.
9. A and B read उज्ज्वल कठिणं, which is incorrect.
10. B read भक्षणार्थे.
11. धनार्थे is omitted in B.

जायते सुभगं तच्च संसारे शोभनं भवेत्।
अनेनैव प्रकारेण जायन्ते धनसम्पदः[1] ।। 68 ।।

पार्वत्युवाच

त्रपौ शुद्धतरे जाते का क्रिया वद मे विभो।
लक्षणार्थं धनसंपत्तेः क्रिया या च पृथक् पृथक् ।। 69 ।।

महादेव उवाच

लक्षणेन[2] च नारीणां भिन्ना भिन्नेन वर्त्तते।
अनेका धनदा विद्या जायते सुभगा महत् ।। 70 ।।
अतीव शोभना सा च गुणसामर्थ्यदायिका।
येन विज्ञानमात्रेण साधके धनमुत्तमम् ।। 71 ।।
जायते नैव सन्देहो धनसामर्थ्यकं महत्।
अतिसौभाग्यसम्पत्तिर्जायते नात्र संशयः ।। 72 ।।
आनयेत् शुद्धं तं वङ्गं गालयेत् टङ्कणैः सह।
पिण्याकसुरसे दिव्ये ढालयेद्यत्नपूर्वकम् ।। 73 ।।
सप्तढालकृते शुद्धे कठिनं रजतोपमम्।
जायते नैव सन्देहः पुञ्जिते रजते शुभे ।। 74 ।।
रजतं जायते शुद्धं संभारं कारयेत् सुधीः।
अन्येष्वेव विधिर्दिव्यो दिव्यरजतसंभवः[3] ।। 75 ।।
आनयेत् शोधितं वङ्गं तदर्द्धे पारदे छलेत्।
निधाय[4] शोभने खल्ले तुर्यांशे दापयेत्ततः ।। 76 ।।
मर्द्दयेत् कन्यकावीर्ये यामं षोड़श यत्नतः।
हंसपद्या रंसे दिव्ये तत्समं मर्द्दयेत् सुधीः ।। 77 ।।
समानाण्डरसेनैव मर्द्दयेद्यत्नपूर्वकम्[5]।
छायाशुष्के च[6] संजाते शोषान्ते[7] बालुके पचेत् ।। 78 ।।

---

1. जायते धनसंपदा, a variant in A and B.
2. A and B read भक्षणेन.
3. दिव्यादिव्यं रजतसंभवा, a variant in A and B.
4. A and B read निधाप्य or निधाप्य्.
5. यत्नपूर्वकः, a variant in A and B.
6. छायाशुष्केन, a variant in A and B.
7. A reads शीश्यांते. B reads सोश्यांते.

अनेनैव प्रकारेण क्रमादग्निं प्रदापयेत्।
प्रहराष्टकेऽष्टकेनैव त्रिवारं हठं दापयेत्।। 79।।

जायते च रसं दिव्यं भक्षणे च सुधासमम्।
नारीणां दापयेद्यत्ने संयोगे नागकेसरैः[1]।। 80।।

तदर्द्धाः सुष्ठु[6] गोधूमाः सहैवं तोलकार्द्धम्।
अतिसौख्यकरं तच्च[2] रक्तिमात्रार्द्धं वर्द्धते।। 81।।

पुंसि अजगुणोपेते यथा च गंधकी गतिः।
नरार्थे भक्षणे यातो जातित्रयसमन्वितः।। 82।।

अनेनैव प्रकारेण क्रमेण वर्द्धयेद्रसम्।
जायते रसजा सिद्धिः नात्र कार्या विचारणा।। 83।।

सुकृत् सुधाकरं वङ्गं गालयेद्यत्नपूर्वकम्।
तज्जलं शोषयत्येव शतांशे रसदापिते।। 84।।

रजतं जायते शुद्धं कलाद्वात्रिंशतः स्फुटम्।
अन्यं वै मेलयत्येव विक्रये शुभदं महत्।। 85।।

अनेनैव प्रकारेण जायते धनसंपदः।
साधयेत् सिद्धिदा विद्या देहसामर्थ्यदायिनी।। 86।।

Tin is to be melted and one-hundredth part its weight of mercury to be amalgamated with it. This [fraudulent substitute for] silver can be used for purpose of exchange, and one can thus amass wealth. 85-86

मेलयेद्भोगभागेन मर्द्दनाद्दरदो विधिः।
पीते भावनया अग्नी[3] सार्द्धवृद्धेन सिद्धयेत्।। 87।।

शुद्धताम्रे प्रदातव्यं शतांशे हाटकोत्तमम्।
जायते नात्र सन्देहो यथा जाम्बूनदोद्भवम्।। 88।।

* * * * *

Pure gold is to be alloyed with one hundred times its

---

1. A reads नागकेसरै.
2. A and B read ते च, which is incorrect.
3. A and B read रग्नौ, which is incorrect.

weight of copper and this [imitation] gold will resemble native gold. 88

पार्वत्युवाच

ताम्रं च तुत्थजं दिव्यं जायते केन विक्रियाम्।
वद विश्वेश्वर शम्भो सिद्धिं ताम्रविधानकीम्[1] ।। 64 ।।

महादेव उवाच

शृणु शुद्धक्रियां दिव्यां तुत्थताम्रस्य सम्भवाम्।
येन विज्ञानमात्रेण साधयेद्रसकुम्पिकाम्।। 65 ।।

साधयेत् पञ्चामृतं दिव्यं महासौख्यप्रदायकम्।
ज्ञात्वा कालेन लीयेत संसारसुखभाजने।। 66 ।।

आनयेत्तुत्थकं दिव्यं पीतगन्धकसम्भवम्।
दापयेच्छोभने खल्ले अर्कक्षीरेण भावयेत्।। 67 ।।

मर्द्दयेद्यत्नपूर्वेण याममात्रमखण्डितम्[2]।
तेनैव धातुयोगेन सुवर्णं[3] सुलभतां व्रजेत्।। 68 ।।

एतादृशी परा धातुर्वर्त्तते धातुर्मध्यमा।
सुलभा शुभदा प्रोक्ता मम सान्निध्यगा सदा।। 69 ।।

पार्वत्युवाच

ये गुणा नागजे ताम्रे वद विश्वस्य वल्लभ।
प्रयोजनवस्तु शक्यं चतुरेणैव चोदितम्।।95 ।।

महादेव उवाच

अनेका गुणा[4] नागेषु पूर्वमुक्ता हि[5] पारदे।
किञ्चित् शेषतरा ये च शृणु देवि प्रयत्नतः ।। 96 ।।

---

1. सिद्धा ताम्रविधानकं, an incorrect variant in A and B.
2. A reads याममात्रमखंडितः, B reads याममात्रे मखण्डितः
3. A and B read सुवर्णः.
4. अनेकान् गुणान्, a variant in A and B.
5. मुक्तानि पारदे, an incorrect variant in A and B.

नागस्य सम्भवं ताम्रं मध्ये मेलापनं कृतम्।
विभागे तु कृते तत्र[1] जायते कुम्पिका शुभा।। 97।।
तन्मध्ये गालयेन्नागं त्रिवारं यत्नपूर्वकम्।
जायते निर्मलं स्वर्णम् उदितं चैव कुम्पिके।। 98।।
विभागे साम्प्रतं दत्त्वा अमृतं जायते शुभम्।
उक्तकार्यकराश्चैते नान्यथा वचनं मम।। 99।।
तस्मात् साधयते दिव्यं भूनागोद्भवजं शुभम्।
अनेका साधयेद्विद्या रससामर्थ्यदायिका।। 100।।

Colophon in A

।। इति श्रीरुद्रयामले उमामहेश्वरसंवादे सुवर्णकल्पे सुवर्णप्रशंसा नामाध्यायः।।

Colophon in B

।। इति श्रीरुद्रयामले उमामहेश्वरसंवादे धातुमंजर्यां सुवर्णप्रशंसा समाप्ता।।

Lead and copper being alloyed together will give rise to gold. 97-98

Colophon : Here ends the Chapter on the extolling of gold in Rudrayāmala.

Chemistry in Suvarṇatantra or Svarṇatantra

सुवर्णतन्त्रात् वा स्वर्णतन्त्रात्

उद्धृता श्लोकाः

A=MS. from Benares.[2]
B=MS. from Ramnākālī's maṭha, Dacca.[3]

श्रीगणेशाय नमः।

श्रीराम उवाच

देवदेव महादेव ऋद्धिबुद्धिफलप्रद।
पूर्वं संसूचिता ऋद्धी रसायनपरा परा।। 1।।

1. A and B read ते च, which is incorrect.
2. The name of this MS. is सुवर्णतंत्रम्.
3. The name of that MS. is स्वर्णतन्त्रम्.

यस्याः साधनमात्रेण स्वाराट्तुल्यो नरो भवेत्।
तां सिद्धिं वद मे देव यदि त्वं भक्तवत्सलः।। 2।।

पूर्वं तु कथितं देव रत्नतन्त्रं त्वया मम।
गुटिकाः कथिताः पूर्वं सहस्रद्वितयं शिव।। 3।।

पारदाः कथिताः पूर्वं षट्शतं मृतिरूपकाः।
धातूनामष्ट कल्पास्तु पूर्वमेव प्रकाशिताः।। 4।।

धातुयोगाख्यकल्पस्तु पूर्वमेव प्रकाशितः।
रत्नानां करणे तन्त्रं पूर्वमेव प्रकाशितम्।। 5।।

किन्तु स्वर्णाख्यतन्त्रं तु न मह्यं कथितं प्रभो।
कश्यपेन महेशानाभ्यर्दितोऽस्मि महेश्वर।। 6।।

भूमिदानं मया दत्तम् ऋषये कश्यपाय वै।
कश्यपेन मयि प्रोक्तं भूमिभागं त्यज प्रभो।। 7।।

स्थानार्थं तु महेशान रक्ताब्धिः प्रार्थितो मया।
वाणमात्रं स्थलं तेन दत्तं मम महेश्वर।। 8।।

स्थानं प्राप्तं महेशान भक्षणं मम नास्ति वै।
भक्षणं देहि मे देव यदि पुत्रोऽस्मि शङ्कर।। 9।।

The origin of the tantra is explained in the opening lines. Paraśurāma having given away his worldly possessions to Kaśyapa is rendered destitute, and thus has to invoke the aid of the God Śiva for his very maintenance.

ईश्वर उवाच

शृणु राम प्रवक्ष्यामि रहस्यातिरहस्यकम्।
स्वर्णतन्त्राभिधं तन्त्रं कल्परूपेण कथ्यते।। 10।।

Śiva said : Listen ! I shall now reveal to you the most wonderful mysteries of Svarṇatantra. 10

तत्राद्यं स्वर्णतन्त्रस्य कल्पं शृणु सुपुत्रक।
तैलकन्दाभिधः कन्दः सिद्धकन्दः प्रकीर्त्तितः।। 11।।

कन्दः कमलवत्तस्य पत्राणि कञ्जवच्छिशो।
तथैव तु महत्पत्रं तैलं स्रवति सर्वदा।। 12।।

जलमध्ये सदा पुत्र त्वार्द्र एव प्रतिष्ठते।
विषकन्देति विख्यातो विषाच्च कायनाशनः ।। 13 ।।

तैलस्रावी महाकन्दः परितस्तैलवज्जलम्।
दशहस्तमिते देशे सरते तैलवज्जनम् ।। 14 ।।

महाविषधरः पुत्र तदधो वसति ध्रुवम्।
कन्दाधः कन्दच्छयायां नान्यत्र गच्छति प्रिय ।। 15 ।।

तत्परीक्षाविधानार्थं कन्दे सूचीं प्रवेशयेत्।
सूचीद्रावः क्षणात् पुत्र तत्कन्दन्तु समाहरेत् ।।16 ।।

तत्कन्दं तु समादाय शुद्धसूतं खलेत्त्रिधा।
मूषायां निक्षिपेत् तन्तु तत्तैलं तत्र निक्षिपेत् ।। 17 ।।
दीप्ताग्निं तु महाराम वंशाङ्गारेण दापयेत्।
तत्क्षणान्मृतिमायाति लक्षवेधी भवेत् सुत ।। 18 ।।

A kind of oil is exuded from the bulbous root of a plant; all around it within a radius of ten cubits oily water is exuded and a venomous snake lives under it. If you want to test the properties of the bulb, you should thrust a needle into it, and the needle at once dissolves. Having procured this bulb, rub it with mercury in a mortar and add the oil and subject the mixture to heat in a crucible. The mercury is at once killed and acquires the property of converting one hundred thousand times its own weight of the base metal into gold.[1] 14-18

ततः प्रभक्षयेद्राम क्षुन्निद्राहारको ध्रुवम्।
तालं शुद्धं समानीय तत्तैलेन खलेत् सुत ।। 19 ।।

सप्तधा प्रत्यहं राम त्वेवं विंशद्दिनं ध्रुवम्।
हरितालो मृतिमेति निर्धूमो जायते ध्रुवम् ।। 20 ।।

अग्नी पुत्र ततो दद्यान्निर्धूमो जायते सुत।
तत्तालं चाष्टधातौ तु दद्याद्द्रावे कृते सति ।। 21 ।।

सर्ववेधी भवेदेव शतिविद्धो भवेत् सुत।
तत्तैलं तु समादाय ताम्रद्रावे विनिक्षिपेत् ।। 22 ।।

1. लक्षवेधी भवेत्.

तत्क्षणात्ताम्रवेधः स्यात् दिव्यं भवति काञ्चनम्।
वङ्गे कास्ये यदा दद्यात्तदा रौप्यं भवेत् सुत।। 23।।
ताम्रे लौहे तथा रीत्यां तारे खर्परसूतके।
तत्क्षणात् वेधमायाति दिव्यं भवति काञ्चनम्।। 24।।[1]

* * * * *

Pure orpiment is to be rubbed with this oil for 20 days, and the former is killed thereby and loses its volantiny. The 8 metals in the molten state being treated with this prepared orpiment acquire to power of transmutation. When the above oil is thrown into molten copper, it is turned into gold of beautiful lustre. Tin and bell-metal similarly treated are turned into silver, and copper, iron, brass and silver into gold. 19-24

शंखद्रावस्य भेदान् हि तत्कल्पान् शृणु साम्प्रतम्।
लौहद्रावस्तथा ताम्रद्रावश्चैव द्वितीयकः।। 1।।
शंखद्रावस्तृतीयः स्यात् हन्तालश्च(?) चतुर्थकः।
दन्तद्रावः पञ्चमः स्यात् अम्लवेधी तु मध्यमः।। 2।।
पञ्चानां तु परीक्षा वै कथ्यते शृणु साम्प्रतम्।
लौहसूचीं समादाय लौहद्रावे विनिक्षिपेत्।। 3।।
तत्क्षणाद्द्रवतां याति सा सूची नात्र संशयः[2]।
ताम्रद्रावे तथा सूचीं सन्धिभेदीं विनिक्षिपेत्।। 4।।
सूचीद्रावो याममात्राद्भवत्येव न संशयः।
शंखद्रावे शंखसूची चतुर्यामेन संद्रवेत्।। 5।।
हन्तालोऽधो[3] यदा गच्छेत् दन्ताश्चाम्ला भवन्ति हि।
दन्तद्रावोऽप्यधो गच्छेद्दन्तद्रावो भविष्यति।। 6।।
एवं परीक्षां कृत्वादौ प्रयोगानारभेद्ध्रुवम्।
वज्रमूषां ततः कृत्वा शुद्धसूतं विनिक्षिपेत्।। 7।।

---

1. The above 24 Ślokas are not found in MS. A.
2. The above 7 hemistichs are not found in A.
3. A reads दंताश्चाधो. B reads दन्ताम्लाध्यो. Both the readings seem to be incorrect; since B has हन्तालश्च चतुर्थकः in the 2nd Śloka.

लौहसूचीद्रावरसं तत्र यत्नेन निक्षिपेत्।
तत्राग्निं दापयेद्यत्नात् पुनस्तत्र[1] रसं क्षिपेत्।। 8।।
स सूतो मृततामेति[2] नात्र कार्या विचारणा।
अष्टधातुषु तं सूतं दत्त्वा काञ्चनतां व्रजेत्।। 9।।
तं सूतं भक्षयेद्यो हि सोऽमरत्वमवाप्नुयात्[3]।
तस्य मूत्रपुरीषेषु शुल्वं[4] भवति काञ्चनम्।। 10।।
ताम्रद्रावप्रयोगं वै शृणु यत्नेन साम्प्रतम्।
तद्रसं तु समादाय शुद्धताम्रे विनिक्षिपेत्[5]।। 11।।
तत्ताम्र स्वर्णतां याति भैरवस्य प्रसादतः।। 12।।

Colophon in A—

।। इति श्रीसुवर्णतंत्रे हरपार्वतीसंवादे दुस्पर्शाकल्पः द्वितीयः।।

Colophon in B—

।। इति स्वर्णतन्त्रेन लोहद्रावादिपञ्चफलकल्पो द्वितीयः।।

Śaṅkhadrāvaka (*aquaregia*) is spoken of as an universal solvent ......This solvent with mercury is to be placed in a glazed crucible and subjected to heat ; the mercury, thus killed, can convert the 8 meatals into gold. By partaking of this mercury one becomes immortal ; even his urine and faces can convert copper into gold.[6] 1-10

1. B reads पुनस्तस्य, which is not correct.
2. B has सा सूची मृतिमाप्नोति.
3. A reads स गच्छेदमरत्वतां, which is incorrect.
4. B reads तुल्यं.
5. शुद्धताम्रेषु निक्षिपेत्, a variant in B.
6. This sort of belief was prevalent also at the time of the composition of Rasārṇava.

Knowledge of Gems

A few words on gems may not be considered beyond the scope of the present work especially as the subject has already drawn the attention of some European scholars.[1] We shall, however, limit ourselves, as far as possible, purely to the physical and chemical properties of the precious stones.

It has already been seen that among the 64 branches of arts and sciences recongnised in the ancient 'Kāmasūtra' of Vātsyāyana occur the following :—(a) the testing [of the genuineness or otherwise] of gold and the gems ; (b) knowledge of the colouring of gems, as also of mines and quarries.[2] Varāhamihira (d. 587 A.D.) in his "Bṛhat Saṁhitā" also conveys much useful information on this subject, but as he in his turn refers to experts who preceded him,[3] it may safely be concluded that long before this time it received adequate attention.

The Garuḍapurāṇa under the heading of Ratnaparīkṣā (examination of the gems) devotes several chapters to a detailed and fairly exhaustive description of the gems. This Purāṇa like so many of its class discusses on a variety of topics, mundane and spiritual ; in short, it may be looked upon as a sort of *multum in parvo.*

---

1. Garbe in his "Indischen Mineralien" (Leipzig, 1882) gives the text and the translation of Ch. xii of Narahari's "Rājanighaṇṭu" this work, however, is comparatively modern. Much useful information on the subject with bibliographical references will be found in Professor Yogesa Chandra Ray's treatise in Bengali entitled "Ratnaparīkṣā."
2. *Vide* Vol. i, p. 191.
3. पूर्वाचार्यैः समुद्दिष्टं .

From time immemorial the gems have been valued in India not only for purely decorative and ornamental purposes, but also as agents for warding off perils and the malignant influences of the planets and for securing luck and longevity. Hence, it is that particular attention is paid to the testing of gems, as their virtue increases in proportion as they are faultless and flawless.

The typical characters which were most relied upon for differentiating on species of gem from another were (1) the relative weight, (2) hardness, (3) lustre, transparency and colour, (4) fusibillty, especially when heated in combination with the alkalies. It will be noticed that even the lapidaries and the mineralogists of the present day have to depend largely upon these properties in the classification of the *stones*. We shall now proceed to say a few words under each of the above characters following in the main the authority of Garuḍapurāṇa.

1. Relative Weight

We have purposely used these terms as contradistinguised from "specific gravity": as the latter connotes peculiar associations. The principle of Archemedes being evidently unknown among the Hindus, the measurement of volume could only be roughly approximate. To cite some illustrations. A stone may be of the same species as, and equal in volume to, a padmarāga (oriental ruby), but will differ in weight from the latter.[1]

Empirical and arbitrary scales of weight were adopted as standards. A ruby occupying the volume of guñja (*abrus precatorius*) may be equal in weight to 10.7 and 3 guñjas repectively. A ruby of the volume of a śṛgālakola (a kind of jujube) should weight 12, 8 and 7 guñjas respectively. Again, a ruby of the volume of an āmalakī (*phyllanthus emblica*) should weight 30, 20 and 16 māṣās respectively.[2] In each case

1. तुल्यप्रमाणस्य च तुल्यजातेर्यो वा गुरुत्वेन भवेन्न तुल्यः ।–Garuḍa.
2. We give below the standard, weights as current at the time of Varāhamihira.

the gem which is of greater weight is superior in quality and the value thereof is enhanced in proportion.[1]

In the case of precious metals, however, adulteration was detected by an ingenious device, which may be regarded as a near approach to the principle of Archmedes. Suppose it is intended to test the purity of a specimen of gold. All that one has to do is to draw a wire of this gold and another of the standard gold through the self-same orifice [of corundum]. As these wires are of equal diameters, equal lengths will have equal weights, provided the sample is pure.[2]

2. Hardness

There are many stones which in external appearance resemble a ruby ; the spurious substitute may easily be detected by its difference in softness (low degree of hardness), and low specific gravity : again in case of suspicion the gem is to be rubbed on a whetstone ; if it simply gains in lustre but does not lose in weight (by abrasion), the specimen is to be regarded as genuine (test of a Ruby)[3]. Diamond and

---

8 White mustard	.... =	1 tandula (a grain of paddy)
4 Taṇdulas	.... =	1 guñja (abrus precatorius)
5 Guñjas	.... =	1 Māṣā
6 Māṣās	.... =	1 Karṣa or suvarṇa or tolaka
4 Karṣas	.... =	1 pala

The standard of weight has always varied in different times in different parts of India. 10 guñjas weigh about 18 grains. The seeds often vary in size.

1. Garuḍa lays special stress on the high specific gravity (=4.0) of a genuine ruby. लघुत्वं विजातिलिङ्गं, again गुरुत्वं मणीनां गुणसंग्रहः। While high specific gravity is indicative of the excellence of the gems in general, in the case of diamond it is just the reverse. गुरुता सर्वरत्नानां गौरवाधानकारणम्। वज्रे तत् वैपरीत्येन सूरयः परिचक्षते।। It is not easy to account for this qualifying clause, seeing that diamond has a sp. gr. of 3.5.
2. एकच्छिद्रे समाकृष्टे समखण्डे द्वयोर्यदा।
   धात्वोः सूत्रं मानसमं निर्दुष्टस्य भवेत्तदा।। Śukranīti.
3. This is true in a qualified sense only. After the diamond corundum is the hardest of all known minerals.

corundum alone will scratch ruby and sapphire. Skilful lapidaries often make imitation diamonds of zircon, ruby, cat's-eye, rock-crystal and glass ; hence they are to be tested by calling in the aid of learned experts. All the gems occuring in this world as also the metals will be scratched by a diamond, but the latter will not be similarly affected by the former. .....A diamond alone will scratch a diamond.[1]

To the above test Rājanighaṇṭu adds another, namely, when a diamond is struck with an iron hammer, it is not broken to pieces.[2]

---

1. The ruby and the sapphire are only varieties of corundum. Varāha classes corundum as a species of ruby, the former being dotted over with back and white patches and feeble in lustre [*i.e.* the non-transparent or coarse kind]. It is scarcely necessary to point out that the very word "corundum" is derived from the Hindi *kurand*, a corrupt form of the Sanskrit *kuruvinda*. The statement "corundum will scratch ruby and sapphire" is not strictly accurate. There are degrees of hardness among the different varieties of corundum, the blue sapphire standing first in this respect.
   We give here the original texts of which we have above almost a literal translation.
   "स्नेहप्रभेदो मृदुता लघुत्वं विजातिलिङ्गं खलु सार्वजन्यम्।"
   "अप्रणश्यति सन्देहे शिलायां परिघर्षयेत्।
   धृष्ट्वा योऽत्यन्तशोभावान् परिमाणं न मुञ्चति।
   स ज्ञेयः शुद्धजातिस्तु ज्ञेयाश्चान्ये विजातयः।।"
   "वज्रं वा कुरुविन्दं वा विमुच्यान्येन केनचित्।
   न शक्यं लेखनं कर्त्तुं पद्मरागेन्द्रनीलयोः।"
2. The reading of the text is very corrupt. Garbe gives as many as 12 variants ; that adopted by him seems to be on the whole reliable as also his rendering: "Wenn er mit anderen steinen oder eisernen Hammern geschlagen nicht zerspringt." On account of its brittleness a diamond wil readily be shivered into splinters under the flow of a hamour. The following extract from Bauer and Spencer's Precious may prove of interest.
   "The hardness of diamond has often been confused with its frangilibity or brittleness. It has been supposed, especially in ancient and madieval times that hammer and anvil may be shattered but not diamond which lies between. This statement was made by Pliny, the great naturalist of ancient days, who was killed in 79 A.D. at the first historic eruption of Vesuvius. He proceeded to say further that the fragmentation of a diamond may be effected by subjecting it to a preliminary immersion in the warm blood of a goat, but that even under these circumstances the hammer and anvil will also be broken!

3. Lustre, Transparency, Colour and other optical properties

Difference in lustre (स्नेहप्रभेद) is often regarded as a diagnostic feature of the gems. According to Varāha gems of superior qualities should possess cool (lit. waxy) lustre, be limpid and emit rich rays. Classification according to colour is often enjoined. For instance, a diamond may be perfectly colourless, or it may be yellow, black, red or copper-tinted.[1] Stones which are translucent, opaque and of dull colour are inferior in quality. In the case of diamond special stress is laid on its displaying the colour of the rainbow due to its high dispersive power]. Some gems have only one shade of colour (एकच्छाय), others again two, (द्विच्छाय)[2]

According to Altertus Magnus (1205-1280) the blood is more efficacious if the goat has previously drunk wine or eaten parsely. "Such being the views then held respecting the unbreakable and indestructible character of the diamond, it is easy to understand why the Greek word *adamas*, signifying unconquerable, should have been applied to this stone, although its application to the diamond is singularly inappropriate and inaccurate when its extreme frangibility is considered. Many a doubtful stone has been submitted to the test of the hammer, with the belief that the blow would be resisted only if the stone were a genuine diamond. Probably many beautiful stones have been sacrificed to this old belief. As a matter of fact, diamond is easily fractured, a very moderate blow from a hammer sufficing for the purpose; its perfect cleavage places it among the most brittle of minerals."

1. रक्तपीतं सितं शैरीषं (Varāha); आताम्रं (Garuḍa).
   Yellow diamonds are unknown in India at the present day at any rate. "India can claim for its own all the finely coloured stones of blue, green and red, not however, yellow diamonds, which come mainly from South Africa." (Bauer and Spencer). Crystals of black diamond should not be confounded with the black carbonado.
2. Probably refers to dichroism. As has already been pointed out the red ruby and the blue sapphire are only varieities of corundum, which often occurs in well-developed crystals belonging to the rhombohedral division of hexagonal system. "The dichroism of the ruby affords a means whereby it may be distinguished with certainly from other red stones. such as spinel and the different varieties of garnet, which crystallise in the cubic system, and thus being singly refracting can show no dichroism." (Bauer and Spencer). "The dichorism of the sapphires of Siam is specially well marked."—*Ibid.*

The changes in colour which the ruby and the sapphire undergo under the influence of heat had evidently been carefully observed. This test is to be applied with extreme caution as in case the proper degree of temperature (अग्निमात्रा) is exceeded, the stone may be materilly injured.[1]

4. Fusibility (Liquefaction)

It was well-known that all these the gems with the exception. of diamond would undergo liquefaction when strongly heated with a mixture of the two alkaline carbonates [now familiar in the Laboratory as the "fusion mixture"]. The liquid principal of gems (रत्नजा द्रुति) was a favourite remedy with the Hindu latro-Chemists (*Cf.* Vol. i, p. 103). As most of the precious stones belong to the corundum and the spinel group, they easily yield to treatment with sodium and potassium carbonates. In the case of the pearls it is recommended that they should be enclosed inside lemons and placed in a mass of paddy grains [for the sake of the warmth].

General and Concluding Remarks

The crystalline character of the minerals is referred to in one or two places. All the old writers agree in describing the diamond as having 6 angles, 8 faces and 12 edges.[2] The octahedral form of the crystal is common in the diamond occurring in the neighbourhood of Panna in Bundel Khand.

1. यावन्तं चंक्रमेदग्निं पद्मरागोपयोगतः (?) ।
इन्द्रनीलमणिस्तस्मात् क्रमेत सुमहत्तरम् ।।
तथापि न परीक्षार्थं गुणानामभिवृद्धये ।
मणिरग्नौ समाधेयः कथञ्चिदपि कश्चन ।।
अग्निमात्रा(ऽ)परिज्ञाने दाहदाषैश्च दूषितः ।
सोऽनर्थाय भवेद्भर्त्तुः कर्त्तुः कारयितुस्तथा । Garuḍa.
Cf. "The red ruby, at a high temperature, is colourless: on cooling it first becomes green, after which it gradually assumes its original find red colour." (Bauer and Spencer). "The blue colour of the sapphire disappears on heating."—*Ibid*.
2. कोट्यः पार्श्वानि धाराश्च षड्ष्टौ द्वादशेति । Garuḍa.

Rājanighaṇṭu further adds that the stone is often endowed with numerous faces and edges.[1] According to Rasaratna-samuccaya the female diamond is flattened and rounded. whilst the neuter is rounded and obtuse-angled.[2] Vaikrānta is also (probably a mineral of the spinel group, *vide* Vol. i, p. 83) described as having 8 faces and 6 angles.

It is sometimes asserted that the phosphorescence of diamond was first observed in 1663 A.D. by the celebrated Robert Boyle, Bhoja, (11th century A.D) however, mentions this property.[3] Faults in precious stones due to isolated grain of minute gravels etc., embedded in the mass now technically called "enclosures", are also referred to.[4]

It may be noted here that what is generally known as the Indian or Tavernier's rule for determining the value of diamond can be traced back to a remote date. Śukranīti speaks of it as the "rule of squares,"[5] while Varāha, himself

1. "In the most cubes of diamond, however, each edge is replaced by two faces,......the twenty-four faces thus derived would, if produced or enlarged". Sufficiently, give rise to the form known as the four-faced cube, or tetrakis-hexahedrom. (Bauer and Spencer).
2. The rendering of the description of diamond in R.R.S. (*vide* Vol. i, p. 100) is based upon a misreading. The correct reading is that adopted by the Poona edition, namely, अष्टास्रं वाष्टफलकं षट्कोणं "a diamond has 8 faces and 6 angles." The author evidently uses अस्र and फलक in the same sense; there cannot be any confusion here, as distinct mention is made of the 6 angles. As a rule अस्र and कोण are used synonymously. Thus in Varāha वज्र is described as षड्स्रि ; the commentator Bhaṭṭotpala explains it षड्भिरस्रिभिर्युक्तं षट्कोणमित्यर्थः। "Rounded" (वर्त्तुल): *Cf.* Bauer and Spencer. "The faces of diamond crystals differ from those of most other crystallised minerals, in that they are as a rule, much curved and rounded instead of being perfectly plane. as is usually the case." p. 119. "Obtuse-angled" (कुण्ठकोणाग्रं) in the crystallographic terminology should be rendered as "with corners truncated off".
3. अन्धकारे च दीप्यते (phosphoresces in the dark).,
4. कलुषा मन्दद्युतयो लेखाकीर्णाः सधातवः खण्डाः।
दुर्विद्धा न मनोज्ञाः सशर्कराश्चेति मणिदोषाः।। Barāha.
Here सधातवः and सशर्कराः no doubt refer to the minute pebles (gravels) enclosed in the body of the stones.
5. यथा गुरुतरं वज्रं तन्मुल्यं रत्तिवर्गतः *i.e.* if a diamond weighing 1 *rati* he worth x, one of the weight of 4 *ratis* will be worth 2 xx.

a great mathematician, gives practically the same rule, though somewhat on an elaborate scale.[1]

It would be unfair to compare the progress made by the Hindus in this branch of knowledge in the 6th century A.D. and perhaps earlier with that of the present day. Reference to the crystalline forms of diamond can be found in the writings of Keppler, Steno and Robert Boyle in the seventeenth century. Scientific crystallography originated with Romede l'Ile and the abbe Haüy at the end of the eighteenth century. The various physical and chemical properties of the minerals culled here bespeak considerable powers of keen obesrvation.[2]

Note on the Method of Preparing Caustic Alkali

In the first volume of our History the detailed description of preparing caustic alkali from the ashes of plants has been quoted at length from the Suśruta and we referred to it "as a proof of the high degree of perfection in scientific pharmacy achieved by the Hindu at an early age."[3] M. Berthelot has no less been struck with the originality of this process; indeed, he goes so far as to suggest that this portion in the Suśruta is evidently modern and in fact a later addition derived from contact with the European chemists.[4]

The process of rendering mild alkali caustic by the addition of lime to be found also in Cakrapāṇi (*c.* 1050 A.D.)

---

1. सितसर्षपाष्टकं तण्डुलो भवेत्तण्डुलैस्तु विंशत्या।
   तुलितस्य द्वे लक्षे मूल्यं द्विद्व्युनिते चैतत्॥
   पादत्र्यंशार्धोनं त्रिभागपञ्चांशषोड़शांशाश्च।
   भागाश्च पञ्चविंश: शतिक: साहस्त्रिकश्चेति॥
   i.e. if a diamond weighing 20 taṇḍulas (p. 33 *ante*) be worth 2 lacs (2,00,000), one weighing 5 taṇḍulas will be worth only (5^2x4=100)th part of 2 lacs or 2,000.
2. Concerning Ratnaparikṣā *cf.* also "Les Lapidaires Indiens" by M. F Finot in the "Bibliotheque de l'Ecole des Hautes Etudes."
3. *Vide* Vol. i, Intro c.
4. Journal des Svants. Jan. 1903, p. 42.

and Vāgbhaṭa. who must have preceded the former by several centuries, and it is well known that both these authors are deeply indebted to the Suśruta. There is no thus the remotest chance of inspiration from the European chemists.

We have evidence from quite an unexpected and independent that the cauterization of bad wounds by means of caustic alkali was an established custom in ancient India. Thus, we read in "The Questions of king Milinda"—

"And when the inflammation had gone down, and the wound had become sweet, suppose he were then to cut into it with a lancet, and burn it with caustic. And when he had cauterized it, suppose he were to prescribe an alkaline wash.....Now tell me, O king ! would it be out of cruelty that the surgeon....thus cut with the lancet and cauterized with the stick of caustic."[1]

The Tantrists, the Rosicrucians and the Seekers after Truth

The East not been always East and the West, West. Modern Europe in materialistic, but she has not disdained in the past to "plunge in thought." This reminds us of the truth of the poet's words ; "one touch of nature makes the whole world kin." Ancient Greece had her orphic mysteries. They were developed into a religious and ethical system by Pythagoras. In the schools of Alexandria the neo-Pythagoreans and the neo-Platonists further added to it the Vedantic doctrine of a world-soul.

"With the dissolution of the neo-Platonic school in the sixth

1. "Sacred Books of the East" Vol. XXXV, p. 168. Although Milinda has been indentified with the Bactrain Greek king Menander (c. 140-110 B.C.) it is not to be supposed that any Greeks influence can be traced in it. In this ethical romance, king Milinda and the Buddhist Bhikṣu Nāgasena have been put forward simply "as men of straw" to quote Rhys Davids. The work in reality conveys the teachings of the older Buddhists (Hīnayānists) with the background of Sāṁkhya and Vedānata Metaphysics.

century. Greek philosophy disappears, it is true, as a distinct phenomenon from the theatre of history, but it continues to exists in combination with foreign elements in the service of a new form of culture in Science of the Middle Ages and of modern times." (Zeller).

Says Waite :—

"The traditions of the neo-Platonic philosophy, with its elaborate theurgical system, were to some extent prepetuated through the whole period of the Middle Ages, for besides the orthodox theology of the great Latin Church and amidst the clamour of scholastic philosophy, we find the secret theosophy of the magician, the Kabbalist, and the alchemical adept borrowing, directly on indirectly, from this prolific fountain of exalted mysticism. The traces of its influence are discoverable in Augustine, in Albertus Magnus, in St. Thomas, the angel of the schools, and in other shining lights of western Christendom, while the metaphysical principles of Johannes Scotus Erigena, even so early as the close of the ninth century, were an actual revival of this philosophy."[1]

History repeats itself. In India the Yoga doctrines of Patañjali paved the way for the origin of Tāntric mysticism and it associated itself with magic and alchemy. Some twelve centuries later, this phase in the stage of the human progress found exponents in the soil of Europe. As Waite says :—

"In an age of progress, of doubt, and of great intellectual activity, it is singular to remark the almost prevalence of mysticism in one or other of its manifold phases, and the close of the sixteenth century beheld spreading over the whole of Germany and passing thence into Denmark, France, England and Italy, a mighty school of mysticism in the great multitude of magicians, alchemists, etc., who directly indirectly were followers of the renowned Paracelsus.[2]

1. History of the Rosicrucians,– p. 27. Regarding the sect of the Cabalists see also Lecky's "Rationalism in Europe." Vol. i, p. 42. ed., 1900.
2. See also Kopp's "Die Alchmie in ält. u. nue. Zeit," ii, pp. I-146, on the "Bekanntschaft mit Geheimnissen der Magie, Theosophie, Alchemie u. A."

Even the lower strata of society in England did not escape the contagion. Sir William Temple observes : "I have had several servants far gone in divinity, others in poetry, have known in the families of some friends, a keeper deep in the Rosicrucian mysteries."

There are those who sneer at the attempts of the alchemists to convert the baser metals into gold and are at the pains to insinuate that the motives of the adepts were sordid. It may be that there have been here and there men— vulgar charlatans—who have pursued and *art* merely from worldly considerations. The high priests of alchemy, however have in every age and clime been honest seekers after truth. Had there been no desire implanted in the human breasts to pry into the hidden and occult mysteries of nature there would haver been no science. Whence come we and whither are we distined to go ? What is this substance made up of, and what are its ultimate constituents, and what will its combination with this or that lead to ? These are queries, the solution of which or rather the very attempts to solve which mark the birth of philosophy. Patañjali anḍ Nāgārjuna in India and Paracelsus in Europe stand forth as the prominent representative of this spirit of inquiry. They have been dreamers, mystics and naturalists all combined in one. Such geniuses have always chafed as the restaints and limitations imposed or frail man and hence their fervid yearnings to peep into regions beyond the ken of our gross senses. As Emerson puts it :—

> "The privilege of this class is an access to the secrets and structure of nature, by some higher method than by experience. In common parlance, what one man is said to learn by experience, a man extraordinary sagacity is said, without experience, to divine.[1]

Plotinus heaved a sigh at the soul with its inflinite possiblities being caged in a frail and corruptible body. The

1. "Swedenborg; or, the mystic."

pursuit of alchemy by the esoteric Tāntrist is easily explained. To him it was only a means to an end. "It (mercury) is alone that can make the body undecaying and immortal.[1] We find echoes of the same sentiment in the authoritative literature of the Rosicrucians.

Says Waite :—

" Among the concourse of inquirers, and the clamour of supposed and pretended discovers, these rose gradually into deserved prominence an advanced school of illumination, who, employing the terminology of the *turba philosophorum*, under the pretence of alchemical pursuits appear to have concealed a more exalted aim....The student is directed by these writers from the persuit of material gold to the discovery of incorruptible and purely spirtual treasures.

...Physical transmutation, the one and supreme end of the practical alchemist, sinks into complete insignificance ; nevertheless, it is performed by the adept and is a landmark in his sublime progress." l. c. p. 32.

The sage and seer of Hoenheim had his wrestlings. Should he pursue knowledge for its own sake or for what it bringeth ? His evil genius hold out the bait thus :—

"Know not for knowing's sake
But to become a star men for ever ;
Know for the gain it gets, the praise it brings,
The wonder it inspires, the love it breeds."

The spritual as last triumphs over the base in man. Paracelsus has his revelation, and he bursts forth :—

"Truth is within ourselves : it takes no rise
From outward things, where' er you belive,
There is an inmost centre in us all
Where truth abides in fulness : and around
Well upon wall, the gross flesh hems it in,
The perfect clear perception—which is truth."[2]

1. *Vide* "History of Hindu Chemistry" Vol. i, Intro lxxvii and xcv–xcvi.
2. Browing's "Paracelsus."

In the above delineation the poet has done no more than bare justice to the inward longings of an idea alchemist, who is only an honest seeker after truth. The spiritual and thought-world has always transgressed all barriers of time and space. Kaṇāda, Patañjali and Nāgārjuna of India and Heraclitus, Empedocles and Plato and the rest— they all form a holy fraternity ; aye, these prophets and oracles of the intellect belong to a "lofty and sequestered class," "the high-priesthood of the pure reason, the *Trismegisti,* the expounders of the principles of thought from age to age. When at long intervals we turn over their abstrues pages, wonderful seems, the calm and grand air of these few, these great spiritual lords, who have walked in the world,—these of the old religion,—dwelling in a worhship which makes the sanctities of Christianity look *parvenues* and popular ; for 'persuasion is in soul, but necessity is in intellect."[1]

1. Emerson : on "Intellect."

The Metals and their Loss in Weight after Calcination

Very vague ideas prevailed among the Hindu philosophers on the constitutions of a metal and its fate after calcination. We are reminded of the diversity of opinion in Europe before the time of Lavoisier. A metal like every material substance was supposed to be formed of one or more of the five bhūtas ; in fact the earthy of saline element was taken to be predominant in it,[1] and hence it lent itself so readily to calcination.[2] It should be noted here that by "bhūta" was understood not so much an *element* in the modern connotation of the terms as certain qualities of individual substance. Even mercury was regarded as composed of the five bhūtas.[3] When a metal was submitted to calcination some of its component bhūtas escaped, leaving the earthy or saline portion behind ; hence the ash (calx) must weigh less than the original metal itself. There is, however, no direct reference to be met with anywhere as regarded as quantitative aspect of the question so clearly established by Lavoisier.

---

1. पार्थिवाः सुवर्णरजतमणिमुक्तामनः शिलामृतकपालादयः Ch. 1., 24. Again in Ch. XLVI (Sūtra) 345 *et seq*. the metals are included among the saline ingredients.
2. *Cf.*"For some maintain that gold is solid light, or, at least that the chief ingredient is light, which is rendered solid by mixture with some particles of earth. *Were it mere earth it might be calcined by fire strongly urged.*–Vol. i, p. 9.
3. पञ्चभूतात्मकः सूतस्तिष्ठत्येव सदाशिवः। इति रसार्णवे।

The porportion of loss which a metal undergoes on ignition was made the basis for determining its purity and we find in the law-books an arbitrary scale fixed. Thus according to Yājñavalkya, "gold is unaffected by fire, while one hundred parts by weight of silver tin and copper lose (lit. undergo decay to the extent of) two, eight and fifteen parts respectively.[1]" This is to be taken as referring not to the roasting (killing) of a metal but simply as allowance made to goldsmiths and other artisans for loss incurred in the shape of dross etc. when it is melted in a crucible.

That copper, lead, tin and other base metals (हीनधातव:) when repeatedly calcined are completely reduced to their calces was well-known as will be seen throughout the body of the texts. The passage from Rasendracūḍāmaṇi deserves some notice. Lead of the weight of 30 palas is to be calcined till its weight is reduced to a karṣa (= ¼ pala). Even if it be calcined now one thousand times, if will not undergo further reduction in weight. Our author evidently had his experience of lead derived from argentiferous galena, which would leave a residue in the shape of "button" of silver after the lead had been cupelled off.[2]

---

1. अग्नौ सुवर्णमक्षीणं द्विपलं रजते शते।
   अष्टौ त्रपुणि सीसे च ताम्रे पञ्चदशायसि।।
2. Similar confusion prevailed in Europe up till the end of the 17th century. The following extract from Robert Boyle's "Concerning the Unsuccessfulness of Experiments" will make the point clear.

   "Having, upon occasion, had the chemistry not long since to visit some mines of lead and other metals, I find that there is a great difference, and discernible even to the eye, betwixt the several ores ; for intance of lead, some of which I can show you so like steel, and so unlike common lead-ore, that the workmen upon that account are pleased to call it steel-ore. which being of more difficult fusion than ordinary, they are wont to mix it with order ore, which they call firm-ore, to facilitate the melting of it. And I likewise took notice of an ore, which for its aptness to vitrify, and serve the potters to glaze their earthen vessles, the miners call pottern-ore, and sell it (at least where I saw it digged up) dearer than other ore, from which it differs both visibly enough and as the workmen affirm in divers other (and those

Antimony

Although there is no direct recongnition of antimony as a distinct metal, some sort of allusion to it as a variety of lead is not wanting. The reason why antimony was often confounded with lead is that at first sight stibnite may be mistaken for galena. The mineral sauvīrāñjana or nilāñjana was thus indiscriminately applied both to the sulphides of antimony and of head (*cf.* vol. I. pp. 93-95) That nīlāñjana contained a new metal was often supected though its nature was not properly understood. Thus in Rasendracūḍāmaṇi we have "nīlāñjana, mined with tīkṣṇa (cast iron), and heated several times yield a superior kind of lead. Which is readily fusible and is of mild black colour.[1]

less obvious) qualities : and yet all these ores, after fusion, do pass indiscriminately under the name and notion of lead. In which therefore it is no wonder, that severer inquiries find a great deal of disparity. I remember I did not long since cause some lead-or to be tried, which being the most promising that ever I saw, made me suppose it might contain some considerable quantity of silver : but though it proved so rich in lead, as to yield after the rate of seventy pound to the hundred, yet one of the most export artists in Europe could not extract one grain of silver out of it ; whereas the lead of very many mines, being skilfully examined, will leave behind it, upon the test, a proportion of pure silver. And though his quantity of silver be not considerable enough to make such mines as yield it pass for silver mines (or, as we are wont to call them mines-royal) because the silver will not quit the cost of extracting it ; yet such mines though they pass but for lead mines with the metalist, may appear to be mixt mines to the naturalist, who may meet with divers experiments, wherein the little silver that is in them, may make their lead operate differently from that of those ores, which are wholly destitute of silver."—Edition of 1772, p. 323.

1. *Vide* Sans. texts, p. 52. The identical distich occurs also in R.R.S. *Cf.* Vol. i, p. 119. *Cf.* also "But to detain you no longer on this subject, give me only leave to strengthen the paradox I have proposed by the authority of that great and candid chymist *Basilius Valentinus*, who speaking of antimony, after he hath told us, that there are several kinds of it, and especially two ; the one more mercurial and of a golden property, witnessed by the shining streaks or beams it abounds with; the other more full of sulphur, but destitute of the gold nature that

The Preparation known as Svarṇasindūra or Makaradhvaja

As the "law of definite proportion" was unknown to the ancients, tedious processes were resorted to for the complete conversion of mercury into its sulphide. "Makaradhvaja" even at the present day is a favourite remedy with the Hindu physicians of the Āyurvedic school.[1] Let us follow the recipe as given in Rasendracintāmaṇi ; "when quicksilver is killed with an equal weight of purified sulphur, it becomes hundred times more efficacious ; when it is killed with twice its weight of sulphur is cures leprosy ; when it is killed with thrice its weight of sulphur it cures mental langour ; when it is killed with four times its weight of sulphur it removes grey hair and wrinkles ; when it is killed with five times its weight of sulphur it cures consumption ; and when it is killed with six times its weight of sulphur it is a panancea for all the ills that flash is heir to.[2] Now 25 parts by weight of mercury can only taken up (*i.e.* combine with) 4 parts by weight of sulphur; the excess of sulphur simply sublimes off unchanged. In the

---

enriches the former ; adds, that there is such a different goodness betwixt the several sorts of antimony as there is betwixt the several sorts of flesh or fish, which, though agreeing in name and if, you please, in nature, do exceedingly differ in point of goodness, which brings into my mind the great difference which I have found, even visible to the eye, betwixt the several sorts of Antimony." Boyle's work—l.c. p. 324.

1. If we take the historical meaning of the term "Āyurvedic" the above preparation cannot lay claim to this distinction, for it was unknown at the time of the Caraka and the Suśruta and was introduced into the Materia Medica during the Tāntric period.

2. तुल्ये तु गन्धके जीर्णे शुद्धाच्छतगुणो रसः
द्विगुणे गन्धके जीर्णे सर्वकुष्ठहरः परः।।
त्रिगुणो गन्धके जीर्णे सर्वजाड्यविनाशनः।
चतुर्गुणे तत्र जीर्णे वलीपलितनाशनः।।
गन्धे पञ्चगुणे जीर्णे क्षये क्षयहरो रुजः।
षड्गुणो गन्धके जीर्णे सर्वरोगहरो रसः।। र० चि०

actual preparation of svarṇasindūra (Lit. vermilion with gold) the mercury is first made into an amalgam with thin gold leaves and then repeatedly rubbed in a mortar with sulphur. During the process of sublimation (उद्ध्र्वपातन) the excess of sulphur volatilises off as we have already explained and the gold in fine particles remains behind as a residue. The lustrous, crystalline, reddish brown sublimate, which collects nearest the source of heat,[1] has the formula HgS. As there is an eroneous impression about the composition of this drug even among educated people in this country we give below the results of analysis of a sample of "mercury killed with six times its weight of sulphur along with gold."[2]

Identification of Metals by their Colouration of Flames

Prominent attention should be drawn to the following extract from Rasārṇava (Vol. 1. p. 68)

"Copper yields a blue flame * * * that of the Tin is pigeon-coloured ; that of the Lead is pale-tinted." * * *

We are not aware of similar tests being applied anywhere at such an early period as a qualitative test for metals.

The Age of Bhikṣu Govinda the author of Rasahṛdaya

There is a belief current in some parts of the Deccan that our Govinda is no other than the celebrated teacher of Śaṅka-rācārya.[3] If there be any historical foundation for the belief

1. Gopālakṛṣṇa very properly lays particles stress upon this point.
स्फोटयित्वा तु मुक्ताभमूद्र्ध्वलग्नं बलिं त्यजेत्।
अधःस्थं रससिन्दूरं सर्वरोगेषु योजयेत्।। इति रसेन्द्रसारसंग्रहः।
2. Percentage of sulphur found=13.89 ; the calculated amount being 13.79; not a trace of gold could be detected.
3. The following Ślokas are cited from Śaṅkaradigvijaya in support of the above :—
तस्योपदर्शितवतश्चरणौ गुहाया द्वारे न्यपूजयदुपेत्य स शंकरार्यः।
आचार इत्युपदिदेश स तत्र तस्मै गोविन्दपादगुरवे म गुरुर्यतीनाम्।।

Govinda should no doubt be placed in the 8th century A.D. Apart altogether from the question whether at such an early date the progress of chemical knowledge as revealed in Rasahṛdaya has been attained in India, the colophon at the end of the work would tell against such an hypothesis (Intro. liii). The author distinctly states that he wrote his book under the auspices of the king of the Kirātas and even throws out a hint that he was of the Buddhist persuasion. We have no valid reasons to believe that Śaṅkara, the sturdy champion of Brahmanical faith and the mighty dialecticians, whose activity proved the ruin of Buddhism in India should have sat at the feet of a Guru of the opposite creed.

व्यासः पराशरसुतः किल सत्यवत्यां तस्यात्मजः शुकमुनिः प्रथितानुभावः।
तच्छिष्यतामुपगतः किल गौड़पादो गोविन्दनाथमुनिरस्य च शिष्यभूतः।।
शुश्राव तस्य निकटे किल शास्त्रजालं
श्रीमच्छंकरदिग्विजयः, पंचमसर्गः, verse 101 and 105-6।
आनन्दाश्रमसंस्कृतग्रन्थावलिः-ग्रन्थाङ्क 22।

In the above Ślokas Govindanātha Muni is stated to be the preceptor of Śaṅkara, the former being himself the pupil of Gouḍapāda, who in his trun was a pupil of Śukadeva, son of Vyāsa, the well-known compiler of the Vedas. The Govinda nātha of the text has been confounded by the commentator, Dhanapati Śūri with the Buddhist Govinda. It is scarcely necessary to take serious notice of the chronological perspective of the author of the Saṅkaradigvijaya.

The Mechanical, Physical and Chemical Theories of the Ancient Hindus

(By Principal B.N. Seal)

I propose in this paper to give a synoptic view of the mechanical, physical and chemical theories of the ancient Hindus. A chronological survey, even if the materials for it were available, would be here of little account. The origins of Hindu natural philosophy in the speculations of the Brāhmaṇas and the Upaniṣads, or in the mythology of the Purāṇas, however interesting from the standpoint of *'cultur-geschichte,'* do not come within the scope of the present exposition, which relates to the results of systematic thought as directed to the phenomena and processes of Nature. I have therefore cofined myself to an account of natural philosophy as expounded in the principal systems of Hindu thought. The Sāṅkhya-Pātañjala system account for the Universe on principles of cosmic evolution the Vaiśeṣika-Nyāya lays down the methodology of science, and elaborates the concepts of mechanics, physics and chemistry. The Vedānta, the Pūrva-Mīmāṁsā, and in a less degree, the Bauddha, the Jaina, and the Cārvāka systems, make incidental contributions on points of special interest, but their value in this regard is critical and negative. The principal authorities followed in this account,—the Vyāsa Bhāṣya on Patañjali sūtras, the Saṁhitā of Caraka the Bhāṣya to Praśastapāda the Vārtika of Udyotakara, and the Vṛhat Saṁhitā of Varāhamihira—all centre round the Hindu Renaissance, the beginnings of the anti-Buddhist reaction, in the fourth, fifth and sixth centuries of the Christian era. Whenever I have made use of later authors, *e.g.* Kumārila Śaṅkara, Śrīdhara, Vācaspati, Udayana, Bhāskara Jayanta, Varavara, Raghunātha, Vijñānabhikṣu etc., I have taken care to see (except where the opposite is expressly mentioned) that no idea is surrepti-tiously introduced which is not explicity contained in the earlier authors.

The Sāṅkhya-Pātañjala System.

This system possesses a unique interest in the history of thought as embodying the earliest clear and comprehensive account of the process of cosmic evolution, viewed not as a mere metaphysical speculation but as a scientific principle, based on the conservation, the transformation, and the dissipation of Energy.

Prakṛti—the Ultimate Ground

The manifested world is traced in the Sāṅkhya to an unmanifested ground, Prakṛti, which is conceived as formless and undifferentiated. limitless and ubiquitous, indestrucible and undecaying, ungrounded and uncontrolled without beginning and without end. But the unity of Prakṛti is a mere abstraction ; it is in reality and undifferentiated manifold, an indeterminate infinite continuum of infinitesimal Reals. These Reals, termed Guṇas, may by another abstraction be classed under three heads, (1) Sattva, the Essence with manifests itself in a phenomenon. and which is characterised by this tendancy to manifestation, the Essence, in other words, which serves as the medium for the reflection of Intelligence, (2) Rajas, Energy, that which is efficient in a phenomenon, and is characterised by a tendancy to do work, or overcome resistance, and (3) Tamas, mass or inertia, which counteracts the tendency of Rajas to do work, and of Sattva to conscious manifestation.

The ultimate factors of the Universe, then, are : (1)

Essence, or intelligence-stuff, (2) Energy, and (3) Matter, characterised by mass of inertia.

These Guṇas are conceived to be Reals, substantive entities—not however as self-subsistent or independent entities (प्रधान) but as interdependent moment in every Real or substantive Existence.

Even Energy is substantive in this sense. The infinitesimals of Energy do not possess inertia or gravity, and are not therefore material, but they possess quantum and extensity (परिमाण–परिच्छिन्नत्व).

The very nature of Energy is to do work, to overcome. resistance (रजश्चलम् उपष्टम्भकं), to produce motion. All Energy is therefore ultimately kinetic. Even potential Energy (अनुद्भूतवृत्तिशक्ति) is only the Energy of motion in imperceptible forms.

The original constituents and their interaction. Every phenomenon. It has been explained, consists of a three-fold *arche,* intelligible Essence, Energy and Mass. In intimate union these enter into things as essential constitutive factors. The essence of a thing (sattva) is that by which it manifests itself to intelligence, and nothing exists without such manifestation in the Universe of Consciousness (समष्टिबुद्धि). But the Essence is only one of three moments. It does not possess mass or gravity, it neither offers resistance, nor does work. Next there is the element of Tamas, mass, inertia, matter-stuff, which offers resistance to motion as well as to conscious reflection. (तमः गुरु वरणकम्).

But the intelligence-stuff and the matter-stuff cannot do any work, and are devoid of productive activity in themselves. All work comes from Rajas, the principles of Energy, which overcomes the resistance of matter, and supplies even Intelligence with the Energy which it requires for its own work of conscious regulation and adaptation.

(अन्यान्याभ्याम् अहङ्काराभ्यां स्वकार्योपजनने राजसाहङ्कारः सहकारी भवति। लोकाचार्य तत्त्वत्रय-अचित्प्रकरण।)

The Guṇas are always uniting separating, uniting again. (अन्योन्यमिथुनाः सर्वे, नैषामादिसम्प्रयोगे वियोगे वा उपलभ्यते). Everything in the world results from their peculiar arrangement and combination. Varying quantities of Essence, Energy and Mass, in varied groupings, act on one another, and through their mutual interaction and interdependance evolve from the indenfinite or qualitatively indeterminate to the definite or qualitatively determinate. (एते गुणाः परस्परोपरक्तप्रविभागाः संयोगविभागधर्म्माणः इतरेतरोपाश्रयेण उपार्ज्जितमूर्त्तयः।–व्यासभाष्य). But though cooperating to produce the world of effects, these divers moments with divers tendencies never coalesce. In the phenomenal product whatever Energy there is due to the element of Rajas, and Rajas alone ; all matter, resistance, stability is due to Tamas, and all conscious manifestation to Sattva. (परस्पराङ्गाङ्गित्वेऽपि असंभिन्नशक्तिप्रविभागा–व्यासभाष्य। अन्योन्याङ्गाङ्गिभावेन उत्पादितेऽपि द्रव्ये प्रकाशगुणः सत्वस्यैव क्रियागुणः रजसएव स्थितिगुणस्तमसएव–विज्ञानभिक्षु योगवार्त्तिक on व्यासभाष्य, *ibid*—नैषां संकीर्यन्ते–वाचस्पतिमिश्र तत्ववैशारदी on व्यासभाष्य *ibid.*).

The nature of the interaction is peculiar. In order that there may be evolution with transformation of Energy, there must be a disturbance of equilibrium, a preponderance of either Energy or Mass-resistance, or Essence over the other moments. The particular Guṇa which happens to be predominant in any phenomenon, becomes manifest in that phenomenon and the others become latent, though their presence is inferred by their effect. For example, in any meterial system at rest, the Mass is patent, the Energy latent, and the conscious manifestation sub-latent. In a moving body, the Rajas, Energy is predominant (kinetic), while the Mass or rather the Resistance it offers is overcome. In the volitional consciousness accompanied with movement, the transformation of Energy (or work done by Rajas) goes hand in hand with the predominance of the conscious manifestation, while the matter-stuff or Mass, though latent, is to be inferred from the resistance overcome. (प्रधानवेलायाम् उपदर्शितसन्निधाना गुणत्वेऽपि च व्यापारमात्रेण प्रधानान्तर्नीतानुमितास्मिता: –व्यासभाष्य).

The starting point. The starting point in the cosmic history is a condition of equilibrium or equipoise consisting in a uniform diffusion of the Reals. The tendencies to conscious manifestation as well the powers of doing work are exactly counterbalanced by the resistance of the inertia or Mass. The process of cosmic evolution (परिणाम) is under arrest. (सत्त्वरजस्तमसां साम्यावस्था प्रकृति:–Pravacana-Sūtra 61, Chap. I. साम्यावस्था अन्यूनानतिरिक्तावस्था अन्यूनाधिकभावेन असंहतावस्था अकार्यावस्था इति निष्कर्ष:–विज्ञानभिक्षु, *ibid.*)

Beginning of Evolution. The transcendental (non-mechanical) influence of the Puruṣa (the Absolute) puts an end to this arrest, and initiates the process of creation. Evolution begins with the disturbance of the original equilibrium. How this is mechanically brought about is not very clear. A modern expounder of the Sāṅkhya supposes that the particles of Sattva, Rajas and Tamas possess a natural offinity for other particles of their own class, and that when the transcendental influence of the Puruṣa ends the state of arrest, the affinity comes into play, breaks up the uniform diffusion, and leads to unequal aggregation. and therefore, to the relative preponderance of one or more of the three Guṇas over the others. Thus commences formative combination among the Reals, and consequent productive activity. (संहननम् आरम्भकसंयोग: न्यूनाधिकभावेन अन्योन्यं संयोगविशेष:। Chap. I. Sūtra 66 प्रवचनभाष्य–विज्ञानभिक्षु। पृथिवीव्यक्तीनां पृथिवीत्वेनैव सत्त्वव्यक्तीनाम् एकजातीयतया एकता। स्वजातीयोपष्टम्भादिना वृद्धिह्रासादिकं च युक्तम् ibid. Sūtra 128, Chapt. I.)

Formation of wholes or systems—collocation of Reals :

Creative transformation accompanied with evolution of motion (परिस्पन्द) and work done by Energy (क्रिया) cannot take place without a peculiar collocation of the Reals (Guṇas). To form wholes or systems (समुदय), it is essential that one Guṇa should for the moment be preponderant, and the others co-operant. And this cannot be without an unequal aggregation which overthrows the original equilibrium (गुणवैषम्य, न्यूनाधिकभाव),–in other words, without unequal forces or stresses coming into play in different parts of the system. (गुणविमर्द–वाचस्पति।) व्यक्तं सक्रियं परिस्पन्दवत्...समुदयः समवायः। स च गुणानां गुणप्रधानभावमन्तरेण न सम्भवति। न गुणप्रधानभावो वैषम्यं विना। न च वैषम्यमुपमदद्वर्योपमर्दकभावादृते। वाचस्पति, कौमुदी on Kārikā 16.)

The Formula of Evolution—Differentiation in Integration.—Evolution (परिमाण) in its formal aspect is defined as differentiation in the integrated (संसृष्टविवेक). In other words, the process of Evolution consists in the development of the differentiated (वैषम्य) within the undifferentiated (साम्यावस्था), of the determinate (विशेष) within the indeterminate (अविशेष), of the coherent (युतसिद्ध) within the incoherent (अयुतसिद्ध). The evolutionary series is subject to a definite law which it cannot overstep (परिणामक्रमनियम). The order of succession is not from the whole to parts, nor from parts to the whole—but ever from a relatively less differentiated, less determinate, less coherent whole to a relatively more differentiated, more determinate, more coherent whole. That the process of differentiation evolves out of homogeneity separate or unrelated parts, which are then integrated into a whole, and that this whole again breaks up by fresh differentiation into isolated factors for a subsequent redintegration, and so on *ad infinitum*, is a fundamental misconception of *the course of material Evolution.* That the antithesis stands over against the thesis, and that the synthesis supervenes and impose unity *ab extra* on these two independent and mutually hostile moments is the same redical misconception as regards *the dialectical form* of cosmic development. On the Sāṅkhya view, increasing differentiation within the evolving whole, so that by this two-fold process what was an incoherent indeterminate homogenous whole evolves into a coherent determinate heterogeneous whole.

The different stadia in the order of cosmic Evolution are characterised as follows —

(1) The inconceivable, the unknowable, the formless, of which no character can be predicated (अलिङ्ग), including Prakṛti, on the Reals in a state of equilibrium.

(2) The knowable, the empirical universe, cosmic matter of Experience, things as matter of stuff of consciousness (लिङ्ग),—comprising Mahat, the intelligible Essence of the cosmos, evolved by differentiation and integration within the formless, characterless, inconceivable Prakṛti.

(3) Individuated but still indeterminate stuff bifurcating into two series—subject experience and Object-experience,—comprising on the one hand the indeterminate unity of apperception on the empirical Ego, as the co-ordinating principle of the Subject-series (अस्मिता), and on the other hand, the indeterminate material potencies, the subtile vehicles of potential Energy (तन्मात्र, सूक्ष्मभूत), the ultimate subtile constituents of the Object-series (the material world). The previous stadium, the cosmic matter of Experience (लिङ्ग, महत्) evolves within itself, by differentiation and integration, an individuated but still indeterminate stuff in two co-ordinated series, Subject and Object.

(4) Determinate stuff (विशेष) evolved within the indeterminate by further differentiation and

integration, *viz.* in the series of Subject-experience, sensory and motor stuff ; and in the Object series, a corresponding atomic matter-stuff actualising the material potencies in the form of specific sensible Energies. The latter includes the different classes of Paramāṇus, the different kinds of atomic constituents of different kinds of gross matter (स्थूलभूत).

(5) Coherent and intergrated matter-stuff, individual substances, characterised by generic and specific properties, which however are not rigidly fixed, but fluent being subject to a three-fold change and constantly evolving, (अयुतसिद्धावयवभेदानुगतः समूहः द्रव्यमिति पतञ्जलिः–व्यासभाष्य, Sūtra 44, Chap. III. सामान्यविशेषसमुदायो द्रव्यं–व्यासभाष्य, *ibid.*, सामान्यविशेषात्मा अन्वयो धर्मी, ibid. धर्मिणां धर्मलक्षणावस्थापरिणामः).

(6) And so the cosmic series moves on in ascending stages of unstable equilibrium (विसदृशपरिणाम) until the reverse course of equilibriation and dissipation of energy (सदृशपरिणाम and साम्यावस्था), which even now constantly accompanies the evolution and transformation of Energy, com-pletes the disintegration of the Universe into its original unmanifested ground the unknowable Prakṛti.

The order of cosmic Evolution according to the व्यासभाष्य (Sūtra 19, Pada II) is shown below, in a tabular form :—

Prakṛti, the unmanifested unknowable ground
(गूढमव्यक्तमलिङ्गं)

Cosmic matter of experience (महत्, लिङ्गं)

Subject series, (अविशेष)
Individuated Indeterminate mind-stuff (unity of apperception, empirical Ego. (अस्मिता)

Determinate mind-stuff (विशेष) sensory and motor psychoses. etc. (ज्ञानेन्द्रिय, कर्मेन्द्रिय and मनः)

Object series (अविशेष)
Individuated but indeterminate matter-stuff (subtile material potencies (तन्मात्र)

Determinate matter-stuff (विशेष) atomic and molecular constituents of gross matter (परमाणु–स्थूलभूत)

Coherent and integrated matter-stuff (अयुतसिद्धावयवः समूहः सङ्घातः द्रव्यं)
Individual substances, with generic and specific characters subject to constant change or evolution *e.g.* inorganic objects composed of atoms or molecules (परमाणु), vegetable organisms (वृक्ष) animal organisms (शरीर) (अयुतसिद्धावयवः सङ्घातः शरीरं वृक्षः परमाणुरिति–व्यासभाष्य, Sūtra 44, Pada III).

तत्र तत् संसृष्टं विविच्यते क्रमानतिवृत्तेः तथा षड्विशेषाः लिङ्गमात्रे संसृष्टा विविच्यन्ते, परिणामक्रमनियमात्, तथा तेषु अविशेषेषु भूतेन्द्रियाणि संसृष्टानि विविच्यन्ते, न विशेषेभ्यः परं तत्त्वान्तरमस्ति तेषान्तु धर्मलक्षणावस्थापरिणामा व्याख्यायिष्यन्ते। (व्यायभाष्य, Sūtra 19 , Pada II)

N.B.:—The usual order given in the Sāṅkhya compendiums is as follows :—Prakṛti, Mahat, Ahaṅkāra,—and then the bifurcation, *viz.* 11 organs sensory motor and common sensori-motor from Rājasic Ahaṅkāra, and

Tanmātras from Tāmasic Ahaṅkāra,—and finally the Paramāṇus of the Sthūla-Bhūtas.

The conservation of Energy (and of Mass)—the transformation of Energy :—

The Guṇas (Reals), though assuming an infinite diversity of forms and powers, can neither be created not destroyed. The totality of the Mass (Tamas), as well as of Energy (Rajas) remains constant, if we take account both of the manifested and the unmanifested, the actual and the potential. But the individual products of the evolutionary process, the conerete phenomenal modes resulting from the combined action of the original Mass, Energy and Essence, are subject to addition and subtraction growth and decay, which are only due to changes of collocation, and consequent changes of state from the potential to the actual (in other words, from the future to the present and from the present to the past, in a time series,)—changes, which are illusorily ascribed to the Reals themselves. The different collocations of Mass and Energy give birth to the divers powers of things, the various forms of Energy which may be calssedas like and unlike ; indeed the course of Evolution from the Reals conforms to a fixed law, not only as regards the order of succession, but also as regards the appearance (and mutual relations) of like and unlike Energies. And this transformation is constantly going on,—the course of Evolution is not arrested for a moment.

गुणास्तु सर्वधर्मानुपातिन: न प्रत्यस्तमयन्ते नोपजायन्ते व्यक्तिभिरेव अतीताऽनागतव्ययागमवतीभि: गुणान्वयिनीभि: उपजनापायधर्मका इव प्रत्यवभासन्ते (व्यासभाष्य, Sūtra 19, Pada. II.) परिणामिनित्यता गुणानाम्। सर्वमिदं गुणानां सन्निवेशविशेषमात्रम्। (*ibid.*, Sūtra 13. Pada IV.) एते गुणा: तुल्यजातीयातुल्यजातीयशक्तिभेदानुपातिन: (*ibid.*, Sūtra 19. Pada II.) परिणामक्रमनियमात् (व्यासभाष्य, Sūtra 19, Pada II.) क्रमान्यत्वं परिणामान्यत्वे हेतु: (Patañjali Sūtra 15, Pada III.) प्रकृति: परिणमनशीला क्षणमप्यपरिणम्य नावतिष्ठते।

The doctrine of Causation, a corollary from the

conservation and transformation of Energy :—the principle of collocation—the storing-up and the liberation of Energy:—

The Sāṅkhya views of causation follows at once as a corollary from this doctrine of the conservation and transformation of Energy. As the total Energy remains the same, while the world is constantly evolving, cause and effect are only more or less evolved forms of the same ultimate Energy. The sum of effects exists in the sum of causes in a potential (or un-evolved) form. The grouping or collocation alone changes, and this brings on the manifestation of the latent powers of the Guṇas but without creation of anything new. What is called the (material) cause or sum of material causes is only the power which is efficient in the production or rather the vehicle of the power. This power is the unmanifested (or potential) form of the Energy see free (उद्भूतवृत्ति) in the effect. But the concomitant conditions are necessary to call forth the so-called material cause into activity. When the favourable combination or co-operation of concomitants is wanting, there is no manifestation of the effect. The question is—what is the aid which the concomitant conditions render to the determination (and production) of the effect existing in potency in its material cause ? First there is the merely mechanical view as illustrated by some commonplace examples, *e.g.*, manifestation of the figure of the statue in the marble block by the causal efficiency of the sculptor's art, or of the oil in the sesamum by pressing, or of the grain of rice out of the paddy by the process of husking. In these cases the manifestation of an effect is only its passage from potentiality to actuality, a stadium in the process of evolution from possible (future) existence to actual (present) existence : and the concomitant condition (सहकारिशक्ति) or efficient cause (निमित्तकारण), the sculptor's chiselling or pressing, the husking, is a sort of mechanical or instrumental help to this passage or transition. कार्यशक्तिमत्त्वमेव उपादान-कारणत्वम्। सा शक्तिः कार्यस्य अनागतावस्था एव। स एव विशेषः (उत्पत्तेः प्राक् कारणे विशेषः) अस्माभिः कार्यस्य अनागतावस्था इत्युच्यते। (विज्ञानभिक्षुप्रवचनभाष्य, Sūtra, 115, Chap I.)

अभिव्यक्तिः वर्त्तमानावस्था। कारणव्यापारोऽपि कार्यस्य वर्त्तमानलक्षणपरिणाममेव जनयति। यथा शिलामध्यस्थ प्रतिमायाः लैङ्गिकव्यापारेण अभिव्यक्तिमात्रं तिलस्थतैलस्य च निष्पीडनेन, धान्यस्थतण्डुलस्य च अवघातेन। (विज्ञानभिक्षु-प्रवचनभाष्य, Sūtra, 129, Chap. I.)

These mechanical examples of the Kapila Sāṅkhya have the merit of simplicity, but the Patañjali-Sāṅkhya brushes them aside and explains causation on the basis of the conservation and transformation of Energy, advancing it as the liberation of potential Energy existing stored up in a Guṇa collocation, (the sum of materials causes) the liberation following on the action of the proximate efficient cause, or concomitant condition (निमित्तकारण).

The causal operation of concomitant conditions (efficient causes) lies only in this that they supply a physical stimulus which liberates the potential Energy stored-up in a given collocation. Everything in the phenomenal world is but a special collocation of the ultimate Reals (Energy, Mass and Essence). The sun of (material) causes potentially contains the Energy manifested in the sum of efforts ; and in the passage from potency to actualisation, the effectuating condition (the concomitant cause) when it is itself accomplished, is only a step in the evolutionary series, which adds a specific stimulus, and renders determinate that which was previously indeterminate. When the effectuating condition is added to the sum of meterial conditions, in a given collocation. All that happens is that a stimulus is imparted, which removes the arrest, disturbs the relatively stable equilbrium, and brings on a liberation of Energy (उद्भूतवृत्तिशक्ति) together with a fresh collocation (गुणसन्निवेशविशेष). सर्वमिदं गुणानां सन्निवेशविशेषमात्रम् इति परमार्थतो गुणात्मनः (व्यासभाष्य, Sūtra 13, Pada IV.) सतश्च फलस्य निमितं वर्त्तमानीकरणे समर्थ न अपूर्वोपजनने। सिद्धं निमितं नैमित्तिकस्य विशेषानुग्रहणं कुरुते। नापूर्वमुत्पादयति। (व्यासभाष्य, Sūtra 12, Pada IV.)

Describing the production of bodies (organic vehicles)

for individual souls, not of matter of Prakṛti, under the influence of their merit and demerit, as concomitant conditions, Patañjali points out that non-material concomitants like merit and demerit do not supply any moving force or Energy to the sum of material conditions but only remove the arrest (the state of relatively stable equilibrium) in a given collocation, even as the owner of a field removes the barrier in flooding his field from a reservoir of water. This description is intended to represent the super-physical influence of non-material concomitants (or causes) like volition, merit and demerit, etc., but the casual operation of a material concomitant condition is essentially the same—there is the same reservoir of stored-up Energy in a given collocation,—the same condition of arrest of relatively stable equilibrium,—the same liberation of the stored-up potential Energy which flows along the line of least resistace; the only difference being that in the case of material concomitants the stimulus which removes the arrest in physical, instead of being transcendental as in the case of non-material causes like will, merit and demerit, etc.

The Vyāsa-bhāṣya helps us to a clear mental representation of the details of this process, being perhaps the finest example before Newton of the exercise of a Scientific Imagination, and as memorable as any it in the whole history of thought containing as it does the theory of potentials as in nutshell—As the owner of many fields can irrigate, from a field which is already flooded, others of the same or a lower level, without forcing the waters thereto with his hands, and merely by making an opening in the barrier or dyke, on which the waters rush in by their own force;–or further, as the same perosn cannot force these waters, or the earthy matters held in solution therein, into the roots of the rice plants, but only removes the obstructive grasses and weeds, on which the fluids of their own power enter roots; such is the action of an effectuating condition (निमित्त) added to a sum of material causes or conditions.

निमित्तम् अप्रयोजकं प्रवृत्तीनां वरणभेदस्तु ततः क्षेत्रिकवत्। (Patañjali Sūtra,

3, Pada IV.) न हि धर्मादि निमित्तं यथा क्षेत्रिक: केदारात् अपां पूरणात् केदारं पिप्लावयिषु: समं निम्नं वा निम्नतरं वा नाप: पाणिनापकर्षति आवरणं तु आसां भिनत्ति तस्मिन् भिन्ने स्वयमेवाप: केदारान्तरम् आप्लावयन्ति।

Chain of Causation—fixed order

The order of Evolution with the transformation of the Energies follows a definite law. The unalterable chain of causes and effects in the phenomenal world illustrates this fixed order. But though the cosmic order is one and fixed. It comprehends divers series arising from different combinations of the original Guṇas, which constitute subordinate or particular laws of cause and effect.(क्रमान्यत्वं परिणामान्यत्वे हेतु:, Patañjali Sūtra 15, Pada III.)

What we call the qualities of things are only modes of Energy acting in those collocations. योग्यतावच्छिन्ना धर्मिण: शक्तिरेव धर्म:-स च फलप्रसवभेदानुमितसद्भाव:। (व्यासभाष्य, Sūtra, 14, Pada III.) ते खल्वमी धर्मा वर्त्तमाना व्यक्तात्मान:, अतीताऽनागता: सूक्ष्मात्मान: सर्वमिदं गुणानां सन्निवेशमात्रमिति परमार्थतो गुणात्मान:। (व्यासभाष्य, Sūtra 3, Pada IV.) And these various Energies are sometimes actual (kinetic), sometimes potential rising to actuality, and sometimes sublatent, subsiding from actuality into sublatency. In fact, the original Energy is one and ubiquitous, and everything therefore exists in everything else, *potentailiter* (सर्वं सर्वात्मकमिति), without prejudice to the generic and specific differences of things (जात्यनुच्छेदेन सर्वं सर्वात्मकं). Inorganic matter, vegetable organisms, and animal organisms are essentially and ultimately one (जलभूम्यो: पारिणामिकं रसादिवैश्वरूप्यं स्थावरेषु दृष्टं तथा स्थावराणां जङ्गमेषु जङ्गमानां स्थावरेषु) so for as Mass and Energy are concerned, but the varied forms of Energy and the generic and specific qualities (or properties) of things, which are but modes of Energy, follow a definite unalterable law in the order of their appearance and succession, under conditions of space, time, mode and causality, and hence all effects do not manifest themselves at once देशकालाकारनिमित्तापबन्धात् न खलु समानकालम् आत्मनामभिव्यक्ति: (*Ibid.*, Sūtra 14, Pada III.) यो यस्य धर्मस्य समनन्तरो धर्म: स तस्य क्रम:। पिण्ड: प्रच्यवते घट उपजायते इति धर्मपरिणामक्रम:। लक्षणापरिणामक्रमक: घटस्य अनागतभावात् वर्त्तमानभावक्रम:, तथा पिण्डस्य वर्त्तमानभावात् अतीतभावक्रम:। (व्यासभाष्य, Sūtra 15, Pada III.)

Time space, and the Causal Series :—

A Tanmātra (infra-atomic particle of subtile matter) is conceived by our understanding to stand in three relations—(1) position in space (देशावच्छिन्न), (2) position in the Time-series (कालावच्छिन्न) and (3) position in the causal series (निमित्तावच्छिन्न.).

These three relations are the work of the intuitive stage of knowledge as opposed to the conceptual (निर्विकल्पप्रज्ञा as opposed to सविकल्पप्रज्ञा.) But this is not the pure relationless intuition of Reality (निर्विचारा निर्विकल्पप्रज्ञा) which may be termed intelletual intuition, but the intuition that imposes its forms on the Real substrate (सविचारा निर्विकल्पप्रज्ञा) or in other words mepirical intuition. तत्र भूतसूक्ष्मेषु अभिव्यक्तधर्मकेषु देशकालनिमित्ता-नुभवावच्छिन्नेषु या समापत्ति: स सविचारा इत्युच्यते। तत्रापि एकबुद्धिनिर्ग्राह्यमेव उदितधर्मविशिष्टं भूतसूक्ष्ममालम्बनीभूतं समाधिप्रज्ञायामुपतिष्ठते। (व्यासभाष्य Sūtra 44, Pada I.)

Infinite Time is a non-entity objectively considered, being only a construction of the understanding (बुद्धिनिर्माण) based on the relation of antecedence and suquence in which the members of the phenomenal series are intuited to stand to one another. These phenomenal changes as intuited by us in the empirical consiousness fall into a series, which the Understanding conceives, as order in Time. The Time-series, then, is a *schema* of the Understanding for representing the course of Evolution. The *schema* of the Understanding supervenes on the phenomenal world as under in time, and hence in the empirical consciousness the Time-series appears to have an objective reality, and to form a continuum. As there is an ultimate and irreducible unit of extensive quantity (परिमाण) in the Guṇas or infinitesimal Reals of Prakṛti, which are without constituent parts, so the moment may be conceived as the ultimate and irreducible unit of this time-continuum as represented in the empirical conciousness. A moment therefore cannot be thought of as containing any parts standing in the relation of antecedence and sequence. If change is represented by the Time-series, a moment as the unit of time may be supposed to represent the unit of change. Now all physical change may be reduced to the motion of atoms in Space, and we may therefore define the moment as representing the ultimate unit of such change—*viz.*, the (instantaneous) transit of an atom (or rather a Tanmātra) from one point in space to the next succeeding point. Even an atom has constituent parts (the Tanmātras), and hence an atom must take more than one moment to change its position. The motion of that which is absolutely simple and without parts from one point in Space to the next must be instantaneous, and conceived as the absolute unit of change (and therefore of time, क्षण). If this is held to be an irreducible absolute unit, it will follow that what we represent as the time-continuum is really discrete. Time is of one dimension. Two moments cannot co-exist, Neither does any series of moments exist in reality. Order in Time is nothing but the relation of antecedence and sequence, between the moment that is, and the moment that went just before. But only one moment, the present, exists. The future and the past have no meaning apart from potential and sublatent phenomena. One kind of transformation, to which a thing is subject, is that it changes from the potential to the actual, and from the actual to the sublatent. This may be called the change of mark (लक्षणा-परिणाम) as opposed to change of quality (धर्मपरिणाम), and the change due to duration or lapse of time (अवस्थापरिणाम). The present is the mark of actuality,—the future, the mark of potentiality—and the past, of sublatency—in a phenomenon.

Only one single moment is actual, and the whole Universe evolves in that one single moment. The rest is but potential or sublatent.

क्षणतत्क्रमयोर्नास्ति वस्तुसमाहार इति बुद्धिसमाहारः मुहूर्त्ताहोरात्रादयः। स खल्वयं कालः वस्तुशून्यो बुद्धिनिर्माणः शब्दज्ञानानुपाती लौकिकानां व्युत्थितदर्शनानां वस्तुस्वरूप इव अवभासते। क्षणस्तु वस्तुपतितक्रमावलम्बी। क्रमश्च क्षणानन्तर्यात्मा। तं कालविदः काल इति आचक्षते योगिनः। न च द्वौ क्षणौ सह भवतः, क्रमश्च न द्वयोः सहभुरवोरसम्भवात् पूर्वस्मादुत्तरभाविनो यदानन्तर्ये क्षणस्य स क्रमः। तस्मात् वर्त्तमान एवैकः क्षणः न पूर्वोत्तरक्षणाः सन्तीति तस्मान्नास्ति तत्समाहारः। ये तु भूतभाविनः क्षणाः ते परिणामान्विता व्याख्येयाः। तेनैकेन क्षणेन कृत्स्नो लोकः परिणाममनुभवति, तत्क्षणारूढ़ाः खल्वमी धर्माः। यथाऽपकर्षपर्यन्तं द्रव्यं परमाणुः एवं परमापकर्षपर्यन्तः कालः क्षणः। यावता वा समयेन चलितः परमाणुः पूर्वदेशं जह्यात् उत्तरदेशम् उपसम्पद्येत स कालः क्षणः, तत्प्रवाहाविच्छेदस्तु क्रमः। (व्यासभाष्य, Sūtra 52, Pada III.)

Vijñāna-bhikṣu points out that this does not amount to a denial of Time. It means that time has no real (or objective) existence apart from the 'moment'. But the latter is real, being indentical with the unit of change in phenomena (गुणपरिणामस्य क्षणत्ववचनात्). But even this is real only for our empirical (relative) consciousness (व्युत्थितदर्शनं) which intuits the relation of antecedence and sequence into the evolving Reals (Guṇas), in the stage of 'empirical intuition' (सविचारा निर्विकल्पप्रज्ञा). The intellectual intuition (निर्विचारा, निर्विकल्पप्रज्ञा). on the other hand, apperehends the Reals as they are, without the empirical imported relations of Space. Time and Causality.

Space as extension and Space as position

Space must be distinguished as Deśa (locus, or rather extension) and Dik (relative position). Space (Dik) as the totality of position, or as an order of co-existent points, is wholly relative to the Understanding, like order in time, being constructed on the basis of relations of position intuited by our empirical (or relative) consciousness. But there is this difference between Space-order and Time-order:—there is no unit of Space as position (Dik), though we may conceive a unit of Time, *viz.*, the moment. (क्षण) regarded as the unit of change in the phenomenal or causal series (परमाणुक्रिया or गुणपरिणामस्य क्षणत्वचनात्–योगवार्त्तिक, Sūtra 51, Pada III.). Spatial position (Dik) results only from the different relations in which the all-pervasive Ākāśa stands to the various finite (or bounded) objects. On the other hand, Space as extension or locus of a finite body, Deśa, (देश स्थित्याधारः), has an ultimate unit, being analysable into the infinitesimal extensive quantity inherent in the Reals (Guṇas) of Prakṛti.(गुणप्रकृतेरणुपरिणामः–योगवार्त्तिक। एतेन नित्या दिगपि अप्रामाणिकी व्याख्याता, सामान्यतो दिग्व्यवहारभावात्। पूर्वादिव्यवहारस्य दिगुपाधिभिरेव सम्भवात्। सामान्यतः कालदिग्व्यवहारसत्त्वेऽपि आकाशादेव तदुपपत्तेश्च। कालाच्च दिश्ययं विशेषः यत् कालः क्षणरूप इष्यते, दिक् तु सर्वथैव नेष्यते।

The Casual series

The relation of Cause and Effect has been already explained. It only remains to add that the category of causality is mediated through the *schema* of order in Time. The Empirical intuition first superimposes relations of antecedence and sequence on changing phenomena (the evolving Guṇas or Reals), and the Understanding out of these relations creats order in Time. The Empirical intuition then intuites the phenomenal series of transformations of Energy in this Time-order and in so doing, imports the relation of cause and effect into the course of Nature. (कार्यकारणभावादीनां क्षणघटितत्वात्–योगवार्त्तिक, Sūtra, 51, Pada III.)

The dissipation of Energy (and of Mass)—their dissolution into the formless Prakṛti :—Cosmic Evolution (परिणाम) is a two-fold process, creative as well as destructive, dissimilative as well as assimilative, katabolic as well as anabolic (अनुलोमसर्ग and विलोमसर्ग, विसदृशपरिणाम and सदृशपरिणाम). In one aspect, there is the aggregation (unequal aggregation) of Mass and Energy, with consequent transformation of

Energy, resulting in the creation of inorganic as well as organic matter, and the genesis of worlds. The successive steps of this process may be described as (1) unequal aggregation with storing-up of Energy in a certain collocation, under a state of arrest (*i.e.* in a state of relatively stable equilibrium), (2) a stimulus removing the arrest, and disturbing the equilibrium, and (3) liberation of the Energy, moving on to a fresh collocation, fresh aggregation, arrest and equilibrium. The process of the world thus moves on from equilibrium to equilibrium, and the result of that process is the development of a coherent determinate heterogeneous whole (लिङ्ग अविशेष विशेष, अयुतसिद्धावयवसमूह, युतसिद्धावयवसमूह) in what is essentially in incoherent indeterminate homogeneous whole (अलिङ्ग).

But there is a second aspect of this evolutionary process. Unequal aggregations are unstable, there is a constant tendency in things to go back to the original stable equilibrium, the state of uniform equal diffusion of Reals. This process is called the resolution of like to like (सदृशपरिणाम), consisting in assimilation and dissipation, and being the exact opposite of the process of "differentiation in the intergrated" which he evolved the Cosmos. The collocations of Mass, Energy, and Essence are always breaking up, and the Energy as well as the mass, however slowly, however imperceptibly, are being dissipated, *i.e.* dissolved into the original formless Prakṛti a state of permanent equilibrium and arrest, from which there is and can be no return, except under the tanscendental influence of the Absolute at the commencement of a new creative cycle. Not that there is a destruction of the Mass or Energy, but a dissipation or dissolution into a condition of equal unifrom diffusion from which there is no return. This is not the phenomenon of kinetic Energy disappearing and becoming potential or sublatent, for in such cases there is restitution or reconversion by natural means. When this reverse current of assimilation (and dissipation) prevails over the current of dissimilation (and integration), the Universe will disintegrate more and more, until is disappears in the formless Prakṛti, its unknowable source and ground. (*Vide* तत्त्वत्रय and कौमुदी).

The Evolution of matter (तान्मात्रिक सृष्टि) :-

The ultimate constitution of Matter is a question of the profoundest interest in the Sāṅkhya-Pātañjala system. Three stages clearly stand out in the genesis of Matter :— (1) the original infinitesimal units of Mass or inertia, absolutely homogeneous and ubiquitous, on which Energy does work, when the original equilibrium comes to an end (भूतादि-तामसाहङ्कार), (2) the infra-atomic unit potentials, charged with different kinds of Energy, which result from the action of Energy on the original units of Mass, (तन्मात्र), and (3) the five different classes of atoms, the minutest divisions of which gross matter is capable, but which are themselves complex Tanmātric systems (स्थूलभूतपरमाणुं.)

The first stadium Bhūtādi is absolutely homogeneous and absolutely inert, being devoid of all physical and chemical characters (रूपादिभिरसंयुतं) except quantum or mass (परिच्छिन्नत्व, परिमाण); and this admits neither of addition nor of subtraction, can neither be created not destroyed. The second stadium Tanmātra represent subtile matter, vibratory, impingent, radiant, etc., instinct with potential energy. These potentials arise from the unequal aggregation of the original mass-units in different proportions and collocations with an unequal distribution of the original Energy (Rajas). (तन्मात्र रूपादेः किं कारणम् इति चेत् स्वकारणद्रव्याणां न्यूनाधिकभावेन अन्योऽन्यं प्रति संयोगविशेष एव-स्वजातीयोपष्टम्भादिना वृद्धिह्रासादिकं च युक्तं).

The Tanmātras possess something more than quantum or Mass and Energy. They possess physical character, some of them penetrability (आकाशदान), others powers of impact is pressure, others radiant heat, others again capability is viscous and cohesive attraction. In intimate relation to these physical characters they also possess the potentials of the energies represented by sound, touch, colour, taste and smell, but being subtile matter they are devoid of the peculiar forms (विशेष) whichy these potentials assume in particles of gross matter like the atoms and their aggregates. In other words, the potentials lodged in subtile matter must undergo peculiar transformations by new groupings or collocations to be classed among sensory stimuli—gross matter being supposed to be matter endued with properties of the class of sensory stimuli, though in the minutest particles thereof the sensory stimuli may be infra-sensible (अतीन्द्रिय but not अनुद्भूत). (तस्मिंस्तस्मिंस्तु तन्मात्राः ते च अविशेषिणः। ते च पदार्थाः शान्तघोरमूढ़ाख्यैः स्थूलगतशब्दादिविशेषैः शून्या एकरूपत्वात्। तथाच शान्तादिविशेषशून्यदादिमत्त्वमेव भूतानां शब्दादितन्मात्रत्वम्।)

The Tanmātras, then, are infra-atomic particles charged with specific potential energies—first, the potential of the sound-stimulus is lodged in one class of particles, Tanmātras which possess the physical energy of vibration (परिस्पन्द), and serve to form the redicle of the ether atom (आकाशपरमाणु), then the potential of the tactile stimulus is lodged in another class of Tanmātras, particles which possess the physical energy of impact or mechanical pressure in addition to that of vibration and serve to form the redicle of the gas atom (Vāyu Paramāṇu) ; next, the potential of the colour stimulus is lodged in a third class of Tanmātras, particles which are charged with the energy of radiant heat and light in addition to those of impact the vibration and serve to form the nucleus of the light and heat corpuscle ; then the potential of the tastestimulus is lodged in other Tanmātras, particles which possess the energy of viscous attraction, in addition to those of heat, impact and vibration, and which afterwards develop into the atom of water, and lastly, the potential of the smell-stimulus is lodged in a further class of Tanmātras, particles which are charged with the energy of cohesive attraction, in addition to those of viscous attraction, heat, impact and vibration and which serve to form the redicle of the earth-atom.

षड्विशेषाः तद्यथा शब्दतन्मात्रं स्पर्शतन्मात्रं रूपतन्मात्रं रसतन्मात्रं गन्धतन्मात्रञ्च इत्येकद्वित्रिचतुस्पञ्चलक्षणाः शब्दादयः पञ्चाविशेषाः। व्यासभाष्य, Sūtra 19, Pada II. अहङ्कारात् शब्दतन्मात्रं ततश्चाहङ्कारसहकृतात् शब्दतन्मात्रात् शब्दस्पर्शगुणं स्पर्शतन्मात्रम्। एवंक्रमेण एकैकगुणवृद्ध्या तन्मात्राणि उत्पद्यन्ते (प्रवचनभाष्य–Sūtra 62, Chap. I.)

शब्दादीनां मूर्त्तिसमानजातीयानाम् एकः परिणामः पृथिवीपरमाणुः तन्मात्रावयवः। भूतान्तरेषु अपि स्नेहौष्ण्यप्रणामित्वाऽकाशदानानि उपादाय सामान्यम् एकविकारारम्भः समाधेयः। (व्यासभाष्य, Sūtra 14, Pada IV.) तद्यथा। गन्धतन्मात्रं वर्ज्जयित्वा चतुस्तन्मात्राणां स्नेहजातीयानाम् एकः परिणामः जलपरमाणुः तेषां च महाजलादिः, एवं गन्धरसौ वज्जयित्वा औष्णाजातीयानां त्रितन्मात्राणां तेजोऽणुः तेभ्यः महातेज आदिः। एवं गन्धरसरूपाणा वर्ज्जनात् द्वाभ्यां वायवणुः। तेभ्यः महावायुवादिः। एवं शब्दतन्मात्रादहङ्कारांशसहकृतात् आकाशाणुः तेभ्यः महाकाशादिः।

विज्ञानभिक्षु remark:— अच दर्शने अयं सिद्धान्तः शब्दादितन्मात्रपञ्चके काठिन्यस्नेहादिव्यङ्ग्या पृथिवीत्वादिजातयः सन्ति। (योगवार्त्तिक, Sūtra, 14, Pada IV.)

Before explaining the gensis of atoms, it is necessary to say something about Ākāśa, which is the link between the infra-atomic particles (Tanmātras), and atoms (Paramāṇus). Ākāśa correspounds in some respects to the ether of the physicists, and in others to what may be called proto-atom (protyle). In one aspect Ākāśa is all-presive (विभु), and devoid of the property of impenetrability which characterises even the infra-atomic potential units (Tanmātras). In another aspect, Ākāśa is described as having originated out of the mass or inertia in Prakṛti (Bhūtādi) when the latter became charged with the first potential vibration (the sound-potential). Vijñāna-bhikṣu in the Yoga-Vārttika boldly tackles the difficulty. Ākāśa, he explains, has two forms, original and derivative, non-atomic and atomic The original Ākāśa is the undifferentaited formless Tamas (mass in Prakṛti, matter-rudiment—Bhūtādi). Which is devoid of all potentials, and is merely the all-pervasive seat or vehicle of the ubiquitous original Energy (Rajas). This Ākāśa must not be confounded with vacuum, which is merely nagative (आवरणाभाव—un-occupiedness), though it must be conceived as all-pervasive, occupying the same space as the various froms of gross matter (समानदेशकम्–अवकाशस्वरूपम् आकाशं–योगवार्त्तिक), and therefore devoid of the property of impenetrability (मूर्त्तान्तरासमानदेशत्वं–योगवार्त्तिक) which characterises atomic matter. But when the original equilibrium (साम्यावस्था) comes to an end, unequal aggregation form collocations is different groups and proportions of the three Guṇas. (न्यूनाधिकभावेन अन्योऽन्यं संयोगविशेष संहननं–विज्ञानभिक्षु, प्रवचनभाष्य and योगवार्त्तिक). The transformation of Energy now begins—working on a collocation of mass (with Essence), (गुणान्तरोपसमृद्धं सदंशतः–विज्ञानभिक्षु, योगवार्त्तिक); it first gives rise to the sound potential (शब्दजननयोग्यार्थरूपेण परिणमति–विज्ञानभिक्षु) and the atomic Ākāśa (proto-atom, protyle) is but an integration of the original unit of mass charged with this vibration-potential. This vibratory (or rather rotary) ether-atom (आकाशाणु) is integrated, limited (परिच्छिन्न) and as such cannot occupy the same space with other (subsequently integrated) atoms. But this proto-atomic integratoin of Ākāśa (कार्याकाश) is formed everywhere, and itself residing in the ubiquitous non-atomic Ākāśa (कारणाकाश–अवकाशरूपमाकाशं) forms the universal medium in which air or gas atoms, light and heat corpuscles, and other atoms move and float about. (वायोरावणं। यदि हि अवकाशरूपम् आकाशं न स्यात् तदा मूर्त्तद्रव्येषु स्थाल्यादिषु अन्तस्तेजआदिप्रवेशो न स्यात्–विज्ञानभिक्षु, योगवार्त्तिक Sūtra, 40, Pada III. आकाशं हि कार्यकारण-रूपेण द्विविधम्। तव कारणाकाशं तमोगुणविशेषतयैव व्यवह्रियते। तदानीम् आकाशताव्यञ्जकशब्दादिगुणविशेषाभावात्। तच्च कारणाकाशं गुणान्तरोसमृद्धं सदंशतः आदौ शब्दजननयोग्यार्थरूपेण परिणमते। ततश्च अणुसंघातेन पृथिवीव देव महाभूताकाशम् अहङ्कारापेक्षया परिच्छिन्नं वायोरावरणम् उत्पद्यते (विज्ञानभिक्षु, योगवार्त्तिक, Sūtra 40, Pada III.)

The gensis of the infra-atomic unit-potentials (Tanmātras) and of the atoms.

The subject of the gensis and the structure of the Tanmātras and the Paramāṇus was a fascinating one to these ancient thinkers, and a wide divergence of view prevailed. I will here notice several typical views :—those of the Viṣṇu Purāṇa, Parāśara, Patañjali and a certain School of Vedāntists reported in the तत्त्वनिरूपण.

I. A famous passage in the Viṣṇu Purāṇa explains the genesis and the structure of the Tanmātras and of the Bhūtas (Paramāṇus) in the following manner :—

The first Tanmātra originated from the rudiment-matter (Bhūtadi), the individuated but still undeterminate potential-less Mass in Prakṛti, under the action of Energy (उपष्टम्भत: परिणमयन्ति- प्रवचनभाष्य, Sūtra 62 Ch. I.) by a process of disintegration and emanation (विकुर्वाण:-विष्णुपुराण- एकविकाराम्भ:-व्यासभाष्य:, Sūtra 14, Pada IV. विकरणभाव:, विकीर्णातास्वभावा व्यापिता इति यावत्, योगवार्त्तिक on Patañjali's Sūtra.— मनोजवित्वं विकरणभाव: etc.) in the menstruum, or surrounding medium of the unindividuated Cosmic Mass (Mahat) (महतावृत:).

This first subtile matter, the first result of Mass-disintegration and Energy-transformation, is charged with the sound-potential, the potential of vibration or oscillation (परिस्पन्द). It is called the sound-potential (शब्द तन्मात्र).

This is typical of the genesis (and structure) of the other Tanmātras (kinds of subtile matter). In each of the remaining cases, an atomic Mass charged with actual specific energies (भूतपरमाणु) disintegrates and emanates, and thus evolves a form ot subtile matter (a kind of Tanmātra) under the action of Energy, and always in the same menstruum or sorrounding medium—that of Bhūtādi, the super-subtile. Each kind of subtile matter becomes charged with a new potential in addition to the potentials already evolved. The gensis of an atom a Bhūta-Paramāṇu, is a quite different process. Here the unit potential (Tanmātra) receives an accretion of Mass, and by a sort of condensation and collocation evolves an atom (Bhūta-Paramāṇu).

The genesis and structure of the Tanmātras and the Bhūta Paramāṇus are worked out below :—

1. The super-subtile individuated Mass (rudiment-matter (Bhūtādi) under the action of the original Energy (रजः) disintegrates and emanates (विकुरुते) in the menstruum or surrounding medium (समावरण) in Mahat, cosmic super-subtile Mass, and evolves a form of subtile matter (तन्मात्र), which becomes charged with the sound-potential (vibration-potential, परिस्पन्द), and is called the unit of sound-potential (शब्दतन्मात्र).
2. This subtile matter, the Mass, charged with sound-potential, receives an accretion of Mass from the rudiment matter (Bhūtādi) any by condensation and collocation evolves the Ākāśa Bhūta, the atomic Ākāśa, the proto-atom charged with the specific energy of the sound stimulus (actual vibratory motion.) (स एव भूतादिः शब्द तन्मात्रात् स्वद्वितीयात् शब्दगुणकम् आकाशं ससर्ज्ज। तथाच अहङ्कारशब्दतन्मात्राभ्यां मिलित्वा आकाशं सृज्यते–योगवार्त्तिक on the Viṣṇu Purāṇa passage, Sūtra, 14, Pada IV).
3. This proto-atom, the atomic Ākāśa, charged with its actual specific energy, again disintegrates and emanates, under the action of the original Energy, and in the menstruum of the rudiment-matter (supersubtile Mass) and thus evolves another kind of subtile matter (Tanmātra) which becomes charged with the touch potential (the potential of impact or mechanical pressure (प्रणामित्व, वहनशीलत्व) in addition to the sound-potential (vibration-

potential—परिस्पन्द) and is called the unit of touch-potential (स्पर्शतन्मात्र).

4. Next, this subtile matter, the mass charged with touch (and sound) potential ; *i.e.* with the potentials of vibration and impact, receives an accretion of mass again from the rudiment-matter (Bhūtādi) and by condensation and collocation, evolves the Vāyu Bhūta, a kind of gaseous matter or air of which the atoms are charged with the actual specific energy of the touch stimulus, *i.e.* with actual energy of impact in addition to the actual energy of vibratory motion.

5. Next, the atom of Vāyu, so charged with the actual specific energy of impact and vibration again disintegrates and emanates, under the action of the original Energy, and in the same menstruum or surrounding medium of the rediment matter (super-subtile Mass—Bhūtādi) and thus evolves another kind of subtile matter (Tanmātra), which becomes charged with the heat-potential (औष्ण्यं–रूपं –heat and light-potential) in addition to the impact potential and the vibration-potential, and is called the unit of colour-potential (रूपतन्मात्र).

6. Now this subtile matter, this radiant matter, charged with light-and-heat-potential, and also with impact and vibration potential, receives an accretion of Mass again from the rudiment-matter (Bhūtādi), and by condensation and collocation evolves the Tejas Bhūta, the light-and heat-corpuscle, which is changed with the specific Energy of the colour-stimulus, *i.e.* radiates actual heat and light (विकिरण) in addition to manifesting the energy of impact (impingency) and of vibration (or osciallation).

7. Next, this atom or light-and-heat-corpuscle disintegrates, and emanates as before a form of subtile matter charged with the taste-potential (रसतन्मात्र),

in addition to the three potentials already generated, and also with the physical potential of viscous attraction.

8. This subtile matter charged with the taste-potential and with the potential of viscous attraction condenses and collocates as before into the water-atom which manifests the actual specific energies of viscous attraction and the taste-stimulus.
9. The viscous water-atom charged with the actual specific Energy of the taste-stimulus disintegrates, and emanates as before a form of subtile matter charged with the smell-potential in addition to the four potentials already generated and aslo with the potential of cohesive attraction.
10. This subtile matter charged with the smell-potential and with the potential of cohesive attraction condenses and collocates as before into the earth-atom, which manifests the actual specific Energies of cohesive attraction and the smell-stimulus.

Vijñāna-bhikṣu in the Yoga-Vārttika briefly summarises the Viṣṇu Purāṇa process as follows :—

Bhūtādi as radicle in conjunction with Mahat produces the sound potential, which as radicle in conjunction with Bhūtādi produces Ākāśa, which as radicle in cojunction with Bhūtādi produces the touch-potential, which as radicle in conjunction with Bhūtādi produces Vāyu, which as radicle in conjunction with Bhūtādi produces the colour potential, which as radicle in conjunction with Bhūtādi produces Tejas and so on.

In this brief summary he does not bring out the force of विकुर्वाण: (the disintegrating process), and the distinction between the genesis of subtile and gross matter (Tanmātra and Bhūta).

यथा प्रधानेन महान्, महता च तथावृतः।
भूतादिस्तु विकुर्वाणः सर्गं तन्मात्रिकं ततः।
ससर्ज शब्दतन्मात्रात् आकाशं शब्दलक्षणं।
शब्दमात्रं तदाकाशं भूतादिः स समावृणोत्।
आकाशस्तु विकुर्वाणः स्पर्शमात्रं ससर्ज ह।
बलवान् अभवत् वायुः तस्य स्पर्शो गुणो मतः।(विष्णुपुराण)

II. A famous passage in Parāśara takes another view of the gensis and structure of the Tanmātras and the Bhūta-Paramāṇus. Kṛṣṇāpāda, in the तत्त्वत्रयविवरण, represents the scheme as follows :—

The Tanmātras originate from one another in one linear series, and each Bhūta originates in a separate line from its own Tanmātra :—

भूतादि–(Bhūtādi).

शब्दतन्मात्र–*Sound-Tanmātra*—as a redicle or centre surrounded or encircled by Bhūtādi generates Ākāśa.

स्पर्शतन्मात्र–*Touch-Tanmātra*—as a radicle or centre encircled by Sound-Tanmātra with Ākāśa-atom as a help generates Vāyu—

रूपतन्मात्र–*Colour Tanmātra*—as a radicle or centre encircled by Touch-Tanmātra with Vāyu-atom as a help generates Tejas—

रसतन्मात्र—*Taste-Tanmātra*—as a radicle or centre encircled by colour-Tanmātra with Tejas-atom as a help generates Ap—

गन्धतन्मात्र–*Smell-Tanmātra*—as a radicle or centre encircled by Taste-Tanmātra with Ap-atom as a help generates Pṛthivī.

The genesis of a Bhūta-Paramāṇu (atom) from the subtile matter of a Tanmātra is not here so simple as in the view of the Viṣṇu Purāṇa. The latter speaks of condensation and collection, but in the passage under reference a Tanmātra is supposed to act as a redicle, as the centre of a system, surrounded or encircled by Tanmātras of the immediately higher order in the medium of its own Bhūta.

Thus an atom of Ākāśa has the following structure :—

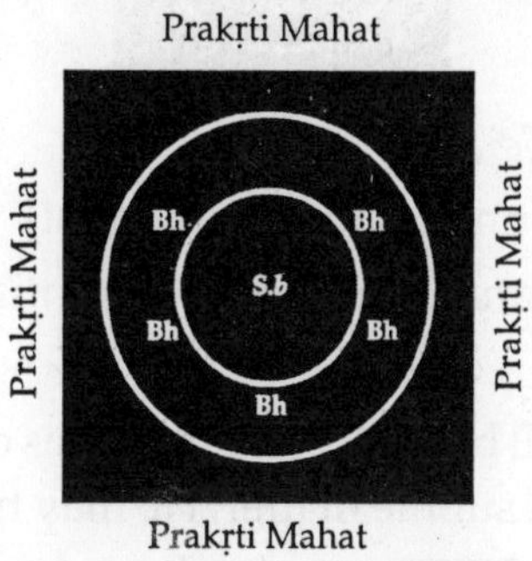

An atom of Ākāśa

Sb=Śabda-Tanmātra (vibration-potential)

Bh=(Bhūtādi)

An atom of Vāyu is constituted as follows—

This takes place within the surrounding medium of Ākāśa.

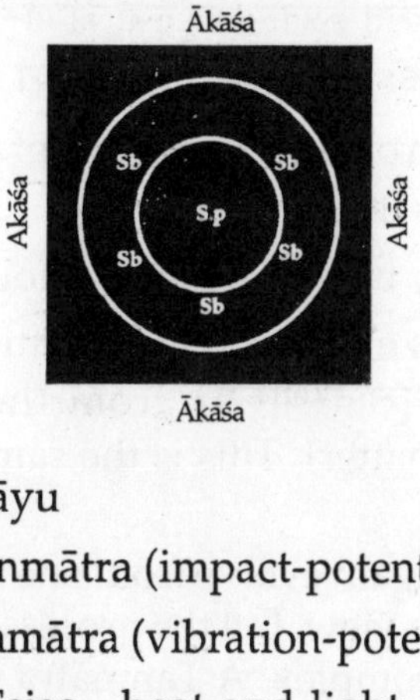

An atom of Vāyu

Sp=Sparśa-Tanmātra (impact-potential)

Sb=Śabda-Tanmātra (vibration-potential)

An atom of Tejas—heat-and-light corpuscle—has the following structure :—

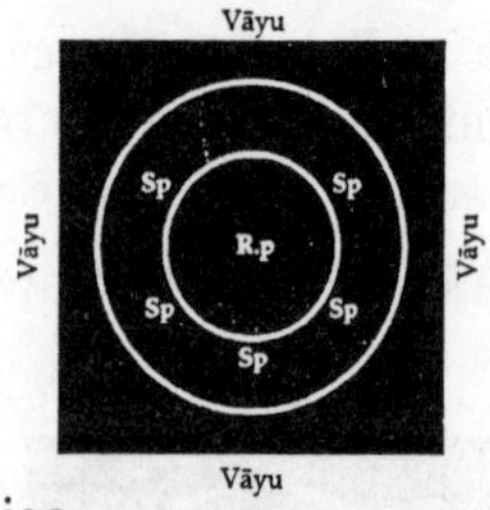

An atom of Tejas

Rp=Rūpa-Tanmātra (light-potential)

Sp=Sparśa-Tanmātra (impact-potential)

and so on.

A chemist will be disposed to push his chemical processes into the region of subtile matter. He may translate आवरण as a menstruum, and सहायक as a catalytic agent. In this case, an atom of Vāyu will be cosidered as generated from the impact-Tanmātra as a radicle, in the menstruum of vibration-Tanmātra, with Ākāśa-atoms as a catalytic agent.

अत्रायं क्रमः–भूतादेः शब्दतन्मात्रं जायते, शब्दतन्मात्रं भूतादिरावृणोति, तत आकाशं जायते, ततः अस्मात् शब्दतन्मात्रात् स्पर्शतन्मात्रं जायते, स्पर्शतन्मात्रं शब्दतन्मात्रामावृणोति, एवं शब्दतन्मात्रावृतात् आकाशसहायकात् स्पर्शतन्मात्रात् वायुर्जायते, ततः अस्मात् स्पर्शतन्मात्रात् रूपतन्मात्रं जायते, रूपतन्मात्रं स्पर्शतन्मात्रमावृणोति, एवं स्पर्शतन्मात्रवृतात् वायुसहायकात् रूपतन्मात्रात् तेजो जायते–and so on. अयं क्रमः तत्त्वत्रयविवरणे कृष्णपादैरुक्तः। वरवरमुनि, तत्त्वत्रयभाष्य (अचित्प्रकरण).

A slight variation of the above views of ascribed to certain school of Vedāntists in the तत्त्वनिरूपण.

The scheme may be represented as follows:—

A Bhūta-atom is evolved by integration (condensation and collocation, स्थूलावस्थारूप) from the corresponding Tanmātra (subtile matter). This is the same view as that of the Viṣṇu Purāṇa.

The Tanmātras again evolve from one another in a lineal series as in Parāśara's view. But the process of this generation is somewhat more complex. A Tanmātra first disintegrates and emanates in a surrounding medium (a menstruum) of the Tanmātra just preceding it in order of genesis and with the help of its own Sthūla Bhūta as a sort a catalytic, generates the Tanmātra next in order ; *e.g.*, the infra-atomic impact particles (स्पर्शतन्मात्र) disintegrate or emanate, in a sorrounding atmosphere of the vibratory subtile matter (शब्दतन्मात्र) and then with the help of their own atomic integration Vāyu, gas, generate the Tanmatra next in order, the subtile matter of radiant light-and-heat (तेजः).

III. Patañjali view, as expounded in the व्यासभाष्य and योगवार्त्तिक is as follows :—

(a) The order is genesis of various forms of subtile matter (potentials):—

(1) Bhūtādi, the rudiment-matter, original Mass, acted on by Rajas, Energy, produces the sound-potential (vibration-potential) अन्यान्याभ्याम् अहङ्काराभ्यां स्वकार्यजनने राजसाहङ्कारः सहकारीभवति।

(2) This subtile vibration-potential, as a radicle, with accretion of rudiment-matter (Bhūtādi) condensing and collocating, and acted on by Rajas, generates the subtile touch-potential (impact-potential) which is impingent as well as vibratory (oscillating).

(3) This subtile impact-potential again, as a radicle, with accretion of rudiment-matter (Bhūtādi) condensing and collocating, and acted on by Rajas, generates the subtile light-and-heat potential (रूपतन्मात्र) which radiates light-and-heat, in addition to being impingent and vibratory.

(4) Next, the light-and-heat potential, as a radicle, with accretion of rudiment-matter (Bhūtādi) condensing and collocating as before, generates the subtile taste-potential, which is charged with the potential of the taste-energy, and of viscous attraction, in addition to being vibratory, impingent and radiant.

(5) Lastly, the subtile taste-potential as a radicle, with accretion of rudiment-matter as before, condensing and collocating, generates the subtile smell-potential, which is charged with the potential of the smell-energy, and also of cohesive attraction, in addition to being vibratory, impingent and radiant.

The order of gensis of the Bhūta Paramāṇus, (forms of atomic matter).

The five classes of atoms are generated as follows :—

(1) The sound-potential, subtile matter, with accretion of rudiment-matter (Bhūtādi) generates the Ākāśa-atom.

(2) The touch-potentials combine with the vibratory particles (sound-potential) to generates the Vāyu-atom.

(3) The light-and-heat potentials combine with touch-potentials and sound-potentials (*i.e.* with impact particles and vibratory particles) to produce the Tejas-atom.

(4) The taste-potentials combine with light-and-heat potentials, touch-potentials and sound-potentials (*i.e.* with radiant, impingent and vibratory particles) or gene-rate the Ap-atom and

(5) The smell-potentials combine with the preceding potentials (*i.e.* with particles of touch-energy and with radinat, impingent and vibratory particles) to generate the earth-atom.

The Ākāśa-atom possesses penetrability, the Vāyu-atom impact or mechanical pressure, the Tejas-atom, radiant heat-and-light, the Ap-atom, viscous attraction, and the Earth-atom, cohesive attraction.

Vijñāna-bhikṣu in one passage gives the following scheme of the genesis of the Bhūtas :—

A radicle of sound-potential with rudiment-matter gives Ākāśa-atom (Bhūtādi), a radicle of touch-potential with Ākāśa-atom gives Vāyu-atom, a radicle of light-and-heat potential with Vāyu-atom gives Tejas-atom, a radicle of taste-potential with Tejas-atom gives Ap-atom, and a radicle of smell-potential with Ap-atom gives Earth-atom. On this view, an atom

of Ākāśa= Bh (Sb)
of Vāyu= {Bh(Sb)} (Sp)
of Tejas= {Bh (Sb)} (Sp) Rp

where Bh=भूतादि, Sb=शब्दतन्मात्र, Sp=स्पर्शतन्मात्र, Rp=रूपतन्मात्र, and so on.

Bhūtās and Paramāṇus—Cosmo-gensis and its successive stages.

The 'five Bhūtas' stand for a classification of substances of the basis of their generic properties resulting, as the Sāṅkhyas hold, from the structural type of their constituent atoms—a classification more physical than chemical, or properly speaking chemico-physical, unlike the purely chemical classification of the so-called elements of modern chemistry. A Paramāṇu, again, is a type of atoms corresponding to each Bhūta class, and indeed one and the same kind of Paramāṇu may comprehend atoms of different masses, if only these should agree in their structural type.

Cosmo-gensis—a bird's eye view:—

Out of the all-pervasive rudiment-matter (Bhūtādi) appeared Ākāṣa (ether), first as a Tanmātra (subtile matter) charged with the potential energy of sound (vibration-potential), and then as an atomic integration of a mono-Tanmātric structure (the Ākāśa-atom—आकाशाणु) also ubiquitous and all-enveloping. In the next stage we find a new kind of Tanmātras, systems of the infra-atomic vibratory particles, so arranged as to manifest a new forms of energy, that of impact or machanical pressure and these Tanmātras (स्पर्शतन्मात्राणि) combining with the vibration-potentials (Ākāśa Tanmātra) produced a new kind of atom, the di-Tanmātric Vāyu-atom, which by aggregation formed a gaseous envelope composed of impinging (driving) vibratory particles (Vāyu). Next appeared the third class of Tanmātras, infra-atomic system of the impinging vibratory particles, which by their collocation developed a new form of Energy—the energy of radiant heat-and-light. These Tanmātras (रूपतन्मात्राणि) combining with the potentials (Tanmātras) of vibration and impact, produced a new kind of atom—the tri Tanmātric Tejas-atom, the light-and-heat corpuscle, which by aggregation enveloped the gaseous world in huge flames. In the next stage we have the fourth class of Tanmātras, new and complex infra-atomic systems of the radiant impinging vibratory particles, which evolved the energy of viscous attraction as well as the potential Energy concerned in the taste-stimulus. These Tanmātras (रसतन्मात्राणि) combining with the three previoius ones, gave rise to another class of atoms, the tetra-Tanmātric Ap-atom, and flaming gases were thus precipated into cosmic masses of viscous fluid matters (Ap.). Finally appeared the fifth class of Tanmātras infra-atomic systems of the viscous radiant impinging vibratory particles which developed new forms of Energy—the Energy of cofesive attraction, as well as the potential energy concerned in the stimulus of smell. These Tanmātras (गन्धतन्मात्राणि) uniting with the other four kinds of infra-atomic subtile particles, formed another class of atoms, the penta-Tanmātric Earth-atom. Thus the viscous fluid matters, were condensed and transformed into the Earth-Bhūta, comprising the majority of the so-called elements of chemistry.

The Purāṇas, in their own fanciful way, conceive that, in the course of cosmic evolution, each succeding Bhūta appeared within an outer envolpes of its immediate predecessor, with a total mass (or volume ?) a tenth less than that of the latter.

वियत् तत्राप्येकदेशगतो वायुः प्रकल्पितः।
चिन्तयेत् वह्निमप्येवं मरुतो न्यूनवर्त्तिनं।।

ब्रह्माण्डावरणेष्वेषां न्यूनाधिक-विचारणा।
वायोर्दशांशतो न्यूनो वह्निर्वायौ प्रकल्पितः।।
पुराणोक्तं तारतम्यं दशांशैर्भूतपञ्चके।।
(पञ्चदशी-भूतविवेक) ślokas 72, 81 and 82)

Examples of the different Bhūtas :—

1. *Ākāśa*. This is ubiquitous.
2. *Vāyu*. Various substances composed of di-Tanmātric atoms—kind of Vāyu—must have been formed in the geseous envelope in the second stage of cosmic evolution, out of the proto-atoms of Ākāśa. But they have either suffered a fresh transformation into substances of a more complex atomic structure, or have dissipated into the mono-Tanmātric Ākāśa out of which they took their rise. The one familiar example now surviving is atmosphric air Water-vapour (वाष्प) is but water (Ap), and smoke, fumes, etc. but earth-particles in gaseous diffusion.
3. *Tejas:*—Various classes of Tejas corpuscles,—substance with tri-Tanmātric atomic structure, *i.e.* two grades subtler than the ordinary elements of chemistry (which are of a penta-Tanmātric structure),—are even now known. (तेजो भीमादिभेदेन बहुविधं। भीमं दिव्यं औदर्यं आकरजं-वरवरमुनि,–तत्त्वत्रयभाष्य)।

First, there is fire, or the light-and-heat emitted by the burning log of wood or lamp (अग्निः–दीपः–भीम तेजः) Now, it is important to note that the flame of a burning log of wood (इन्धन) or an oil-lamps is not pure Tejas, a pure mass of light-and-heat corpuscles. There is chemical union with Earth-particles (particles of the hard penta-Tanmātric substance) acted on by Energy ; and then the Tejas corpuscles, light-and-heat particles which are latent (absorded) therein, come forth as flame (पार्थिवीपष्टम्भेन तदनुगतात् तेजसः अग्निर्भवति–विज्ञानभिक्षु–प्रवचनभाष्य, Sūtra 110, Chap. V.) Then there is the light of the sun and the stars (दिव्यं सौरं आदित्यादि) which are flaming masses of molten viscous matters (जलमात्रेन्धनं तेज: दिव्यं तच्च सूर्यादि-तत्त्वत्रयभाष्य, अचित्प्रकरण) or of molten earthy matters (सूर्यादीनि सर्वाणि तेजांसि आर्थिवद्रव्यसङ्गेनव अवस्थितानि, विज्ञानभिक्षु, प्रवचनभाष्य, Sūtra 13, Chap. III.) There is also the lightning, which liberates a kind of Tejas latent in the aqueous particles and vapours, under the action of Energy, in the same way as an ordinary fire liberates the Tejas latent in the wood or other fuel. Next there are the stores of animal heat derived from the break-up of the nutritive material (औदर्यं). Lastly there comes the puculiar form of the Tejas Energy (radiant Energy stored up in the metalliferous ores and igneous rocks which have been formed in the subterranean heat. Here Earthly matters are mixed up, but the radiant Energy predominates in the composition of the metals (तेजआधिक्येन तैजसादिता सुवर्णादीनां-विज्ञानभिक्षु, प्रवचनभाष्यं, Sūtra 19, Chap. III.) Aniruddha, a late Sāṅkhya commentator, notes in reference to 'igneous bodies' that the greater part of their mass is derived from the Earth-Bhūta, though the Tejas particles determine the peculiar chemical combination, which produces them; and this must also be his view of the composition of the metals. तत्र अपि (तैजसे शरीरे अपि) बहुतरपार्थिवावयवावष्टम्भकत्वं, अल्पत्वे-चानुपभोगात्। (अनिरुद्धवृत्ति-Sūtra, 112, Chap. V.)

4. Ap. This viscous fluid of a tetra-Tanmātric structure has but one pure example, *viz.*, water, though the various organic acids, the juices of fruits and the saps of plants, are supposed to be transformations of watery radicles combined with different kinds of earthy accretions.
5. Lastly the Earth-Bhūta, the hard full-formed matter, with its penta-Tanmātric atoms, comprises by far the majority of the so-called chemical elements.

The question is—how does one and the same Bhūta, of the same formal structure, comprise different kinds of elements, with different atomic masses, and different characteristic properties ? And the answer is not far to seek. The properties of a thing are only the energies that are manifested in the particular collocations of the three Guṇas—Mass, Energy and Essence; and a tri-Tanmātric, or a penta-Tanmātric atom, *i.e.* an atom composed of three or of five kinds of Tanmātras may differ from another of the same class, in respect of the number of constituent Tanmātras of any particular kind, as also of their collocation or grouping, and therefore in mass as well as in generic and specific characters.

The Sāṅkhya-Pātañjala conceives the properties (or energies) of substances to result from the grouping or the quanta of the Tanmātras or the Guṇas themselves, and hence any radical differences in substances of the same Bhūta class must characterise their atoms, though in an infrarsensible form. In the Nyāya-Vaiśeṣika, on the other hand, the atoms of the same Bhūta class are alike in themselves, homogeneous; and the variety of substances comprehended under the same Bhūta, is ascribed merely to the different arragements or groupings of the atoms (व्यूह) and not of their components, for components they have none.

As a typical and familiar instance of the variety of characteristic properties (or energies) that may result from variations in accompaniment or grouping, the Sāṅkhya-Pātañjala points to the various kinds of fruit acids and juices, all originating from one and the same Bhūta (water) with different accretions of earthy matters (भूविकाराः). In the same way, though we speak of only five classes of Tanmātras and atoms, the infinite variety of the world results from the infinitely varied collocations of the three original Guṇas, which underlie Tanmātra and atom alike (सत्त्वरजस्तमसां अपकर्षकाष्ठापन्नः संघातः परमाणुरिति–न्यायवार्त्तिक, उद्योतकर, reporting the Sāṅkhya view : अयुतसिद्धावयवः संघातः परमाणुरिति–व्यासभाष्य, Sūtra 44, Pada III.—परमाणुः सामान्यविशेषात्मा च्युतसिद्धावयवभेदानुगतेः समुदायः-

व्यासभाष्य, Sūtra 44, Pada IV,—योग्यतावच्छिन्ना धर्म्मिणाः शक्तिरेव धर्म्मः (*ibid.*, Sūtra 14, Pada III.) यो धर्मेषु अनुपाती सामान्य-विशेषात्मा सोऽन्वयी धर्मी (ibid., Sūtra 14, Pada III.)

स्यादेतत्–कथं एकरूपाणां गुणानां अनेकरूपा प्रवृत्तिः–इत्यत आह परिणामतः सलिलवत्। यथा वारिदविभुक्तं उदकं एकरसं अपि भत्तद्भूविकारान् आसाद्य नारिकेल-तासी-तिल्वचिरविल्व-तिन्दुकामलक-कपित्थ फलरसतया परिणामात् मधुराम्लवणतिक्त कटुकषायतया विकल्पते, एव मेकैकणगुणसम्भवात् प्रधानं गुणं आश्रित्य अप्रधानगुणाः परिणामभेदान् प्रवर्त्तयन्ति। तदिदमुक्तं प्रतिप्रति गुणाश्रयविशेषात्, एकैकगुणश्रयेण यो विशेषस्तस्मात् इत्यर्थः।

(वाचस्पति–कौमुदी–On Kārikā 16.)

If we take a unit of rudiment-matter (Bhūtādi) for the unit of mass (*cf.* the mass of an electron with a charge of motion etc.) and represent the first Tanmātra by t_1, such units, and if further, t_2 , t_3, t_4, t_5 units of mass (Bhūtādi) be successively added to each accretion, to form a fresh Tanmātra, then the second, third, fourth and fifth Tanmātras will respectively contain t_1+t_2, $t_1+t_2+t_3$, $t_1+t_2+t_3+t_4$, and $t_1+t_2+t_3+t_4+t_5$ units of mass.

Also the Vāyu atom (bi-Tanmātric system) will contain $t_2+(t_2+t_2)$, *i.e.* $2t_1+t_2$ units of mass; the Tejas-atom (tri-Tanmātric system) will contain $t_1+(t_1+t_2)+(t_2+t_2+t_3)$, *i.e.* $3t_1+2t_2+t_2$ units; the Ap-atom (tetra-Tanmātric system) $t_2+(t_2+t_2)+(t_1+t_2+t_3)+(t_1+t_2+t_3+t_4)$, *i.e.* $4t_1+3t_2+2t_3+t_4$ units ; and the Earth-atom (penta Tanmātric system), $t_1+(t_1+t_2)+.....+(t_1+t_2+t_3+t_4+t_5)$, *i.e.* $5t_1+4t_2+3t_3+2t_4+t_5$ units.

If t units of mass be added to the first Tanmātra to form the atom of Ākāśa, the latter will contain t_1+t units of mass.

In other words the number representing the mass-units (Tamas) in the different classes of atoms (gross matter) will form an ascending series, *viz.*, t_1+t, $2t_1+t_2$, $3t_1+2t_2+t_3$, $4t_1+3t_2+2t_3+t_4$, and $5t_1+4t_2+3t_3+2t_4+t_5$.

Now if a follower of the Sāṅkhya-Pātañjala were asked to account for different among Paramāṇus of the same Bhūta class, he would perhaps suppose t_1 to vary from α_1 to β_1 t_2 from α_2 to β_2 t_3 from α_3 to β_3 t_4 from α_4 to β_4 and t_5 from α_5 to β_5.

Therefore, the mass-units contained in the Vāyu atoms of the different possible Vāyu substances would be represented by $2\alpha_1+t, 2\alpha_1+1+t, 2\alpha_1+2+t, \ldots\ldots 2\beta_1+t$, in A.P. with unity as common difference, there being $2(\beta_1—\alpha_1)+1$possible Vāyu substances.

The mass-units contained in the Tejas-atoms of the different possible Tejas substances would be represented by the series, $3\alpha_2+2\alpha_2+\alpha_3$, $3\alpha_1+2\alpha_2+\alpha 3+1, \ldots\ldots 3\beta_1+2\beta_2+\beta_2$, increasing in A.P. by units as common difference, then being $3(\beta_1-\alpha_3)+2(\beta_1-\alpha_2)+(\beta_3—\alpha_3)+1$, Tejas substances possible.

The mass-units contained in the Ap-atoms of the different possible Ap-substances would form the series $4\alpha_1+3\alpha_2+2\alpha_2+\alpha_4$, $4\alpha_1+3\alpha_2+2\alpha_3+\ \alpha 4+_1, \ldots\ldots 4\beta_1+3\beta_2+2\beta_3+\beta_4$, there being $4(\beta_1—\alpha_1)+3(\beta_2—\alpha_2)+2(\beta_3—\alpha_4)+3$ Ap-substances possible.

The mass-units contained in the Earth-atoms of the different possible Earth-substances would form the series in A.P.,$5\alpha_1+4\alpha_2+3\alpha_3+2\alpha_4+x5\alpha_I, \ldots\ldots 5\alpha_1+4\alpha_2+3\alpha_3+2x4+\alpha 5\alpha_I, \ldots\ldots 5B_1+4B_2+3B_2+2B_4+B_5$, $5(B_1—\alpha_1)+4(B_2—\alpha_2)+3(B_2—\alpha_3)+2(B_4—\alpha_4)+(B_5—\alpha_5)+1$ Earth substances possible.

Size (परिमाण).–As to size or volume, the Sāṅkhya accepts only two kinds—the infinitesimal, which is also without parts (अणु–निरवयव) and the non-infinitesimal, which consists of parts (महत्–सावयव). The latter varies from the excessively small (the so-called Aṇus, Tanmātras and Paramāṇus) to the indefinitely great (परममहत्–विभु,–*e.g.*—Ākāśa).

The Guṇas alone are infinitesimal, with the exception of those ubiquitous ones that evolve into Ākāśa-atoms and Mind-stuff (आकाशाणु, and अन्तःकरण):–all the rest of the evolved products (whether subtile or gross matter) are non-infinitesimal.

Vijñānabhikṣu notes that all the Guṇas (Reals) cannot be ubiquitous. If this were the case, that distrubance of equilibrium, that unequal aggregation with unequal stress and strain, with which cosmic evolution begins, would be impossible. The Guṇas, which give rise to Ākāśa and Mind-stuff, must be held to be ubiquitous, and this will suffice for the ubiquity of Prakṛti. (न च अस्मदर्शने गुणाख्यप्रकृतेः विभुत्वमेवावगम्यते इति वाच्यम्। अन्तः करणाकाशहेतुगुणानां विभुत्वेन तदुपपत्ते। सर्वगुणानां विभुत्वे सति आद्यपरिणामहेतुक्षोभसंयोगाद्यसंभवात् (योगवार्त्तिक, Sūtra 51, Pada III.)

Chemical analysis and synthesis—Elements and Compounds.

What then is the equivalent in the Sāṅkhya Pātañjala of the distinction between a chemical element and a chemical compound, or is there none ? Did or did not this elaborate physical analysis and classfication of things lead on to a classification based on chemical analysis and chemical synthesis ? These are questions of singular interest, the answer to which will disclose some new points of view from which the ancient Hindu thinkers approached the problems of chemical physics and physical chemistry.

Aggregates (समूहाः) may, in regard to their structure, be divided into two classes, (1) those of which the parts are in intimate union and fusion, being lost in the whole (अयुतसिद्धा-वयवाः समूहाः) ; and (2) mechanical aggregates, or collocations of distinct and independent parts (युतसिद्धावयवाः समूहाः).

A substance is an aggregate of the former kind, and may be divided into two classes, (1) the Bhūtas and their 'isomeric' modifications (भूत, भूतभेद and भूतविकार); and (2) chemical compounds (मिलितद्रव्य संहतभूतार्थ). Chemical compounds, again may be subdivided into two classes, (1) those composed of atoms of the same Bhūta class, *i.e.* of different isomeric modifications of the same Bhūta and (2) thsoe composed of atems of different Bhūta classes. In the first case, there is contact between 'isomeric' atoms (सजातीय-संयोग), in the second case between heterogeneous or 'plymeric' atoms

(विजातीय-संयोग). The first contact leads to intimate union (सङ्गः,–सङ्गाख्यः यः संयोगविशेषः तेनैव द्रव्याणां विकारो भवति–विज्ञानभिक्षु प्रवचनभाष्य): the isomeric atoms by a peculiar liberation of Energy (सजातीयोपष्टम्भ)–the action of similars on similars) are attracted towards one another, and being riveted as it were, form the so-called material cause (उपादान कारण) of the compound product. The second kind of contact (that between unlike or 'polymeric' atoms of heterogeneous Bhūtas) begins with a liberation of Energy (उपष्टम्भ), which breaks up each of the Bhūtas, and taking particles (or atoms) of one as nuceei or radicles groups particles of the rest round these radicles in a comparatively free or unattched condition. In this case, one Bhūta, that which serves to furnish the radicles, not necessarily that which is numerically of quantitively predominant, gets the name of material cause (उपादान कारण), and the others, which by their collocation cause the liberation of Energy (उपष्टम्भ, अवष्टम्भ or विष्टम्भ), are called efficient causes (निमित्त कारण)–एवमेकैकगुणसम्भवात् प्रधान गुणं आश्रित्य अप्रधानगुणाः परिणामभेदान् प्रवर्त्तयन्ति–वाचस्पति, कौमुदी on Kārikā 16. The illustrations given (*viz.*, the Rasas as modifications of Ap, water with Earth-accretions) show that this process applies not only on the Guṇas, but also to the Bhūtas. यथा आकाशादेकरसं सलिलं पतितं नानारूपात संश्लेषात भिद्यते–(गौड़पाद on Kārikā 17). तत्र अपि (तैजसे शरीरे अपि) बहुतरपार्थिवावयवाष्टम्भात्, अल्पत्वे च अनुपभोगात् (अनिरुद्ध on Sūtra 12, Chap. V.), जातिसाङ्कर्यस्य अस्माकमदोषत्वात् सामग्रीसमवधाने अनेकैरपि इन्द्रियैः एन्दा एकवृत्युत्पादने बाधकं नास्ति। (विज्ञानभिक्षु–प्रवचनभाष्य).

Aniruddha goes so far as to hold that both 'isomeric'and 'polymeric' (or 'heterogenic') combinations are real cases of constitutive contact. (आरम्भक संयोग, *e.g.* भौतिकवायोर्देहारम्भकत्वम्–अनिरुद्ध on Sūtra 113, Chap. V.) But in the later Sāṅkhya-Pātañjala the current teaching denied this—बहूनामुपादानायोगात् (Sūtra 102, Chap. V), when विज्ञानभिक्षु notes—बहूनां भिन्नजातीयानां चोपादानत्वं न दृष्टमिति सजातीयमेवोपादानाम्। इतरच्च भूतचतुष्टयमुपष्टम्भकम्।

But besides these transformations of substance (द्रव्यान्तर-परिणाम) by 'isomeric' or 'heterogenic' process, ceaseless changes go on in the characters, the modality, and the states of substances—changes which are due to the unequal distribution of force (or of stress and strain,—pressure) among the Guṇas, which are in themselves constant. (धर्म्मलक्षणावस्थापरिणामा: न द्रव्यान्तरत:,–यथा एका रेखा शतस्थाने शतं दशस्थाने दश, एकञ्चैकस्थाने, यथा चैकत्वेऽपि स्त्री माता चोच्यते दुहिता च स्वसा चेति,–गुणिनित्यत्वेऽपि गुणानां विमर्दवैचित्र्यात्–व्यासभाष्य, Sūtra 13, Pada III).

"Even as the same figure '1' stands for a hundred in the place of hundred, for ten in the place of ten, and for a unit in the place of unit."[1]

Now the question is—in these mixed substances does

1. This conclusively proves that the decimal notation was familiar to the Hindus when the Vyāsa-Bhāṣya was written, *i.e.* centuries before the first appearance of the notation in the writings of the Arabs or the Greco-Syrian intermediaries. Vācaspati, who comment on the Vyāsa-Bhāṣya, composed his Nyāya-Sūcī-Nibandha in वस्वङ्कवसुवत्सरे *i.e.* Saṁvat 898 or 842 A.D. This cannot be Śaka 898, for apart from the decisive use of यत्सर, which by this time had come to signify the Saṁvat era, Vācaspati's commentator, Udayana, wrote the Lakṣaṇāvalī in Sāka 906—

   तर्काम्बराङ्कप्रमितेष्वतीतेषु शकान्तत: ।
   वर्षेषूदयनश्चक्रे सुबोधां लक्षणावलीं ।।

   and Udayana, who wrote the Pariśuddhi on the Tātparyaṭīkā of Vācaspati, could not have been a contemporary of the latter, as will also appear from the invocation to Sarasvatī in the opening lines of the Pariśuddhi. Vācaspati then preceded Udayana by 142 years, and must have been himself preceded by the author of the Vyāsa-Bhāṣya by a longer interval still for Vācaspati ascribes the Bhāṣya to Veda-Vyāsa himself (वेदव्यासेन भाषिते भाष्ये व्याख्या विधास्यते). The internal evidence also point to the conclusion that the Bhāṣya cannot have been composed later than the sixth century,—*cf.* the quotations from Pañcaśikha, Vārṣagaṇya, and the Śaṣṭhi-Tantra-Śāstra, without a single reference to Īśvara-Kṛṣṇa which is decisive. I may add that I remember to have come across passages, of a similar import in Buddhist and other writings of a still earlier date.

the fusion take place by Paramāṇus or by larger masses (or lumps)? Now a Paramāṇu is defined to be the smallest portion of any substance which exhibits the charcteristics qualities of that substance, in other words, it stands for the smallest homogeneous portion of any substance. It is not without parts and therefore not indivisible. It is subject to disintegration. In a Bhūta or its isomeric modification, the Paramāṇu, the smallest homogeneous component particle, is unmixed, and therefore correspounds to the atom of modern chemistry. In a mixed substance (मिलितद्रव्य, संहतभूतार्थ whether it is an 'isomeric' or a 'polymeric' compound. the qualities are due to the mixture, and therefore its Paramāṇu, the smallest homogeneous particles possessing its characteristic qualities, must result from the mixture of the Paramāṇus (in smaller or larger numbers as the case may be) of the component substances. The Paramāṇu of a mixed substance therfore corresponds to what we now call molecule. (अपकर्षपर्यन्तं द्रव्यं परमाणुरिति–व्यासभाष्य। लोष्ट्रस्य हि प्रविभज्यमानस्य यस्मिन्नवयवे अल्पत्वातारतम्यं व्यवतिष्ठते सः अपकर्षपर्यन्तरपरमाणुः–वाचस्पति–तत्त्ववैशारदी–Sūtra 52, Pada III.—नाणुनित्यता न निर्भागत्वं। पृथिवीपरमाणुः जलपरमाणुः रित्यादिव्यवहारस्तु पृथिव्यादीनां अपकर्षकाष्ठाभिप्रायेणैव। विज्ञानभिक्षुः,–प्रवचनभाष्य, Sūtra 88, Chap. V.)—That the Paramāṇus form molecules (द्व्यणुक) in forming substances, is acknowledge by the Sāṅkhyas as will appear from Gauḍapāda,—तथा अन्योन्याश्रयाश्च द्व्यणुकवत्गुणाः। (गौड़पाद, or Kārikā 12). Even the Vaiśeṣikas, with their prejudice against 'polymeric' or 'heterogenic' combination, acknowledge that in 'polymeric' compounds the different Bhūta-substances unite by their Paramāṇus (or atoms), though they rigidly insists that in such cases only one atom should be regarded as the 'radicle' (उपादान or द्रव्यारम्भक कारण) and the others as co-efficient causes (निमित्त कारण–उपष्टम्भक, *e.g.* प्रशस्तपाद–एवं समुतपन्नेषु चतुर्षु महाभूतेषु, महेश्वरस्याभिध्यानमात्रात् तैजसेभ्योऽणुभ्यः पार्थिवपरमाणु सहितेभ्यो महदण्डमुत्पद्यते (प्रशस्तपादभाष्य—where Śrīdhara notes पार्थिवा अवयवा

उपष्टम्भका: and Udayana तैजसानां परमाणूनां पार्थिवाणुसाहित्येन अण्डस्य हिरणतायत्वं विवक्षितं)।

It is only in the medieval Sāṅkhya-Pātañjala that under the influence of the Nyāya-Vaiśeṣika doctrine a radical difference was conceived to exist between the structure (or constitution) of a molecule composed of 'isomeric' atoms, and that of one composed of heterogeneous (or "polymeric") atoms. In the former case, that was delivered to take place intimate union (सङ्ग), in the latter case, only a grouping of comparatively free or loosely attached atoms round a radicle atom (व्यूहन), with liberation of Energy (उपष्टम्भ,–अवष्टम्भ–or विष्टम्भ) and the setting up of unequal stress and strain (गुणवैषम्यविमर्द–ईश्वरकृष्ण). At the same time, it was of course admitted that this distinction does not apply to the forms of subtile matter (Tanmātra—सूक्ष्मभूत) which could unite in intimate fusion, wheter homogeneous or heterogeneous. For example, the subtile body (सूक्ष्मशरीर) which is supposed to be the seat or vehicle of the conscious principle is acknowledged by Vijñāna-bhikṣu to be penta-Bhautic (पञ्चभूतात्मक); in the other words, all the five Tanmātras serve as material causes, though the gross body (the animal organism) is stated to be only a "polymeric" compound with the Earth-Bhūta as radicle or base. अधिष्ठानशरीरं च सूक्ष्मं पञ्चभूतात्मकं वक्ष्यते। तन्मात्रकार्यं यत् भूतपञ्चकं लिङ्गाधिष्ठानशरीरम् (प्रवचनभाष्य–Sūtras 11 and 12, Chap. III.—स्थूलशरीरं पार्थिवमेव अन्यानि च भूतानि उपष्टम्भकानि (*ibid*. Sūtra 19. Chap. III.)

But in the original Sāṅkhya-Pātañjala it appears that the production of a new substance by mixture of unlike Bhūtas (विजातीयसंयोग) was conceded as freely as in the Vedānta, and was conceived as nowise differing from the formation of a compound of atoms of the same Bhūta class. The Sāṅkhya analysis of all change into transformations of Energy due to collocations of unchanging Guṇas, in other words the prevailing chemico-physical (or physical) point of view, naturally recognises on distinction between collocations of 'isomeric' and those of heterogeneous (or 'polymeric') atoms. At bottom they are all collocations of the Guṇas. Even Vijñāna-bhikṣu who, as one of the latest expounders of the Sāṅkhya-Pātañjala, has been most affected by the Vaiśeṣika prejudice against 'polymeric' combination (विजातीयसंयोग) urges that the qualities of a compound substance are not necessarily the result of similar qualities in the component elements. सजातीय-कारणगुणस्यैव कार्यगुणारम्भकता इति तु तेषां (तार्किकाणां) अपि न नियमः।) Elsewhere he explains that far from the vital activity being independent it is originated and maintained by the combined operation or fusion of the different sensory and motor reflexes of the living organism, and notes that the united operation of mixed (or miscellaneous) causes, where there is a fit collocation of matter, offers no difficulty to a follower of the Sāṅkhya-Patañjala :—(जातिपाङ्कर्यस्य अस्माकमदोषत्वात्, सामग्रीसमवधाने सति अनेकैरपि इन्द्रियैः एकदा एकवृत्युतपादने बाधकं नास्ति–प्रवचनभाष्य, Sūtra 32 Chap. II.). In the middle of the ninth century, we find Vācasapati instancing some 'heterogenic' or 'polymeric' compositions as typical examples of evolutionary change (परिणाम) and unhesitatingly accepting the substantive character of the products. In the Kaumudī he describes the various acids and juices of fruits as modifications of the same original water in the presence of different kinds of earthy accretions. The process is peculiar. The water-particle (or atom) serves as a radicle or centre of a system, and the different kinds of the Earth-Bhūta centering round this as a nucleus become the seat of the forces, which bring in the development of new energies (and new qualities *e.g.* tastes) in the water. (यथा वारिदविमुक्तं उदकं एकरसमपि त त्तद्भूतविकारान् आसाद्य नारिकेल-ताली-विल्व-कपित्थादि-रसतया परिणामात् मधु-राम्लवणतिक्तकटुकषायतया विकल्पते। इति एकैकगुणसम्भवात् प्रधानगुण आश्रित्य अप्रधानगुणाः परिणामभेदान् प्रवर्त्तयन्ति (कौमुदी on Kārikā 16). Vācaspati also points out that different substances may be transformed into one and same substance (*e.g.* the production of salt by the cow, the horse, the buffalo and the elephant, thrown into the salt factory of Sambara in Rajputana or of the flame of a candle by the combination of wick, oil and fire. (परिणामैकत्वं बहूनामपि एकः परिणामः दृष्टः। तद्यथा गवाटश्व-महिषमातङ्गानां रुमानिक्षिप्तानां एकलवणत्वजातीयलक्षणाः परिणामः, वर्त्तितैलाननानां च प्रदीप इति।-तत्त्ववैशारदी on व्यासभाष्य, Pada IV. Sūtra 14.)

Earlier still, *i.e.*, not later the sixth century, the Vyāsa-Bhāṣya, noting that inorganic matter, vegetable substances and animal substances do not differ from one another essentially in respect of their potential energies and ultimate constituents, points out that various bi-Bhautic chemical compounds of water and earth substances, in the shape of saps, acids and juices, are found in plants in their different parts. (जलभूम्योः पारिणामिकं रसादिवैश्वरूप्यं स्थावरेषु दृष्टम्). In other words, bi-Bhautic compounds are here placed in the same category as 'isomeric' compounds of substances of the same Bhūta class, for here the particles of both the Bhūtas are regarded as forming the matter (material cause—उपादानकारण) of the smallest homogeneous portions of the compounds substances.

N.B.—Vācaspati naturally interprets this to mean separate modifications of the two Bhūtas.

The view of the earlier-Sāṅkhyas that atoms of different Bhūtas may chemically combine to form molecules of compound substances as much as atoms of different modes of the same Bhūta comes out clearly in Utpala's brief reference to the Sāṅkhya system in his commentary on Varāhamihira's Vṛhat Saṁhitā. (एभ्यः पञ्चभ्यः तन्मात्रेभ्यः) पञ्चमहत्-भूतानि भवन्ति। तेभ्यः शरीरिणां शरीराणि। यतः पञ्चमहाभूतमयानि शरीराणि (Utapala, Chatp. I, Śloka 7).

Chemistry in the Medical Schools of Ancient India

As a matter of fact, long before the fifth century, probably, as early as the first century A.D., the prevailing schools of medicine and surgery which were based on the Sāṅkhya teaching with a methodology derived from the Nyāya-Vaiśeṣika doctrine (*cf.* Caraka, Sārīrasthāna, Chap. 1, Vimānasthāna, Chap. VIII—also Śuśruta, Śārīrasthāna, Chap. 1) had founded an elaborate theory of inorganic compounds, which equally admitted iso-Bhautic and hetero-Bhautic combinations. Like the Vedāntists, Caraka and Suśruta held that each of the gross Bhūtas (Mahābhūtas) is a peculiar ultra-chemical compound of five original subtile Bhūtas. In this sense, every substance is penta-Bhautic, but for purposes of chemical analysis and synthesis, *i.e.* considered with reference to the Mahābhūtas, all substances in their chemical constitution, belong to one or other of the following classes : Mono-Bhautic, bi-Bhautic, tri-Bhautic, tetra-Bhautic, and penta-Bhautic. Compounds of diffferent Bhūtas, again, may combine to form more complex substances, and these in their turn, higher compounds still, and so on in progressive transformation, as is more specially the case with organic substances and products.

Physical characters of the Bhūtas : The prevailing physical characters of the different Bhūtas and their isomeric modes are enumerated as follows :

Earth-substances—Heavy, rough, hard, inert, dense, opaque, exciting, the sense of smell.

Ap-substances—Liquid, viscous, cold, soft, slippery, fluid, exciting the sense of taste.

Tejas-substances—Hot, penetrative, subtle, light, dry, clear, rerefied, and luminous.

Vāyu-substances—Light cold, dry, transparent rerefied, impingent.

Akāśa-substances—imponderable (or light), rarefied, elastic, capable of sound (vibration).

(गुरु खर कठिन मन्द स्थिराविषदसान्द्र-स्थूल गन्धगुणबहुलानि पार्थिवानि। द्रव स्निग्ध शीत मन्द मृदु पिच्छिल सररसगुणबहुलानि आप्यानि। उष्ण तीक्ष्णसूक्ष्म-लघुरूक्षविषदरूपगुणबहुलानि आग्नेयानि। लघु शीत रूक्ष खर विषद सूक्ष्म स्पर्शगुणबहुलानि वायव्यानि। मृदु लघु सूक्ष्म श्लक्ष्ण शब्दगुणबहुलानि आकाशात्मकानि। Caraka, Śārīrasthāna, Chap. 26 ; compare Suśruta, Sūtrasthāna, Chap. 41).

Caraka points out that the primary qualities of specific physical charcters of the five Bhūtas are tactile qualities, *i.e.* sensible to touch, *e.g.* hardness (or roughness) for Earth, liquidity (or yielding to pressure) for Ap, impelling or moving force (pressure) for Vāyu, heat for Tejas, and Vacuum (non-resistance, penetrability) for Ākāśa.

(खरद्रवचलोष्णत्वं भूजलानिलतेजसाम्,
आकाशस्याप्रतीघातो दृष्टं लिङ्गं यथाक्रमम्।
लक्षणं सर्वमेवैतत् स्पर्शनेन्द्रियगोचरः।

Caraka-Śārīrasthāna, Chap. I)

(*cf.* the elaborate enumeration of physical characters quoted in Vijñāna-bhikṣu, Yoga-Vārttika, Sūtra 42, Pada I;– also Varavara's commentary on Tattva-traya, Acit-prakaraṇa).

The Mahā-Bhūtas—Mechanical Mixtures

Suśruta-notes that each of the gross Bhūtas (Mahābhūtas) is found mixed up with the other Bhūtas ;—*e.g.*, the Mahābhūta Ākāśa is the receptacle (or vehicle) of air, heat-and-light, and water vapour ; the Mahābhūta Vāyu, of water-vapour, light-and heat, and even fine particles of Earth held in suspension ; the Mahābhūta Tejas, of earth-particles in the shape of smoke, and also water-vapour.

(अन्योऽन्यानुप्रविष्टानि सर्वान्येतानि निर्दिशेत्–Suśruta Śārīrasthāna Chap. I.) अन्ये पुनः अन्यथा व्याचक्षते। आकाशे पवन-दहन-तोयानि। वायौ तेजोऽम्बुनी। भूरपि अणुताविशेषेण। भूमिरपि धूमादिरूपेण तेजसि। तेजोद्रव्ये पानीयमपि।–Dalavaṇa on Suśruta, *loc. cit.*)

Mono-Bhautic Earth-substances

Caraka and Suśruta regard the following as Earth-substances—Gold, the five Lohas (silver, copper, lead, iron and tin) and their rust, arsenic, orpiment, various mineral earths and salts, sand, precious stones, (Caraka, Sūtrasthāna, Chap. I. *Cf.* also Suśruta, Sūtrasthāna, Chap. I पार्थिवाः सुवर्णरजतमणिमुक्ता मनःशिलामृत्कपालादयः। सुवर्णस्य इह पार्थित्वमेवाङ्गीक्रियते गुरुत्वकाठिन्यस्थैर्यादिहेतुभिः। सूत्रे आदिग्रहणात् लोहमलसिकतासुधा हरिताल लवण गैरिकरसाञ्जन प्रभृतीनां–Dalavaṇa Suśruta, *loc. cit.*

The salts include common salt, saltpetre etc. Suśruta mentions the alkalis, borax, natron, Yavakṣhāra (carbonate of potash) etc.

A. Searing of hard rocks to enable them to be cut (or pulvised) (शिलादारण).

Sprinkle on the rock taken red-hot from the fire of Palāśa and Tiṇḍuka wood (Butea Frondosa and Disopyros Embryopteris) :—(a) diluted milk, or (b) a solution of wood ashes the ashes of the *Mokṣaka* mixed with those of reeds), or (c) a decoction of (the fruit) of the jujube (Zizyphus Jujuba) kept standing for 7 nights in a mixture of whey, vinegar and spirits, in which Kulattha (Dorichos Uniflrous of Biflorus) has been steeped, or (d) a solution of the ashed of the Neema bark and leaves (Azadirachta Indica), the sesame pod, the resinous fruit of the (Sinospora Cordifolia), with cow's urine. Repeat the process seven times (in the last case. six times).

(Varāhamihira, Vṛhat-saṁhitā, Chap. 33, Slokas 112-117).

Of these Earth-substances, some were known to be compounds, *e.g.* the chemical salts of the metals, collyrium etc. Suśruta describes of preparation of the metallic salts. The leaves of the metals were pasted over with the salts, and then roasted (अयस्कृति) (Cikitsāsthāna, Chap. 10). These metallic salts are therefore mono-Bhautic Earth-compounds. Suśruta also gives the preparation of mild and caustic alkalis. (Sūtrasthāna, Chap. 11).

B. Hardening of steel (शस्त्रपान).

(1) Plunge the steel red-hot into a solution of plantain-ashes in whey, kept standing for 24 hours ;— then sharpen on the lathe.

(2) Make a paste with the juice of the Arka (Calotropis Gigantea), the gelatine from the horn of the sheep, and the dung of the pigeon and the mouse : apply it to the steel after rubbing the latter well with (sesame) oil. Plunge the steel, thus treated, into fire; and when it is red-hot, sprinkle on it water, or the milk of the horse (or the camel or the goat), or ghee (clarified butter), or blood, or fat or bile. Then, sharpen on the lathe.

(Virāhamihira, खड्गलक्षणम्, chap. 49, ślokas 23-26).

Ap-substances, simple and compound :—

Suśruta, following Caraka, enumerates various classes of Ap-substances (द्रवद्रव्य) as follows :— waters, acids, milk, curbs, butters, oils (vegetable as well as animal) fats, honeys, molasses, alcoholic liquors, urines etc.

Pure Ap (Mahābhūta) is tasteless and the six tastes are developed when the Mahābhūta Ap enters into combination, mechanical or chemical, with other Mahābhūtas. Suśruta notes that varoius kinds of Earths are dissolved in the waters of different localities, and where the particles so dissolved are predominary Earthy, the water tastes acid or salt,—where predominantly watery, the resulting taste is sweet,—where the Earth particles are mixed up with Tejas, the water tastes pungent or bitter etc. Such is the case with mechanical mixtures. In the case of bi-Bhautic or tri-Bhautic compounds Caraka mentions that subtances with Mahābhūta Ap predominating in their composition taste sweet ; with Mahābhūtas Earth and Tejas predominating, acid ; with Mahābhūtas Ap and Tejas predominating, salt ; with Mahābhūtas Vāyu and Tejas predominating, pungent ; with Mahābhūtas Vāyu and Ākāśa predominating, bitter ; and with Mahābhūtas Vāyu and Earth predominating, astringent. (Caraka, Sūtrasthāna. Chap. 26,—*cf.* Suśruta Sūtrasthāna, Chap. 42).

In fact with the exception of Suśruta's waters, which are mechanical mixtures, or rather solutions, all these Ap-substances are organic products and, as such , penta-Bhautic, *i.e.* compounded of all the five Mahābhūtas, and the particular 'taste' which is developed depends on the relative proportion of the Mahābhūtas, and the predominance of one or more of them in the penta-Bhautic compound in question.

Qualities of Compounds

The isomeric modes of each Mahābhūta have specific colours, tastes etc. due to their structure, *i.e.* the arrangement of their atoms, and the physico-chemical characters of compounds whether of the same or of different Mahābhūtas result from the collocation in unequal proportion of the different forces latent in the atoms of the component substances. Caraka adds that the varied forms (textures) and colours of organic substances, whether vegetable or animal, are derived in the same way. एवमेतेषां, रसानां षट्त्वमुपपन्नं न्यूनातिरेकविशेषात् महाभूतानाम्। भूतानामिव स्थावरजङ्गमानां नानावर्णाकृतिविशेषाः।

Suśruta ignores Caraka's distinction between Mahābhūta and subtile Bhūta, and views every substances as in reality penta-Bhautic, and it is only the relative predominance of a particular Bhūta or Bhūtas predominance of a particular Bhūta or Bhūtas in any substance that determines its class. (पृथिव्यप्तेजीवायुवाकाशानां समुदायात् द्रव्याभिनिर्वृत्तिः। उत्कर्षस्तु अभिव्यञ्जकौ भवति इदं पार्थिवं इदं आप्यं इदं तैजसं इदं वायव्यं इदं आकाशीयमिति। Suśruta, Sūtrasthāna, Chapt. 41).

The extant Caraka and Suśruta—succession of medical authorities —

The extant Caraka and Suśruta, the sources of our present information regarding the progress of scientific knowledge in the medical schools of Ancient India. are both redactions of original authorities. The extant Caraka is a redaction by Dṛḍhabal of the genuine Caraka Saṁhitā, which was itself a redaction by Caraka of the original work of Agniveśa, by disciple of Ātreya Punarvasu as distinguished from Kṛṣṇātreya and Bhikṣu Ātreya, also well-known medical authorities. The extant Suśruta is a redaction by Nāgārjuna of an original work (Vṛddha Suśruta) by Suśruta the disciple of Dhanvantari. That Caraka preceded Suśruta is almost certain. Nāgārjuna was probably earlier than Dṛḍhapala. This Nāgārjuna is believed to have been the author of a commentary of Patañjali's Mahābhāṣya (vide Bhojavṛtti and Cakrapāṇi), and was probably indentical with the alchemist Nāgārjuna (Siddha Nāgārjuna), the metallurgist Nāgārjuna (author of a treatise on metallurgy, Lohaśāstra), and the Buddhist Nāgārjuna, author of the Mādhyamika Sūtravṛtti. Caraka and Suśruta continued to receive additions after Dṛḍhabala and Nāgārjuna, and even after Vāgbhaṭa, but the whole of the extant Caraka is probably much earlier than the commentator Cakrapāṇi, and the whole of the extant Suśruta earlier the Dalavana, the commentator, and Mādhava, the author of Rugviniścaya. The extracts in Vāgbhaṭa make it certain that the passage I have quoted or shall quote from the Sūtrasthāna and Śārīrasthāna of Caraka and Suśruta cannot be later than the sixth century of the Christian era.

Preparation of Chemical Compounds

The knowledge of chemical compounds and of their preparation continued to make progress in the Caraka and Suśruta Schools. The great metaleurgist, Patañjali, in his treatise on Metallurgy (लोहशास्त्र) gave elaborate directions for many metallurgic and chemical processes specially the preparation of the metallic salts, alloys and amalgams, and the extraction, purification and assaying of metals. Probably it was Patañjali who discovered the use of the mixtures called Viḍas, which contained aqua regia or other mineral acids in potentia. Unfortunately Patañjali's *magnum opus* appears to have been lost, but extracts from it are frequently found in madieval works on Medicine and Rasāyana, which leave no doubt as to its remarkable scientific value. The metalurgist Nāgārjuna advanced the knowledge of chemical compounds of his preparations of mercury. The Harṣacarita, in the seventh century, relates a fable concerning this Nāgārjuna, and speak of him as a friend and contemporary of Śātavāhana. The relative priority of Patañjali and Nāgārjuna is a vexed question in the history of metallurgy. That Nāgārjuna's Lohaśāstra was earlier than the final redaction of Patañjali will appear from the following circums-tances :— (1) Cakradatta in the summary of Nāgārjuna mentions that the chemical process of testing pure iron must be repeated twice before it can be regarded as decisive, whereas Śivadāsa Sen's extract from Patañjali shows, that the latter directed the process to be repeated seven times; (2) Patañjali in the Abhraka-vidhi (mica operation) adds mercury, which in this particular operation is wanting in Nāgārjuna's is recipe (*cf.* Cakradatta, and Patañjali as reported in the Yogaratnāka-rasamuccaya) ; and (3) Nāgārjuna is quoted in the earlier compilations, Patañjali in the later.

Early in the sixth century, Varāhamihira in the Vṛhat Saṁhitā gives several preparations of cements or powders called Vajra-lepa, "cements strong as the thunderbolt ;" and there was ample use for these in the temple architecture of the Buddhist period, the remains of which bear testimony to the adamantine strength of these metal or rock coments. (प्रासादहर्म्यवलभीलिङ्गप्रतिमासु कुड्यकूपेषु सन्तप्तो दातव्यो वर्षसहस्रायुतस्थायी। Chapter 56. *Ibid.*). Varāhmihira also alludes to the experts in mechinery (यन्त्रविदः, यन्त्रज्ञाः) and the professional experts in the composition of dyes and cosmetics (रागगन्धयुक्तिविदः Ch. 16, also ch. 5). I would also refer to the interesting chapter on perfumery (Ch. 76) where Varāhamihira gives various recipes for artificial imitations of natural flower scents as of the essence of Vakula, Utpala, Campaka, Atimuktaka etc., arranges compound scents in a sort of scale according to the proportions of certain ground essences used in their preparation, and determines by the mathematical calculus of combination (लोष्टकप्रस्तार) the number of variations of the different notes in this scale. To these classes of professional experts were due three of the great Indian discoveries in the chemical arts and manufactures which enabled India to command for more than a thousand years the markets of the East as well as the West and secured to her an easy and universally recognised pre-eminence among the nations of the world in manufactures and exports :—(1) the preparation of fast dyes for textile fabrics by the treatment of natural dyes the Mañjiṣṭhā with alum (तुवरी मंजिष्ठा रागबन्धिनी) and other chemicals (*e.g.* sulphate of iron) also cow-dung, (*cf.* the "cowdung substitute", Roscoe); (2) the extraction of the principle of indigotin from the indigo plant by a process which however crude is essentially an anticipation of modern chemical methods ; and (3) the tempring of steel in a manner worthy of advanced metallugry, a process to which the medicval world owed its Damascus swords. It was this applied chemistry much more than handicraft skill which gave India her premier position in the middle ages and earlier (indeed from Pliny to Tavernier) in exports and manufactures ; for in handicraft skill as in design and workmanship, great as were her merits, India came to be surpassed by her disciples China and Japan.

The Vāsvadattā and the Daśakumāra Carita in the sixth century allude to the preparation of a mass of fixed or coagulated mercury (पारदपिण्ड इव कालधातुवादिनः–वासवदत्ता); a chemical powder, the inhalation of which would bring on deep sleep or stupor (योगचूर्ण–दशकुमारचरित) ; a chemically prepared stick or wick for producing light without fire (योगवर्त्तिका–दशकुमारचरित) ; and a powder which like anaesthetic drugs of curare, paralyses sensory and motor organs (स्तम्भनचूर्णमिव इन्द्रियाणाम्–वासवदत्ता–धीन्द्रियकर्मन्द्रियाणां इति दर्पण:). Vṛnda (circa 950 A.D.) notices the preparation of sulphide of mercury (रसामृतचूर्ण) composed of one part of sulphur, and half its weight of mercury ; and also of caprous sulphide (पर्पटिनाम्न). Cakrapāṇi (circa 1050 A.D.) mentions the preparation of the black sulphide of mercury, "by taking equal parts of mercury and sulphur."

The Rasārṇava (circa 1200 A.D.) notices the colours of a metallic flames, probably after Patañjali ; *e. g.* copper gives blue flame ;—tin, pigeon-coloured ; lead, pale ; iron, tawny ; blue vitriol, red. If may be that the Rasārṇava regards mercury as a penta-Bhautic substance.

The Rasaratna-samuccaya divides the mineral kingdom (Earth-substances, simple and compound) into the following classes:— (1) The eight Rasas, mica, pyrites, bitumen, blue vitriol, calamine, etc. (2) The eight Uparasas (useful in operations of mercury), sulphur, red ochre, green vitriol, alum, orpiment realgar, collyrium, and medicinal earth, to which may be added the eight Sādhāraṇa Rasas, sal-ammoniac, cowrie, cinnabar, rock vermillion, etc. (3) The gems, diamond, emerald, sapphire cat's eye, sun-stone, moon-stone, pearl etc. (4) The metals, gold, silver, iron, copper, lead, in and the alloys—brass and bell-metal. Other Earth-substances are six salts, three alkalis, mineral earths, and several poisons (*cf.* vol. I, pp. 32-98).

Chemical compositions and decompositions—metallurgic processes—In these writings. we frequently come across instances of chemical composition and decomposition, by processes, more or less crude, of calcination, distillation. sublimation, steaming, fixation, etc. (भस्मीकरण, अध:पातन:, उद्ध्र्वपातन, स्वेदन, स्तम्भन etc.) *e.g.* the preparation of perchloride of mercury by taking common salt and mercury (रसकर्पूर–पारदस्य श्वेतभस्मीकरणे चन्द्रिकाकार:) of sulphide of mercury

(हिङ्गुल) by taking sulphur and mercury; of Sindūra from lead (सिन्दूरं नागसम्भवम्–Amarakoṣa 6th cent. A.D.);—of the medicinal compounds, Svarṇasindūra and Rasasindūra with mercury, sulphur and gold, where gold may have been fancied to influence to resulting compound in some mysterious way, either as a 'dynamic' or as a catalytic;—also the extraction by chemical decomposition of mercury from sulphide of mercury (हिङ्गुल,–*vide* रसरत्नसमुच्चय) ; of copper from sulphate of copper (तुत्थ, सस्यक) by heating this substance with one-fourth of its weight of borax (Rasaratna-samuccaya,—*cf.* Bhāvaprakāśa, तुत्थ नाम्रोपधातु किञ्चित् ताम्रेण तद्भवति); of zinc (वङ्गाभ, जसद) for calamine (रसक–Rasārṇava); of copper from pyrites (माक्षिक, विमल, etc.—Rasārṇava, (though the golden pyrites were supposed to be a semi-metal of gold, containing some gold along with the essence of copper)—the purification of mercury by repeated distillation from lead and tin with which was want used to be adulterated in the market. The various metallurgic processes described are—extraction, purification, killing (formation of oxides, chlorides and oxy-chlorides for the most part), calcination, incineration, powdering, solution, distillation, precipitation, rinsing (or washing), drying, steaming, melting, casting, filling, etc. to all which each of the known minerals was successfully subjected by the use of apparatus and regents and the application of heat in different measures (खरपाक–मध्यमपाक–and मृदुपाक),–methods which if often crude, especially from the absence of independent and isolated mineral acids, were yet in several instances remarkably simple and effective, and which, after all by the use of various Viḍas potentially containing mineral acids (aqua regia, sulphuric acid, hydrochloric acid, etc.) virtually accomplished the practical ends kept in view. To these were added several special processes for mercury (*e.g.* fixation), bringing up the number of mercurial operations of nineteen.

It may be noted that the mixtures called Viḍas, which potentially contained dilute mineral acids, were regularly employed not only in killing the metals (forming their oxides and chlorides), but—what is of fundamental importance—for purposes of chemical decomposition of metallic salts, etc. and the extraction and purification of metals.

Organic Compounds

Organic Compounds are either vegetable or animal substances (स्थावर जङ्गम भूत). The molasses, the fermented liquors, the saps and juices of plants, fruit acids, vegetable ashes and alkalis together with the tissues of plants, are vegetable compounds (औद्भिद, स्थावर भूत). Honey milk, curd, butter, fat, bile, urine and other excreta, together with the organs and tissues of animals, are animal substances. Caraka notices vegetable as well as animal oils. The viscous (oily) substances are classed under four heads—butters, oils, fats, and marrows (सर्पिस्तैलं वसा मज्जा स्नेहो दृष्टश्चतुर्विधः). Salt may be either mineral or vegetable salt.

Suśruta divides poisons into two calsses—Vegetable and animal, but several poisons expressly termed mineral poisons (धातुविष) are included in the first class.

All organic substances, whether animal or vegetable, are penta-Bhautic being compounded of greater or less proportions of the five Mahābhūtas.

Taking the human body Caraka finds that the foetus is composed of sixteen organic substances, *viz.* four, composing the Sprem-cell which comes from the male ; four composing the germ-cell which comes from the female ; four added by the transformation of the nutritive material, and finally, four kinds of subtile matter, which together form the vehicle of the conscious principle. As to the four organic substances which compose the sperm-cell, or other four which compose the germ-cell, it is not clear whether in Caraka's view they are also in their turn compounded of less complex organic substances, or whether their constituent elements are inorganic penta-Bhautic compounds.

The tissues that appear in the course of developement of the foetus are further transformations (higher compounds) of these foetal substances. All the component substances of the body are penta-Bhautic compounds, though sometimes they are assigned to the particular Bhūtas which predominate in their composition, *e.g.* bile to Tejas, lymph, chyle, blood fat, urine, sweat and other secretions to Ap, and skin, flesh, bones, nails, hair, etc. to Earth. (शुक्रं वायुवग्निभूम्यवृगुणपादवत्।–गर्भस्य चत्वारि चतुर्विधानि भूतानि मातापितृसम्भवानि आहारजानि आत्मकृतानि चैव Caraka Śārīrasthāna, Chap. II. *vide* also Caraka, Śārīrasthāna, Chap. VII. *cf.* Gaṅgādhara's Jalapakalpataru. पित्तं तृतीयं भूतं शरीरारम्भकतेजः प्रधानपञ्चभूतविकारात्मकं तेजः स्वरूपं अग्निसंज्ञम्। कफश्चतुर्थ भूतमापः शरीरारम्भकसोमप्रधानपञ्चभूतः विकारविशेषः। शरीरं तु पञ्चमहाभूतविकार-समुदायात्मकम्। *Cf.* Caraka, Śārīrasthāna, Chap. IV. मातृजादयोऽपि अस्य महाभूतविकाराः।

Foetal Development (after Suśruta). The ovum fertilised by the sperm-cell and developing under the influence of animal heat forms successive layers and tissues, even as layers of cells and fibres are formed in wood. First are formed seven layers apithelial and dermal (सप्तत्वचः), then follow the several tissues (कलाः) the flesh, the vascular tissue, the fact and marrow, the lymphatic (and glandular) tissue, the intestinal tissues, the biliary and the seminal vessels,—tissues which, are regarded by some as modifications of the original dermal layers of the ovum (*cf.* the layers of the blastoderm and their relation to the tissues in Embryology). The tissues are supposed to be developed successively, one out of another, by chemical action or metabolism (पाक), *e.g.* Chyle is transformed into blood, blood into flesh, flesh into fat, fat into bone, bone into marrow, marrow into sperm-cell. The organs are next formed out of the tissues. The liver gall-bladder (क्लोम), spleen and lungs are referred to the blood ;

the intestines to the blood, lymph and bile ; the kidneys to the blood and fat ; the testicles to the blood, lymph and fat ; the heart to the blood and lymph ; and the tongue to the lymph, blood and flesh. Vāyu, with the accompaniment of animal heat, impels the 'currents' (स्रोतांसि) in the system : Vāyu acting on the flesh gives rise to the masucles, and it is Vāyu again, which, with the essence of fat (or marrow), produces the nerves, arteries and tendons. (तस्य खलु शुक्रशोणितस्य अभिपच्यमानस्य क्षीरस्येव सन्तानिकाः सप्तत्वचो भवन्ति।–कलाः खल्वपि सप्त सम्भवन्ति। यथाहि सारः काष्ठेषु छिद्यमानेषु दृश्यते। तासां प्रथमा मांसवरा। द्वितीया रक्तधरा। तृतीया मेदोधरा। चतुर्थी श्लेष्मधरा। पञ्चमी पुरीषधरा। षष्ठी पित्तधरा। सप्तमी शुक्रधरा।–गमस्य यकृतप्लीहानौ शोणितजौ।–et seq Suśruta, Śārīrasthāna, Chap. IV.—रसाद्रक्तं ततो मांसं मांसान्मेदः प्रजायते। मेदसोऽस्थि ततो मज्जा मज्ञः शुक्रस्य सम्भवः। Sūtrasthāna, Chapt. XIV).

The following parts (tissues and organs) in the foetus are in a special modifications of the four organic substances contributed by the sperm-cell of the male parent ; hair, nails, teeth, bones, nerves, veins arteries. tendons and ligaments, and the sperm-cell the following of the substances derived from the mother,—skin, blood, flesh, fat, the heat, liver and spleen, kidneys, stomach, intestines, etc. (Caraka, Śārīrasthāna. Chap. III. यानि खल्वस्य गर्भस्य पितृजानि पितृतः सम्भवतः सम्भवन्ति, तद्यथा केश-श्मश्रु-नख-लोभ-दन्तास्थिशिरा-स्नायु-धमन्यः शुक्रमिति पितृजानि। यानि खल्वस्य गर्भस्य मातृजानि यानि च अस्य मातृतः सम्भवतः सम्भवन्ति-तद्यथा त्वक् च लोहितश्च मांसश्च भेदश्च हृदयं च यकृत् च प्लीहा वृक्कौ च पुरीषाधानं च आमाशयश्च इति मातृजानि।)

Chemistry of Digestion

The food we eat contains five classes of penta-Bhautic organic compounds. From their radicles or predominant elements, the substances are named Earth-compounds and Ap-compounds, Tejas-compounds, Vāyu-compounds and Ākāśa-compounds. The Earth-compounds supply the hard formed matter of the body the Tejas-compounds give the animal heat (or the metabolic heat), the Vāyu-compounds are the sources of the motor force in the organism, the Ap-compounds furnish the watery parts of the organic fluids and the Ākāśa-compounds contribute to the finer etheric essence which is the vehicle of the conscious life. Roughly speaking, the Earth-compounds answer to the nitrogen compounds in the food, the Tejas-compounds to the hydro-carbons (heat-producing) and the Vāyu-compounds to the carbohydrates (dynamic). The Ap-compounds are the watery parts of food and drink. The flesh, for example, is a tissue composed principally of Earth-compounds, the far of the Earth and Ap-compounds, the bones of Earth, Vāyu and Tejas compounds. "The Tejas compounds predominant in the composition of the blood." For purposes of digestion it is stated that different operations of the metabolic heat (perhaps different digestive fluids are also meant) are required to digest the different substances in the flood.

The course of metabolism is described as follows—

The entire alimentary canal is called Mahāsrotas (the great channel).

The food goes down the gullet by the action of the biomotor force, the Prāṇa-Vāyu.

In the stomach (आमाशय) the food becomes mixed up first with a gelatinous mucus (फेणीभूतं कफं) which has a saccharine taste, and then gets acidulated by the further chemical action of a digestive juice (विदाहादम्लतां गतः–evidently the gastric juice is meant). Then the biomotor force, the Samāna Vāyu, begins to act the drives down the chyme by means of the Grahaṇi Nāḍi to the Pittāśaya (deodenum), lit. the receptacle of the bile and thence to the small intestines (the आमपक्वाशय). In these, the bile (or rather the digestive substance in the bile as opposed to the colouring element) acts on the chyme, and converts the latter into cycle (रस), which has at first, a kaṭu taste (pungency). This chyle contains in a decomposed and metamorphosed condition all the penta-Bhautic organic compouds, *viz.* tissue-producing Earth-compounds, water parts or Ap-compounds, heat-producing Tejas-compounds, force-producing Vāyu-compounds and lastly, finer etheric constituents which serve as the vehicle of consciousness. The essence of chyle (सूक्ष्मभाग) from the small intestines is driven by the biomotor force, the Prāṇa Vāyu, along a Dhamanī trunk (*cf.* the thoracic duct) first to the heart (which is a great receptacle of chyle), and thence to the liver (and the spleen), and in the liver, the colouring substance in the bile acts on the essence of chyle, especially on the Tejas substance therein, and imparts to it a red pigment, trasforming it into blood. But the grosser part to chyle (स्थूलभाग) proceeds along the Dhamanīs, being driven by the biomotor force, the Vyāna Vāyu, all over the body.

When the blood has been formed, the essence of chyle in the blood, acted on by Vāyu (biomotor force) and Māṁsāgni (the flesh forming metabolic heat), forms the flesh-tissue, the Earth-compound of the flood substance especially contributing to this tissue. Of the flesh tissue thus formed, the grosser part goes to feed or replenish the flesh tissue all over the body. The finer essence of flesh in the blood in the chyle, acted on again by Vāyu (biomotor current) and the fat-forming metabolic heat (मेदोऽग्नि) in the menstruum of lymph (कफं समाश्रित्य), receives viscosity and whiteness, and produces the fatty tissue. the Earth-compounds and Ap-compounds of the food especially contributing to the product. This fat in the chyle (or blood), or rather the grosser part of it replenishes the fatty tissue of the body, but the finer essence of fat in the flesh in the blood in the chyle, acted on by Vāyu (biomotor current) and the marrow-forming metabolic heat, in the menstruum of lymph (श्लेष्मणावृत). becomes hard (or crystalline). and forms bone, the Earth, Vāyu and Tejas compounds contributing principally to the product. The essence of the fat fills the hollow channels of the bones, and acted on again by biomotor Vāyu and metabolic heat, becomes transformed into the marrow. The marrow is transformed into the semen, which is conveyed dowen by means of a pair of Dhamanīs (ducts) (द्वे शुक्रवहे), lodged in its receptacles (शुक्रधरा–कृषणौ and dischargeed by means of another pair of ducts (द्वे विसर्गाय). The semen, or rather all the elements in their finer essence, give off Ojas, which returns to the heart, the receptacle of chycle and blood, and again floods the body, and sustains the tissues, thus completing the wheel (or self-returning circle) of metabolism (परिवृत्तिस्तु अक्रवत्. *Cf.* Caraka and Vāgbhaṭa).

It is to be noted that, throughout, the fluid in the cyle (or blood) acts as the menstruum, though occasionally the lymph, which is itself a derivative from the chyle is added as in the case of fatty tissue and the marrow ; and that each preceding element or constituent of the body (धातु–शरीरारम्भकधातु, takes up the proper organic compounds from the food-chyle to form the next element or tissue. Throughout also the chemical changes are due to the metabolic heat which breaks up the compounds and recombines, but the operations and even the vehicles perhaps of this heat are different. For example, these heat-corpuscles in the biliary ducts produce the bile, but the bile secretion is supposed to contain two distinct substances, (1) a digestive fluid in the deodenum (पित्ताशय) which acts on the chyme to produce the chyle, (अन्नस्य पक्तृ पाचकाख्यं पित्तं) ; and (2) a colouring bile-substance in the liver which adds a red pigment to the chycle and transforms it into blood (रञ्जकाख्यं पित्तम्). Besides, there are three other biles, of which the acqueous humour in the eye is supposed to be one (आलोचकपित्तम्) helping in the formation of visual images (रूपग्राहकम्). This is the view of Dhanvantari and his school, but Ātreya holds there is no evidence that the bile really performs the first (digestive) function, for this can be accounted for by the animal heat arising from the working of the whole bodily machine. There are three different hypotheses regarding the course of metabolism and the successive transformations of the chyle (क्षीरदधिन्याय-केदारीकुल्यान्याय-खलेकपोतन्यायेति त्रिधा धातुपोषणक्रमः–Cakrapāṇi, Bhānumatī, Sūtrasthāna, Chap. 14, Śloka 10 ; also his commentary on Caraka, Sūtrasthāna, Chapt. 28), but my account is based upon the second hypothesis which has the preference of Cakrapāṇi (स्वरसः). It may be added as a curiosity that each element of the body (धातु) under the metabolic heat is supposed to give off a finer essence (सूक्ष्मभाग) which serves as the meterial of the next succeeding element, and a dross (मल), which forms some of the excreta in the body, (including the nails, the hair, etc.), besides retaining its own substance (the gross or main part) which is driven along by the Vāyus (biomotor or vital currents) or by the Srotas to its destination in the body. Some idea of circulation appears to have been entertained, for the heat which receives and then sends down the chyle through the Dhamanīs gets it back transformed into blood. and one Ojas also proceeds from the heart and returns to it along with the chyle and blood. (*Cf.* Vāgbhaṭa—ताः हृत्स्थाः शिराः रसात्मकं ओजः अभिवहन्त्यः)। पञ्चभूतात्मके देहे आहारः पाञ्चभौतिकः। विपक्वः पञ्चधा सम्यग्गुणान् स्वानभिवर्द्धयेत्। Suśruta, Sūtrasthāna Chapt. 46. *Cf.* also वियतपवनजाताभ्यां वृद्धिमाप्नोति मारुतः आग्नेयमेव यद्द्रव्यं तेन पित्तमुदीर्यते। *ibid.*, Chap. 41. भौमाप्याग्नेय वायव्याः पञ्चोष्माणः सनाभसाः, पंचाहारगुणान् स्वान् स्वान् पार्थिवादीन् पचन्त्यनु। यथास्वं ते च पुष्णन्ति पक्त्वा भूतगुणान् पृथक् पार्थिवाः पार्थिवानेव शेषाः शेषांश्च देहगान्। अतिरिक्ता गुणा रक्ते वह्लेर्मांसे तु पार्थिवाः। मेदस्यपां भुवश्चास्थि पृथिव्यनिलतेजसाम्। Caraka, quoteed by Dalvana. आदौ षड्रसमप्यन्नं मधुरीभूतमीरयेत, फेणीभूतं कफं यातं विदाहादम्लतां ततः। वायुना समानाख्येन ग्रहणीमभिनीयते। षष्ठी पित्तधरा नाम या कला परिकीर्त्तिता, आमपक्वाशयान्तःस्था ग्रहणी साऽभिधीयते। अग्न्याधिष्ठानमन्नस्य ग्रहणाद्ग्रहणी मता भुक्तमामाशये रुद्धा मा विपाच्य नयत्यधः। बलवत्यबला तवन्नमाममेव विमुञ्चति।.....अन्नस्य पक्तृ पितन्तु पाचकाख्यं पुरेरितम्। दोषधातु–मलादीनामुष् मेत्यात्रेयशासनम्।....तेजोरसानां

सर्वेषामम्बुजानां यदुच्यते, पित्तोष्मणा सरागेण रसो रक्तत्वमृच्छति। वाय्वग्नितेजसा युक्तं रक्तं मांसत्वमृच्छति। श्लेष्माणं च समाश्रित्यं मासं वाय्वग्निसंयुतम्, स्थिरतां प्राप्य शौक्ल्यं च मेदो देहेऽभिजायते। पृथिव्यग्न्यनिलादीनां सङ्घातः श्लेष्मणावृतः, खरत्वं प्रकरोत्यस्य जायतेऽस्थि ततो नृणाम्। करोति तत्र सौषिर्य्यमस्थां मध्ये समीरणः। मेदसा तानि पूर्यन्ते स्नेहो मज्जा ततः स्मृतः। तस्मान्मज्ज्ञश्चन्यः स्नेहः शुक्रं संजायते नमः। Caraka—Dṛḍhabala Saṁhitā, quoted by Aruṇa in his commentray on Vāgbhaṭa. यथा केदारनिषक्तं कुल्याजलं प्रत्यागसन्नां केदारीमाप्लावयति, तथा रस एव प्रथमं रक्तं प्लावयति । तत्र रक्तस्थान-सम्बन्धात् रक्तसादृश्यं रक्तव्यपदेशं च अनुभवति। रक्तंच रक्तसमानेन लोकेनांशनेन अस्त्रं पोषयति। ततो रक्तमाप्लाव्य मांसमाप्लावयति, एवमुत्तरोत्तरधातून् रस एव प्लावयति। (Cakradatta, Bhānumatī). This passage show that the 'venous blood' was conceived to be chyle-essence mixed blood, and that the circulation of the chyle so for as it was held to contribute its quota to the constituent elements and tissues of the body was really supposed to be indentical with the circulation of the blood (नेता रक्तमाप्लाव्य मांसमाप्लावयति). This will be abundantly clear from the following account of the course of the chyle and the blood :—

ततः सारभूतस्याहाररसस्य द्वौ भागौ भवतः। स्थूलः सूक्ष्मश्च....ततः सूक्ष्मौ भागः प्राणावायुना प्रेरितो धमनीमार्गेण शरीरारम्भकस्य रक्तस्य स्थानं यकृत्-प्लीहरूपं गत्वा तेन सह मिलितो भवति। ततः प्राक्तन-रक्तधातौ एव तिष्ठति।....ततः सारभूतस्य आहाररसस्य द्वौ भागौ भवतः। स्थूलः सूक्ष्मश्च। स्थूलो भागो रजकाख्येन पित्तेन रक्तीकृतः शरीरारम्भकं रक्तं पोषयन् व्यानवायुना प्रेरितो धमनीभिः संचरन् सकल-शरीरगतानि रुधिरावणि पुष्णानि। ततः सूक्ष्मो भागः व्यानवायुना प्रेरितो धमनीभिः शिराभिश्च शरीरारम्भकाणि मांसांनि याति। *et seq.*

This finer essence of chyle which nourishes the flesh must also be carried in the blood, on the 'irrigation channel' hypothesis (केदारीकुल्यान्याय). (For diagrams of the central circulation and the sympathetico-spinal nervous system, *vide* Appendix. The machanism of life and the inheritance of specific and other congenital characters are also dealt with in the Appendix.)

Formation of molecular qualities in chemical compounds:

The Caraka school, which, we have seen, was an offshoot of the Sāṅkhya (*cf.* Caraka, Vimānasthāna, Chap. 8, यथा वादित्य: प्रकाशक: तथा सांख्यवचनं प्रकाशकमिति) supplemented the above account of inorganic and organic compounds with a characteristically Sāṅkhya explanation of the formation of molecular qualities by chemical combination. In Caraka's view, the colours, tastes, etc. of the molecules of chemical compounds result from the collocation in unequal proportion and unstable equilibrium of the different forces latent in the atoms (Paramāṇus) themselves. (एवमेतेषां रसानां षट्त्वमुपपन्नं न्यूनातिरेकविशेषात् महाभूतानां। भूतानामिव स्थावरजङ्गमानां नाना वर्णाकृतिविशेषा:। Caraka.)

Chemistry of Colours

As an interesting example of the way in which a follower of Caraka would account for the colours of chemical compounds, I may note the explanation given by the late Gaṅgādhara Kavirāja Kaviratna in the Jalpakalpataru, a commentary on the Caraka Saṁhitā, published at Kolkata in 1869, premising that the Kavirāja's view is pure and genuine Caraka doctrine. Gaṅgādhara begins with a simple statement. The qualities of the atom, he writes, tend to produce similar qualities in the molecule. A molecular quality is therefore, the result of the conjunction or opposition, as the case may be, of the atomic tendencies. When, for example, the five Bhūtas combine to produce an organic compound (the human body), Tejas Ap and Earth tend to produce red, white and black respectively, but in the body (compound substance) the yellow colour may happen to be produced as the result of these tendencies in that particular proportion and collocation. The point to note here is that the molecule forms a fresh collocation, *redistributes the Mass and Energy*, and sets up new forces in the system which coming into play modify the potencies (or tendencies) in component atoms, and thus determine the resultant. This is elaborated into a curious but complete theory of the colours of chemical compounds.

The colours (and other qualities) of a simple substance (an isomeric mode of any Bhūta) are the result of the potencies lodged in that particular collocation of Mass. Energy and Essence. Now when two such substance unite, their colours etc. tend to be produced, but the combination brings on a fresh distribution of Energy. Mass and Essence, and the forces thus set free may powerfully modify or even extinguish the separate tendencies or potencies of the component simple substances. For example, when we prepare a collyrium by mixing equal parts of sulphur and mercury (the black sulphide of mercury) we find the resulting com-pound black. To explain, this it has to be remembered that each of the substances (sulphur and mercury) contains Sattva (Essence), Rajas (Energy) and Tamas (Mass) in different proportions, and that predominent Tamas (Inertia, Mass) always produces black, predominant Sattva (Essence) white and predominant Rajas (Energy) red. Now in the black sulphide of mercury, the white of the mercury tends to produce white, and the yellow of the sulphur yellow ; and if these tendencies were not obstructed, the result would be a mixed colour. But, in the particular collocation in question, the Tamas of the mercury becomes intensive (तीक्ष्ण), and the black of the now intensive Tamas extinguishes the white in the uncompounded mercury, which was due to prevaling Sattva, as well as the yellow of the uncompounded sulphur, which was due to the combined operation of White-producing Sattva and red-producing Rajas. Again, when, with proper apparatus and by the application of heat, we combine mercury and sulphur to produce the red sulphide of mercury, the resulting colour is explained by the fact that in this new collocation the Rajas (Energy)—probably of the mercury, though Gaṅgādhara does note specify—becomes intensive (तीक्ष्ण), and extinguishing both the white-producing Sattva of the mercury and the yellow-producing Sattva-Rajas of the sulphur, imparts a red colour to the compound. In these cases, as also in the formation of red by mixing powdered turmeric with lime, *i.e.*, whenever a new colour is produced in the compound, it is to be explained by the dominance of the Tamas, Rajas of Sattva, or their combinations, and the extinction of the uncompounded tendencies (or potencies) by the forces set free in the new collocation. But there are other cases where the colour of the compound is a mixed colour resulting from the colour of the combining substances, *e.g.*, when sulphide of mercury and calcined tin are mixed, the resulting colour is evidently a mixed one (Pātala, pink), which is easily explained by the colour of the component elements (the red of the sulphide of mercury and the white of the calcined tin). In the same way, in a mechanical aggregate, as in a piece of cloth, the colour is white, where the threads are white.

तत्कार्यारम्भे पृथिव्यादिद्रव्यावस्थागुणा गन्धादयः खलु स्वस्वजातीयं गन्धविशेषादिकं आरभमाणाः यदारभ्यमाणं स्वस्वविजातीयं कार्य स्वारम्भकं कारणञ्च अहङ्कारादिस्थं गुणांश उभयथा कुत्रचित् हत्वा कुत्रचित् न हत्वा स्वजातीयं गुणविशेषं आरभन्ते। यथा पारदगन्धकयोः संयोगे कज्जलीभूतद्रव्ये पारदस्थः शुल्कः शुल्कविशेष आरभमाणः गन्धकस्थपीतः पीतमारभमाणः पारदस्थतीक्ष्णेन तैक्ष्णयमारभमाणेन विरोधिना वध्येते। सर्वत्रैव शौक्ल्यं सत्वगुणयोनिकं तमोगुणयोनिकेन तैक्ष्ण्येन विरोधिना वध्येते। गन्धकस्थपीतस्थौ च लोहितशुक्लौ रजः सत्त्वयोनिकौ वध्येते। तयोः शुक्ललोहितवधे तैक्ष्ण्यसमानयोनिः कृष्णः अभिव्यज्यते। इति पारदगन्धकोभयात्मके कार्ये श्वेतपीतमेलनेन सम्भाव्यं कार्य यद्वर्णं तत् तीक्ष्णगुणेन वध्यते, ततकारेण शुक्ललोहितं हत्वा। एव हिङ्गुले पारदगन्धकाभ्यामारभ्यमाणे यन्त्रेविशेषे वह्निना पच्यमाने रजोबहुलयोनिकेन तीक्ष्णेन सत्त्वतमोयोनियकौ शुक्लकृष्णौ वध्येते। पारदस्थशुक्लगन्धस्थपीतमेलनेन सम्भाव्यं कार्यं वर्णं च वध्यते। रजो योनिकनलौहित्यं च अभिव्यज्यते इति। एवं हरिद्राचूर्णसंयोगेऽपि लोहित्यं अभिव्यज्यते चूर्णस्थतैक्ष्ण्येन हरिद्रास्थ शुल्ककृष्णवधे। अथ अवधे यथा–हिङ्गु लवङ्गभस्मसंयोगे श्वेतलोहित-

वर्णमेव उभयात्मकं भवति पाटलवर्णं, न च तत्र हिङ्गुलस्य लौहित्यं वङ्गभस्मन्थं शौक्ल्यं च केनापि वध्यते। एवं शुल्कसूत्रनिर्मितपटस्य शौक्ल्यमेव।) (Jalpakalpataur, Sūtrasthāna of Caraka, pp. 198-200 ; edition published by Bhabanchandra Basāk, Kolkata, Samvādaratnākara Press, Saṁvat 1925.) This is quite in a line white Caraka's explanation of the tastes of chemical compounds which has been already noticed.

Pariṇāma-vāda versus Ārambha-vāda

Caraka's view of the formation of a new quality or a new substances is based on the Sāṅkhya teaching as to the conservation and transformation of Energy, and brings chemical synthesis in a line with evolutionary change (परिणाम). On this view, a new substance may arise by spontaneous or isomeric change, *i.e.,* by inter-play of Energies within the system of any given substance, in the absence of any action from without. New qualities like new substances are only readjustments of the old, and continual changes are going on by spontaneous disintegration and recombination. Opposed to this evolution view of chemical synthesis of the Nyāya-Vaiśeṣika doctrine of Ārambha-vāda, according to which no change of substance of quality, no effect, in short, can take place except by the action of one component element (substances or quality) on another. A binary molecule, for example, cannot possess any specific quality (विशेषगुण) of a kind not represented in each of the two component atoms. In the cosmic process, no atom can exist free and uncombined with another atom, and every 'specific quality' in a substance can be ultimately analysed into the union of two 'specific qualities' *of the same class* in two ultimate particles which cannot be further divided. A single colour, smell or taste in a single particle, until it can link itself on to another specific quality of *its own class* in a second particle, cannot characterise any substance formed by the union of these particles as material causes. Hence an Earth-atom cannot unite with an Ap atom, to form a new substances of which both the particles must be equally regarded as material cuases. At any rate, such a compound, if effected, would be smell-less, as of the two constituent atoms, only one, *viz.*, the Earth-atom, possesses smell. A compound of Earth and Vāyu would be smell-less, colourless and tasteless, and so on. The Nyāya-Vaiśeṣika does not deny that there may be compounds of different Bhūtas, nor does it deny the causal operation of specific qualities as efficient or energising (dynamic) causes (उपष्टम्भक, निमित्तकारण), but if refuses to place these compounds on the same footing as compounds of isomeric modes of the same Bhūta ; and it accepts the 'material' causality, in such cases of only one of the Bhūtas, regarding the other as co-efficients (निमित्तकारण).

The earlier Sāṅkhya including the medical schools of ancient India brushed all this aside as a distinction without a difference. The Vedantists, as we shall presently see, flouted this doctrine of Ārambha-vāda. The Jainas, in opposing this Vaiśeṣika view of atomic combination, hit upon a solution of the problem of chemical affinity. Others again, found out of a *via media.* They held, as we learn from the reports of Udyotakara in the Nyāya-vārttika, and of Vācaspati Miśra in the Tātparyyaṭikā, that a molecule of the structure EA (one atom of Earth and one of Ap) would exhibit some variety of colour and taste resulting from the joint action of the atoms and of their several colours and tastes. But as in the combination EA only the Earth-atom possesses smell, and the Ap-atom is smell-less, and as moreover no quality in a compound substance can result except from the joint action of the similar (potential) qualities of at least two component elements, it follows that a molecule of the structure EA would not manifest the energy of smell potentially contained in the Earth-atom. Hence, admitting the combination EA for a smell-less compound, the upholders of this view would suppose a molecule of the type E.A. (*i.e.* two atoms of Earth and one of Ap) to explain any bi-Bhautic compound of Earth and Ap (like the plant saps and fruit juices) which exhibits smell in addition to the peculiarities of colour and taste. (*Cf.* Vācaspati's comment on Udyotakara's refutation of this view:—

अयमभिसन्धिः। पृथिव्याप्यपरमाणू तावन्नकं द्व्यणकमारब्धुमर्हतः तयोः रूपरसस्पर्शवत्त्वेन तदारम्भसम्भवे अपि गन्धवत्ता भावप्रसङ्गात। एवं पार्थिवाणु समवेतस्य गन्धस्य एकत्वेन अनारम्भकत्वात् नापि पार्थिवपरमाणु द्वयं एकश्च पाथसीयपरमाणुः इति अणूनामारम्भकत्वे गन्धवत्वोपपत्तिः इति साम्प्रतं। परमाणूनां बहूनां अनारम्भकत्वात्। तथाहि त्रयः परमाणवः न कार्यमारभन्ते।)

Measures of Time and space. Size of atoms

The Siddhānta-Śiromaṇi gives the following measures of Time :—

30 Kṣaṇas (क्षण)=1 day, 2 Ghaṭikās=1 Kṣaṇa,

30 Kalās =1 Ghaṭikā, 30 Kāṣṭhās =1 Kalā,

18 Nimeṣas = 1 Kāṣṭhā, 30 Tatparas

1 Nimeṣa, and 100 Truṭis =1 Tatpara.

This makes a Truṭi of time equal to 1/33750 of a second, which is nearly the measures of the Paramāṇu of time, as given in the Viṣṇupurāṇa (*vide* Bhāskara's Siddhānta-Śiromaṇi—कालमानाध्याय)।

The above measures were in use among the astronomers, but the physicists computed accroding to the following table given both in Udayana's Kiraṇāvali and Śrīdhara's Nyāya-kandali :—30 Muhūtras =1 day (24 hours) 30 Kalās =1 Muhūrta, 30 Kāṣṭhās =1 Kalā, Nimeṣas =1 Kāṣṭhā, 2 Lavas =1 Nimeṣa, 2 Kṣaṇas= 1 Lava.

क्षणाद्वयं लवः प्रोक्तो निमेषस्तु लवद्वयम्,
अष्टादशनिमेषास्तु काष्ठा त्रिंशत्तु ताः कला।
त्रिंशत्कला मुहूर्तः स्यात् त्रिंशद्रात्र्यहणी च ते।

(Udayana, Kiraṇāvali)

This makes 1 Kṣaṇa of the Nyāya-Vaiśeṣika equal to 2/45 of a second. The Nyāya assumes that the unit of physical change (or the time occupied by any single antecedent step in a causal series before the succeeding step is ushered in) is equal to a Kṣaṇa (or 2/45 of a second). The astronomers were familiar with far smaller measures of time. The astronomical Truṭi of time measures about the thirty-four-thousandth part of a second. This is of special value in determining the exact character of Bhāskara's claim to be regarded as the precursor of Newton in the discovery of the principle of Differential Calculus, as well as in it application to astronomical problems and computations. This claim, as I proceed to show, is absolutely established ; it is indeed far stronger than Archimedes's to the conception of a rudimentary process of Integration. Bhāskara in computing the "instantaneous motion" (तात्कालिको गति) of a planet compares its *successive* positions, and regard its motion as constant during the interval (which of course cannot be greater than a Truṭi of time, though it may be indefinitely less). This tātkālika motion is no other than the differentia of the planet's longitude and Bāpudeva Śāstrī claims that both the conception of the instantaneous motion and *the method of determining it* plainly show that Bhāskara was acquainted with the principle of the Differential Calculus. On the data before him. Mr. Spottiswoode remarks that Bāpudeva Śāstrī "overstates the case." Bhāskara "makes no allusion to one of the most essential features of the Differential Calculus, *viz.*, the infinitesimal magnitude of the intervals of time and space therein employed. Nor indeed is anything specifically said about the fact that the method is an approximate one." "With all these reservations" Mr. Spottiswoode continues, "it must be admitted that the formula he establishes and the method of establishing it bear a strong analogy to the corresponding process in modern mathematical astronomy" (*viz.*, the determination of the differential of the planet's longitude, by no means the first step in transcendental analysis or in its application to astronomy). And Mr. Spottiswoode concludes by stating that methematicians in

Europe will be surprised to learn of the existence of such a process in the age of Bhāskara (*circa* 1150 A.D.—born 1114 A.D.) Mr. Spottiswoode's second objection that Bhāskara does not specifically state that the method of the Calculus is only approximative cannot be taken seriously. The conception of limit and the computation of errors came late in the history of the Calculuses of Fluxions and Infinitesimals. For the rest, Bhāskara introduces his computation expressly as a "correction" of Brahmagupta's rough simplification. The first objection (*viz.*, that Bhāskara makes no allusion to the infinitesimal mangnitude of the intervals of space and time employed) would be more to the point, if it were well founded. But it is not, and Mr. Spottiswoode's error was due to the insufficiency of the data supplied to him. As a matter of fact, even Bhāskara's unit, the Truṭi of time (or Paramāṇu). is exceedingly small as the very name implies, being about one thirty-four-thousandth of a second of time. And in the passage in which Bhāskara describes the process, he distinguishes between Sthūla-gati and Sūkṣma-gati (velocity roughly measured, and measured accurately, *i.e.*, by reference to indenfinitely small quantities, for Sūkṣma, as we have seen, has always a reference to the Aṇu, the indefinitely small). Indeed he expressly mentions that the Sthūla-gati takes only Sthūla-kāla (finite time) into consideration, and that the determination of the Tātkālikī Gati (Sūkṣma-gati) must have a reference to the moment (प्रतिक्षणम्), which is an indefinitely small quantity of time being of course smaller than his unit, the Truṭi, (*Cf.* इयं किल स्थूला गतिः अथ सूक्ष्मा तात्कालिकी कथ्यते.....यदा आसन्नस्तिथ्यन्तस्तदा तात्कालिक्या गत्या तिथिसाधनं कर्त्तुं पुज्यते। तथा समीपचालनञ्च। यदा तु दूरतरस्तिथ्यन्तः दूरचालनं या चन्द्रस्य तदा आद्यया स्थूलया कर्त्तुं युज्यते स्थूलकालत्वात्। यतश्चन्द्रगतिः महत्त्वात् प्रतिक्षणं समा न भवति अतस्तदर्शं अयं विशेषोऽभिहितः nothing can be clearer than this conception of 'momentary' motion.—Bhāskara, Siddhānta-Śiromaṇi, Gaṇitādhyāya, Gatisphutiprakaraṇa : *cf.* also प्रतिक्षणं सा न समा महत्त्वतः, *ibid.*, *cf.* also Golādhāya, Tātkālikī-karaṇa-vāsanāprakaraṇa, where Bhāskara points out that the mode of computing adopted by the Acāryya (Brahmagupta) is a rough simplification. The computation of relative motion and idea of resolved components of motion were of course familiar to astronomers.—(*Cf.* एवं सत्याचार्येण लाघवार्थं इष्टघटीसम्वम्विन्यो गतिकला अर्के प्रक्षिप्ताः। *ibid.*) I may add *en passant* that Bhāskara's formula for the computation of a table of Sines also implies his use of the principle of the Differential Calculus.

Measures of weight and capacity. The Amarakoṣa mentions measures of three kinds—weight, length and capacity (मान तुलाऽङ्गुलिप्रस्थिः).

The Kṛṣṇala (Guñja, Raktikā, the black and red berry of the Shrub Abrus Precatorius) was employed as a natural measure of weight, 80 Kṛṣṇala berries on the average weight 105 gains Troy, and this must be taken as the basis of our computation, though in current practice 80 Kṛṣṇalas are taken to be equivalent to 210 gains. One Kṛṣṇala was supposed to weigh as much as 3 medium—sized barley seeds (मध्य यव). one of the later as much as 6 white mustard seeds (गौरसर्षप), one white mustard as much as 3 Rāji mustard seeds (राजि सर्षप), one of these seeds as much as 3 Likṣas and one Likṣu as much as 8 Rajas or Trasareṇus.

We now come to conventional measures, One gold Māṣā was the weight of 5 Kṛṣṇalas of gold, 1 Suvarṇa of Tolā weighed as much as 16 Māṣās, and one Pala as much as 4 Suvarṇas of Tolās. A Pala of gold therefore weights 320 Kṛṣṇalas (Manu, Chapt. VIII, Viṣṇu, Chapt. IV, and Yājñavalkya, Chap. I).

A Māṣā of gold therefore would weigh $6\frac{9}{16}$ grains ; a Tolā, 105 grains, (in current practice it weighs nearly double as I have stated) ; and a Pala, 420 grains Troy.

The measures of silver were the following–1 Silver Māṣa=2 Kṛṣṇal, 7 Dharaṇa=16 Silver Māṣār and 1 Pala =10 Dharaṇas. A Pala of silver would therefore weigh 320

Kṛṣṇalas. In other words, the Pala was a fixed measure of weight, and was equal to about 420 gains Troy. or double this, if we take the Kṛṣṇala of current paractice.

A Pala, which equals 320 Kṛṣṇalas, was subdivided by 4, 16 and 5 successively for gold, and by 10, 16 and 2 successively for silver. A Suvarṇa (or Tolā) of gold corresponds roughly to a Dharaṇa of silver, and a gold Māṣā to a silver Māṣā, but the sizes (or volumes) are not the same, and we must not therefore conclude that gold was supposed to be heavier than silver in the proportion of 5 to 2.

We find that 1296 Trasareṇus equal 1 Kṛṣṇala. A Trasareṇu, as a measure of weight, therefore, is the equivalent of 7/6912 of a grain Troy. or double this according to current measures.

But the Trasareṇu of physics is a different conception. It stands for the *minimum visibile, i.e.,* as the physicists define it, that which is *just discernible* as a glancing particle in the slanting beams of the morning (or afternoon) sun, coming into a dark room through a chink or orfice of a window. This is a measure of size (or rather stimulus limen).

Measures of Capacity. Here the standard was furnished by the Kuḍava (कुडव) a vessel described as 3 Aṅgulis long a 3 Aṅgulis broad, and 1½ Aṅguli deep,—with a cubical capacity of 13½ cubical Aṅgulis. 4 Kuḍavas =1 Praṣtha, 4 Praṣthas =1 Āḍhaka, 4 Āḍhakas=1 Droṇa, and 4 Droṇas= 1 Khārī or Bhāra.

24 Aṅgulis make 1 Hasta, cubit which may be taken to be 18 or 19 inches. A Kuḍava was divided into 4 Palas, and there can be no doubt that originally water of the weight of 4 Palas was found to be actually contained in a vessel of the cubical capacity of a Kuḍava. If we take the ancient cubit to have been 19 inches. the Kuḍava would contain nearly 4 Palas of distilled water at 30° Centigrade. On a cubit of 18 inches, the Kuḍava would contain about 3¾ Palas.

The Kuḍava in current medical practice is supposed to represent a cubical vessel, each side being 4 Aṅgulis. This would give a capacity of 27 cubic inches, if we take the modern cubit of 18 inches. The Kavirājas take a Kuḍava to contain 8 Palas of water, and as 1 Pala=320 Kṛṣṇalas, and 80 Kṛṣṇalas are now taken to be equal to 210 grains Troy, a vessel of a capacity of 27 cubic inches is accordingly supposed to contain about 6720 grains Troy,–which is not very wide of the mark, being about 1 : 3 per cent. short for distilled water at 62° Fahrenheit or 30° Centegrade.

Size of the *minimum visible*; size of an atom.

The supposed thickness of the just discernible mote in the sunbeam called a Paramāṇu in Technology, Śilpa-Śāstra, (and a Trasareṇu in Natural Philosophy follows directly from Varāhmihira's table :—8 Paramāṇus make 1 Rajas (or Rathareṇu,—cp. the Manasara), 8 Rajas make 1 Vālāgra (filament of hair), 8 Vālāgras make 1 Likṣā, 8 Likṣās make 1 Yūkā, 8 Yūkās make 1 Yava, 8 Yavas (the Manasara has 4) make 1 Aṅguli (superior), 24 Aṅgulis make 1 Hasta (cubit. lesser cubit, 18 inches.) The thickness of the *minimum visible* (the finest perceptible mote in the slanting sunbeam) is therefore taken to be 3.2—or 1/349525 of an inch. The volume of a spherical Trasareṇu (or Paramāṇu of the Śilpa-śāstra) would therfore be $4/3\pi\ 3.2^{-63}$ of a cubic inch. It may be here noted that such a Trasareṇu is supposed in the medical schools to contain 30 chemical atoms (Paramāṇus of Natural Philosophy) according to one estimate, or 60 according to another. The size of an atom must then have been conceived to be less than $\pi 3.5^{-1}\ 2^{-62}$ of a cubic inch.

परमाणुरजोबालाअलिक्षयूकं यवोऽङ्गुलं चेति अष्टगुणानि यथोत्तरम् अङ्गुलमेकं भवति संख्या–Varāhamihira. Vṛhatsaṁhitā, Chap. Śloka 2—जालान्तरगते भानो यदणुतरं दर्शनं रजो याति तद्विद्यात् परमाणुं प्रथमं तद्धि प्रमाणानाम्। *ibid.*, Śloka 1,—*Cf.* Utpala, तत्परमाणुं नाम प्रमाणं जानीयात्। प्रमाणानां सर्वेषां तत् प्रथममाद्यं प्रमाणं परमाणुसंज्ञम्। Ch. also, त्रसरेणुस्तु विज्ञेयः त्रिशंता परमाणुभिः)।

The magnitude of a Paramāṇu is called Pārimāṇḍalya (पारिमाण्डल्य) in the Nyāya-Vaiśeṣika, the name suggesting that the Paramāṇus were conceived to be spherical in shape. The Nyāya-Vaiśeṣika calls a Paramāṇu a mere point without any dimensions, but in the Sāṅkhya-Patañjala, a Paramāṇu, though indentifnitely small, had still dimensions, being visible into Tanmātras, which were themselves integrations of Bhūtādi. the diameter of a spherical Paramāṇu must have been coneived to be less than 3.2^{-20} of an inch (*i.e.* less than the conventional Paramāṇu with which linear measures begin), and the volume of a Paramāṇu would therefore, in accordance with Bhāskara's formula, be smaller than $4/3\pi . 3^3 2^{-63}$ or $\pi . 3^2 . 2^{-61}$ of a cubic inch were $\pi . 3927/1250$. The Tanmātras were concieved as smaller still.

That these were conventional measures arbitrarily assumed goes without question, for, of course, the Hindus had no physical data for a mathematical, calculation of these minute quantities. A Valāgra (hair-tip, or finest filament of hair), for example, is taken to be 3.2^{-14} of an inch thick ; *i.e.* less than one five-thousandth fraction of an inch in thickness ; and a fibril or the networks of Dhamanī or Nāḍī (nerve) that supply the pores of the skin (papillac मुखानि रोमकूपप्रतिवद्धानि, Suśruta) was supposed to be about a thousandth part of the finest hair in thickness, or 1/125 of the *minimum visible,'* and therefore about $3.10^{-3} 2^{-14}$ of an inch thick (*cf.* सूक्ष्मनाड़ीप्रचारत: रोम्न: सहस्रभागेनतुल्यासु प्रचरति, Pañcadaśī), but it is evident that these measures were arbitrarily fixed upon, instead of being arrived at by calculation or actual measurement. Indeed, Caraka expressly states, that the number of Śirās and Dhamanīs in the body (three million fifty-six thousand nine hundred) is only a conjectural estimate (अनिर्देश्यमत: परं तर्क्यमेव–Śārīrasthāna, Chapt. VII).

My account of the chemistry of the Sāṅkhya-Pātañjala, and of the affiliated Yoga and medical schools, has anticipated in several points the views of the Vedānta and the Nyāya-Vaiśeṣika. The chemical facts, processes and apparatus are indeed common to all the schools. In the following account of the chemistry of the schools other than the Sāṅkhya-Pātañjala, I will confine myself to the theory of the subject, and even of this I will attempt only the briefest outline.

The Vedāntic System

The Vedāntists believe Māyā to be the material cause (उपादानकारण) of the world. The power of Māyā is power to realise the unreal—to impart practical Reality or mediate existence to that which does not and cannot possess absolute Reality or self-existence. Māyā is at once real and unreal, while the Brahma (Self) is absolute Reality, absolute the intelligence and absolute Bliss. The world evolves out of Māyā (मायापरिणाम), so that Māyā in the Vedānta replaces the Prakṛti of the Sāṅkhya. But Māyā, and by implication the world, originate out of Brahma, not by a process of evolution (परिणाम), but of Vivarta (self-alienation). The self-alienation of the Absolute, acting through Māyā produces in the beginning Ākāśa, one, infinite ubiquitous, imponderable, inert and all-pervasive. The world thus begun goes on evloving, in increasing complexty. The other Sūkṣma Bhūtas, classers of subtile matter, evolve from Ākāśa, in an ascending linear order,—Ākāśa giving off Vāyu, Vāyu giving off Tejas, Tejas giving off Ap, and Ap giving off Earth Ākāśa, one, infinite all-pervasive, has the capacity of sound. Vāyu, subtile gaseous matter, emanates from the universal Ākāśa, and is instinct with the potential of mechanical energy (impact, pressure). (ईरण, प्रणामित्व, वहन, व्यूहन,–वायोश्च सर्वचेष्टाहेतत्वेन सर्वाविना–भूतत्वात्–Vidvanmanorañjinī). Tejas, subtile radiant matter, emanates from Vāyu, and contains in potential the energy of light and heat. Ap, subtile viscous matter, is the transformation of Tejas, and is instinct with the energy that stimulates the nerve of taste, and lastly, Earth, subtile hard matter, which is transformation of Ap, possesses the latent energy of smell.

But the subtile rudiments of matter must be compounded in various ways, to give rise to the gross constituent matter of the world. These forms of gross matter are called Mahābhūtas. There are five kinds of Mahābhūta (gross matter corresponding to the five Sūkṣma Bhūtas (subtile matter, and the process by which a Mahābhūta is produced from the Sūkṣma Bhūtas is called Pañcīkaraṇa (quintuplication). All the five Sūkṣma Bhūtas are present is ingredients, though in different proportions, in each Mahābhūta.

The Mahābhūta Earth, gross earth-matter, is composed of four parts of subtile earth-matter, and one part each of the other forms of subtile matter. The Mahābhūta Vāyu is composed of four parts of subtile gaseous matter and one part each of other forms of subtile matter. And similarly with other Mahābhūtas.

Hence if ak, v, t, ap, e, represent the five forms of subtile matter (Ākāśa, Vāyu, Tejas, Ap and Earth), and AK, V, T, AP, E, stand for the corresponding Mahābhūtas we may represent the constitution of the Mahābhūtas as follows :—

$AK=ak_4.(v_1.t_1.ap_1.e_1)$, ak_4 being the radicle.

$V=v_4.(tak_1.t_1.ap_1.e_1)$, v_4 being the radicle.

$T=t_4.(ak_1.v_1.ap_1.e_1)$, t_4 being the radicle.

$AP=ap_4.(ak_1.v_1.t_1.e_1)$, ap_4 being the radicle.

$E=e_4.(ak_1.v_1.t_1.ap_1)$, e_4 being the radicle.

In forms of gross of compounded matter, the potential energies (or qualities) become actualised. The Mahābhūta Ākāśa manifest sound ; Vāyu sound and mechanical energy ; Tejas, sound, mechanical energy and heat-light ; Ap, the energy of the taste-stimulus in addition ; and finally Earth the energy of the smell-stimulus added to the foregoing.

The Pañcadaśī characterises the different Mahābhūtas by their sound ; *e.g.*, Ākāśa by the echo (hollow booming sound) ; Vāyu (air) by a sibilant sound, hissing, susurration (imitative sumbol, Visi) : Tejas (fire) by a puffing (or roaring) sound (imitative symbol, Bhugubhugu); Ap (water) by a liquid sound (imitative symbol Culu-Culu, gurgle, plash-plash, glut-glut); and finally Earth by a splitting or rattling sound, a crack or a thud (symbol, kad-kada)—Chap. II. Bhūtaviveka, Śloka 3, Pañcadaśī. *Cf.* also Jayanta Mañjarī.)

Others hold that Ākāśa, Ether, never enters as a component part, and is always one and indivisible. In this view, the four Mahābhūtas as—Vāyu Tejas, Ap the Earth alone are supposed to be compounded, and by a process which may be called queternion (*cf.* the Neo-Platonist queternion) :—

$$V= v_3 (t_1 . ap_1 . e_1)$$
$$T=t_3 (v_1 . ap_1 . e_1)$$
$$AP= ap_3 (v_1 . t_1 . e_1)$$
$$E=e_3 (v_1 . t_1 . ap_1)$$

These compound forms, as before, are supposed to exercise their specific energies actively. Others again hold that the Mahābhūtas—Tejas, Ap and Earth alone are compounded by a process named Trivṛt-Karaṇa (triplicatoin). Thus $T=t_2 (ap_1 . e_1)$, $AP=ap_2 (t_1 . e_1)$, $E=e_2 (t_1 . ap_1)$.

The Sūkṣma-bhūtas are forms of homogeneous and continuous matter, without any atomicity of structure ; the Mahābhūtas are composite ; but even these are regarded as continuous, and without any atomic structure. The Vedānta speaks of Aṇu (Paramāṇu) not as an ultimate indivisible discrete constituent of matter. but the smallest coceivable quantum or measure of matter. In the Sāṅkhya doctrine, the atomic structure is ordinarily accepted. The Guṇas are supposed to be परिच्छिन्न and अणु bounded and indefinitely small in size (except the Guṇas giving rise to Ākāśa and Manas which are ultimated अपरिच्छिन्न ; and hence the Tanmātras and Paramāṇus must be conceived to have a discrete structure.

As I have already noted in my account of the genesis of Tanmātras and Paramāṇus various, schools of Vedāntists (*e.g.* Rāmānujists and the followers of Nimbārka) combined, in the orthodox fashion of the Smṛtis and the Purāṇas, the Vedāntic theosophy with the Sāṅkhya cosmology especially as regards Prakṛti and the order of creation and dissolution. For example, the Vedānta-kaustubhaprabhā, fortifying itself with texts from the Viṣṇu Purāṇa and the Subala and Gopāla Upaniṣads, contends that, at the cosmic dissolution (Pralaya), each Mahābhūta merges into the one that preceded it in the order of creation by first disintegrating into its own proper Tanmātric form (तन्मात्रद्वारा), and that the Mahābhūta Ākāśa merges into the original Tanmātras, which then lapse into Bhūtādi, the supersubtile rudiment matter, proto-matter (Chap II, Pada 3, Sūtra 14).

Pariṇāma—Evolutionary Process

When the Mahābhūtas are once formed, the different kinds of substance are derived from them by the Evolutionary process called Pariṇāma (परिणाम transformation). Matter is constantly undergoing change of state. The effect is only the cause in a new collocation (कारणस्यैव संस्थानमात्रं कार्यम्). Change is of two kinds :—

(1) Change by a spontaneous process, without external influence, including isomeric change (स्वाभाविक परिणाम). The Vedāntists believe in spontaneous disintegration and reintegration. Action from without, impressed force *ab extra*, is not, pace the Naiyāyikas, always a condition of change of state (whether of rest or of motion),—not is it necesary that more than one substance should combine to generate another substance or variety of substance (*e.g.* the formation of curds from milk, of ice from water etc.) All this is directed against the Nyāya doctrine (Ārambhavāda).

(2) Change due to combination with other substances (द्रव्यान्तरसंयोग). Such combination may produce (1) a com-

pound substance possessing like qualities with the constituents (समानजातीयोत्पत्ति), or (2) unlike compounds with new qualities, "heteropathic effects" (विजातीयोत्पत्ति). Any new quality thus evolved through (chemical) combination is called Saṁhata-bhūta-dharma (संहतभूतधर्म), *e.g.* the intoxicating power of the fermented rice and molasses, which does not exist in the ingredients taken separately. (मद्यबीजानां प्रत्येकं अवर्त्तमानापि समुदायशक्त्या मदशक्तिः दृश्यते). This Sambhūyakriyā (सम्भूयक्रिया, समुत्थान) corresponds of chemical combination, and the Vedāntists, like the Sāṅkhyas, explain this only as the evolution of the latent energy (शक्ति, अनुद्भूतशक्ति) in a new collocation (संस्थान, अवयव-सन्निवेश). But, unlike the medieval Sāṅkhya, the Vedānta freely recognises the combination of heterogeneous Bhūtas. Thus, Earth, Ap, Tejas and Vāyu freely combine in different proportions and groupings to produce the variety of substances in the world. For example, the animal oragnism is a compound of all the five Bhūtas (पाञ्चभौतिक). It is only morely the concomitant or efficient causes that may be heterogeneous to the material cause, as the Naiyāyikas contend, but the several heterogeneous substances (or Bhūtas) may unite as 'material causes' to produce a new substance.

The Vedāntists resolve all activity, physical vital as well as psychical, into modes of motion, subtile comic motion (परिस्पन्द, सर्वलोकपरिस्पन्द, Śaṅkara ; वायोः परिस्पन्दात्मकत्वात्,-प्राणस्य परिस्पन्दात्मकत्वादेव-यद्दवतं स्थूलं सूक्ष्मं च तत्सर्व मनःस्पन्दितमात्रम्-Śaṅkara) ; but they give a separate substantive existnece to the agents, the vital principal (प्राण) and the mind (मनः) though these are also evolutionary transformations of the Sūkṣma Bhūtas (forms of subtile matter). What is common to the Nyāya. The Vedānta and the Sāṅkhya is that Consciousness or Intelli-gence (चैतन्य) transcends Matter ; but the Naiyāyikas as pluralists hold that vital and psychical activities are also immaterial and cannot be resolved into motion (परिस्पन्द);–the Vedāntists resolve these activities into subtile motion, but ascribe them to a substantive quasi-material Life Principle and Mind, the all-mirroring Intelligence (चैतन्य) alone being immaterial and transcendent; and the Sāṅkhyas accept the substantive existence of the Mind Principle (मनः) as derived co-ordinately with the Sūkṣma Bhūtas or Tanmātras from individualised Prakṛti (Ahaṅkāra), but resolve Life into a mere resultant activity of the bodily organs, *viz.*, the organs of sense and movement, and the psychic principle (मनः).

पञ्चीकरण :–द्विधा विधाय चेकैकं चतुर्द्धा प्रथमं पुनः।

स्वस्वेतरद्वितीयांशैयोजनात् पञ्चपञ्च ते।

यथा त्रिसर्गश्रुतौ सृष्टानां भूतानां स्फुटतरव्यवहारनामरूपस्य करणोपायतया त्रिवृत्करणं श्रुतं तद्वत् भूतपञ्चकसर्गश्रुतौ अपि तथा। तदानीं पञ्चीकरणान्तरं आकाशे शब्दोऽभिव्यज्यते स्फुटतयेति सर्वत्र योजनीयं (विद्वन्मनोरंजिनी)–अस्ति हि शरीरे सर्वेषामपि भूतानां कार्यसंप्रतिपत्तिः अवकाशव्यूहनपवनक्लेदनकाठिन्यानां सर्वजनानुभवसिद्धत्वात्। अतस्तत्कारणतया पञ्चानि भूतानि एकस्मिन् देहे सन्तीति स्थिते तन्तुपटयोरिव अवयवावयवित्वमेव पञ्चभूतदेहयोर्युक्तं। न च स्पर्शशून्यत्वात् एकद्रव्यत्वात् च आकाशस्य आरम्भकत्वानुपपत्तिरिति वाच्यं। आरम्भवादस्य अनङ्गीकारात्। एकस्यापि दुग्धावयविनो दध्यारम्भकत्वदर्शनात्। वस्तुस्तु पञ्चानां भूतानां पञ्चात्मकत्वस्य दर्शितवात् आरम्भवादस्य निराकृतत्वात् च। तस्मात् सिद्धं शरीरं पाञ्चभौतिकमिति। Vidvānmanorañjinī.

न तावत् समानजातीयमेव आरभते न भिन्नजातीयमिति नियमो नास्ति। समवायिकारणे एव समानजातीयत्वांभ्युपगमः न कारणान्तरविषय इति तदपि अनैकान्तिकम्। नापि अनेकमेव आरभते नकमिति नियमोऽस्ति। अणुमनसोराद्यकर्मारम्भाभ्युपगमात्। एकैको हि परमाणुश्चाद्यं स्वकर्मारभते न द्रव्यान्तरः संहत्य इत्यभ्युपगव्यते। द्रव्यारम्भे एव अनेकारम्भकत्वनियम इति चेत् न परिणामाभ्युपगमात्। तदेव तु द्रव्यं विशेषवदवस्थान्तरं आपद्यमानं कार्यं नामाभ्युपगम्यते। तच्च क्वचित् अनेकं परिणमते मृद्बीजाङ्कुरादिभावेन, क्वचिदेकं परिणमते क्षीरादि-दध्यादिभावेन। नेश्वरशासनं अस्ति अनेकं एव कारणं कार्यं जनयति इति। Śaṅkara, Śārīraka-bhāṣya, Sūtra 7, Pada 3. Adhyāya II).

The Atomic Theory of the Buddhists

The Vaibhāṣikas and the Sautrāntikas hold that the Vāyu-atoms are touch-sensibles, having impact or pressure for their characteristic property, and by aggregation form the element Vāyu ; the Tejas-atoms are colour-and-touch-sensibles, having heat for their characteristic, and by aggregation form the Tejas-Bhūta ; the Ap-atoms are taste-colour-and touch-sensibles with a characteristic viscosity, and form the Ap-elements by aggregation ; and finally the Earth-atoms are smell-taste-colour-and-touch-sensibles possessing a characteristic dryness or roughness (खरत्व), and by their aggregation form the Earth-element. The Bhūtas thus originated to form aggregates, which are classed as inorganic substance, organisms and organs. (कामेऽष्टद्रव्यकोऽणु :—Udyotakara's extract from the Buddhist Sūtras, Nyāya Vārttika Chap. I. Āhnika I, Sūtra 14. *cf.* Vācaspati's fuller extract :—कामे अष्टद्रव्यकोऽणु रशब्दः रूपधातुररूपधातुः कामधातुः–also the Buddhist commentary :—कामे कामधातौ अष्टद्रव्यकः अणुः। रूपरसगन्धस्पर्शा इति चत्वारि द्रव्याणि पृथिव्यप् तेजोवायुः इति चत्वारि–Vācaspati. Tātparyyaṭikā, *loc.cit.*—also स्वरस्नेहोष्णोरणस्वभावानि भूतानि–*Vide* Udyotakara, Chap. III, Āhnika 2, Sūtra 12,—*Vide* also Śrīnivāsa's Vedānta-kaustubha. Chap, II. Pada 2. Sūtra 18).

The Atomic Theory of the Jainas

Of the nine categories of the Jainas, that of Ajīva (the not-soul or non-Ego) consists of five entities, four of which are immaterial (अमूर्त्त), *viz.*, merit, demerit, space and Time, and the fifth, material (मूर्त्त, possessing figure). The last is called Pudgala (matter), and this alone is the vehicle of Energy, which is essentially kinetic, *i.e.* of the nature of motion. Everything in the world of non-soul (the non-Ego) is either an entity (द्रव्य), or a change of state in an entity (पर्याय). Pudgala (Matter) and its changes of state (पर्याय). whether of the nature of subtile motion (परिस्पन्द) or of

Evolution (परिणाम), must furnish the *physical* as opposed to the *metaphysical* basis of all our explanations of Nature. Pudgala (Matter) exists in two forms—Aṇu (atom) and Skandha (aggregate). The Jainas begin with an absolutely homogeneous mass of Pudgalas, which, by differentiation (भेद) breaks up into several kinds of atoms qualitatively determined and by differentiation, integration, and differentiation in the integrated (संघातात् भेदात्, संघातभेदात्–Umāsvātī, Chap. V, Sūtra 26) forms aggregates (Skandhas). An Aṇu has no parts, no beginning, middle or end. An Aṇu is not only infinitesimal, but also eternal and ultímate. A Skandha may vary from a binary aggregate (द्व्यणुक) to an infinitum (अनन्ताणुक). A binary Skandha is an aggregate of two Aṇus (Atoms), a tertiary Skandha is formed by the addition of an atom. (Aṇu) to the binary (द्व्यणुक) and so on *ad infinitum*. The ascending grades are (1) what can be numbered (संख्येय), (2) idefinitely large (असंख्येय), (3) infinity of the first order (अनन्त), and (4) infinity of the second order (अनन्तानन्त), and so on.

General Properties of Matter

The specific characters of the Pudgalas (Matter) are of two kinds, (1) those which are found in atoms as well as in aggregates, and (2) those which are found only in aggregates. Qualities of touch, taste, smell and colour same under the first head. The original Pudgalas being homogeneous and indeterminate, all sensible qualities, including the infra-sensible qualities of atoms, are the result of evolution (परिणाम). Every atom thus evolved possesses an infra-sensible (or potential) taste, smell and colour, (one kind of each) and two infra-sensible tactile qualities, *e.g.* a certian degree of roughness or smoothness (or dryness and moistness ?) and of heat or cold. Earth-atoms, Ap-atoms, etc. are but differentiations of the originally homogeneous Pudgalas. The tactile qualities (रुक्ष, स्निग्ध, उष्ण, शीत) appear first, but qualities of taste, smel and colour are involved in the possession of tactile qualties, An aggregate (Skandha), whether binary, tertiary or of a higher order, possess (in addition to touch, taste, smell, and colour) the following physical characters:—(1) sound, (2) atomic linking, or mutual attraction and repulsion of atoms, (3) dimension, small or great. (4) figure, (5) divisibility, (6) opacity and casting of shadows and (7) radiant heat and light.

Sensible Qualities

Tactile qualities are of the following kinds—hardness or softness, heaviness or lightness (degrees of pressure). heat or cold, an roughness or smoothness (or dryness, and viscosity?). Of these, the atoms (Aṇus) possess only temperature, and degrees of roughness or smoothness, but all the four kinds of tactile qualities in different degrees and combinations characterise aggregates of matter from the binary molecule upwards. The Jainas appear to have thought that gravity was developed in molecules as the result of atomic linking. Simple tastes are of five kinds,—bitter, pungent, astringent, acid the sweet. Salt is supposed by some to be resolvable into sweet, while others consider it as a compound taste. Smells are either pleasant or unpleasant. Mallishena notes some elementary varieties of unpleasant smell, *e.g.* the smell of asafoetida, ordure, etc. The simple colours are five— black, blue, red, yellow, and white. Sounds may be classed as loud or faint, bass (thick) or trable (hollow), clang or articulate speech.

The most remarkable contribution of the Jainas to the atomic theory relates to their analysis of atomic linking, or the mutual attraction (or repulsion) of atoms, in the formation of molecules. The question is raised in Umāsvāti's Jaina Sūtras (circa A.D. 40)—What constituents atomic linking? Is mere contact (or juxtaposition) of atoms sufficient to cause linking? No distinction is here made between the forces that bind together atoms of the same Bhūta, and the chemical affinity of one Bhūta to another. The Jainas hold that the different classes of elementary substances (Bhūtas) are all evolved from the same primordial atoms. The infra-atomic forces which lead to the formation of chemical compounds do not therefore differ in kind from those that explain the original linking of atoms to form molecules.

Mere juxtaposition (संयोग) is insufficient; *linking* of atoms or molecules must follow before a compound can be produced. The linking takes place under different conditions. Ordinarily speaking, one particle of matter (पुद्गल) must be negative, and the other positive (विषमगुणयुक्त) ; the two particles must have two peculiar opposite qualities, roughness and smoothness (रक्षत्व and स्निग्धत्व or dryness and viscosity?). to make the linking possible. But no linking takes place where the qualities, though opposed, are very defective or feeble (जघन्यगुण). We have seen that, ordinarily speaking, two homogeneous particles, *i.e.* both positive, or both negative, do not unite. This is the case where the opposed qualities are equal in intensity. But if the strength or intensity of the one is twice as great as that of the other, of exceeds that proportion, then even similar particles may be attracted towards each other. In every, case, change of state in both the particles is supposed to be the result of this linking, and the physical characters of the aggregate depend on the nature of this linking. When particles of equal intensity (negative and positive) modify each, other there is mutual action ; in case of unequal intensity, the higher intensity transformation the lower, it being apparently thought that an influence proceeds from the highter to the lower. All changes in the qulities of atoms depend on this linking. A crude anticipation this, of the ionic theory of chemical combinations, very crude but immensely suggestive, and possibly based on the observed electrification of smooth and rough surfaces as the result of rubbing. The interpretation of रूक्ष and स्निग्ध as dry and viscous (or as vitreous and resinous ?) must be rejected in this connection as untenable. The Tattvārthādhigama of Umāsvātī, which expounds the theory most probably dates back to the first half of the first century A.D.

Cf. Umāsvāti-Tattvārthādhigama, Chap. V.

(अजीवकाया धर्माधर्माकाशपुद्गलाः। द्रव्याणि जीवाश्च। नित्यावस्थितानि अरूपाणि। रूपिणः पुद्गलाः। आकाशादेकद्रव्याणि। निष्क्रियाणि च। पुद्गलजीवास्तु क्रियावन्तः। (क्रियेति गतिकर्माह)। सर्वेषां प्रदेशाः सन्ति अन्यत्र परमाणोः। संख्येयासंख्येयाश्च पुद्गलानां। नाणोः। स्पर्शरसगन्धवर्णवन्तः पुद्गला। तत्र स्पर्शोऽष्टविधः कठिनो मृदुर्गुरुलघुः शीत उष्णः स्निग्धः रूक्ष इति। रसः पञ्चविधस्तिक्तः कटुः कषायोऽम्लोमधुर इति। गन्धो द्विविधः सुरभि-रसुरभिश्च। वर्णः षञ्चविधः कृष्णो नीलो लोहितः पीतः शुक्ल इति।

शब्दबन्धसौक्ष्म्यस्थौल्यसंस्थानभेदतमश्छायातपोद्यातवन्तश्च। तत्र शब्दः षड् विधः ततो विततोः धनः शुशिरो वर्षो भाव इति। अणवः स्कन्धाश्च। उक्तं च। कारणमेव तदन्त्यं सूक्ष्मो नित्यश्च भवति परमाणुः। एकरसगन्धवर्णो द्विस्पर्शः कार्यलिङ्गश्च। तत्राणवोऽबद्धाः स्कन्धास्तु बद्धा एव। स्कन्धास्तावत् संघातभेदेभ्य-उत्पद्यन्ते। संघाताद्भेदात् संघातभेदादिति। द्वयोः परमाण्वोः संघाताद्द्विप्रदेशः। द्विप्रेदेशस्याणोश्च संघातात् त्रिप्रदेशः। एवं संख्येयानमसंख्येयानामनन्तानां अनन्तानन्तानां च प्रदेशानां संघातात् प्रदेशाः। भेदादणुः। भेदादेव परमाणुरुत्पद्यते न संघातादिति। अत्राह। किं संयोगमात्रादेव संघातो भवति। आहोस्विदस्ति कश्चिद्विशेष इति। अत्रोच्यते। सति संयोगे बद्धस्य संघातो भवतीति। स्निग्धरुक्षत्वाद्बन्धः। न जघन्य-गुणानाम्। स्निग्धेरुक्षयोः पुद्गलयोः स्पृष्टयोर्बन्धो भवतीति। जघन्यगुणस्निग्धानां जघन्यगुणरूक्षाणां च परस्परेण बन्धो न भवतीति। उक्तं भवता जघन्यगुणवर्ज्जानां स्निग्धानां रुक्षेण रुक्षाणां च स्निग्धेन सह बन्धो भवतीति। अथ तुल्यगुणयोः किमत्यन्तप्रतिषेध इति। अत्रोच्यते। न। गुणासाम्ये सदृशानां। गुणसाम्येसति सदृशानां बन्धो न भवति। तद्यथा। तुल्यगुणस्निग्धस्य तुल्यगुणस्निग्धेन तुल्यगुणरूक्षस्य तुल्यगुणरूक्षेण इति। अत्राह सदृशग्रहण किमपेक्षते इति। अत्रोच्यते। गुणवैषम्ये सदृशानां बन्धो भवति इति। अत्राह किमविशेषेण गुणवैषम्ये सदृशानां बन्धो भवतीति। अत्रोच्यते। द्व्यधिकादिगुणानां तु द्व्यधिकादिगुणानां सदृशानां बन्धो भवति। द्विगुणाद्यधिकस्निग्धेन। द्विगुणाद्यधिकस्निग्धस्य स्निग्धेन। रूक्षस्यापि द्विगुणाद्यधिकरूक्षस्य रूक्षेण। एकादिगुणाधिकयोस्तु सदृशयोर्बन्धो न भवति। अत्राह परमाणुषु स्कन्धेषु च ये स्पर्शादयो गुणास्तु किं व्यवस्थिताः तेषु आहोस्विदव्यवस्थिता इति। अत्रोच्यते। अव्यवस्थितायुतः।

परिणामात्। अत्राह द्वयोरपि बध्यमानयोर्गुणवत्वे सति कथं परिणामो भवति इति। उच्यते। बन्धे समाधिकौ पारिणामिकौ। बन्धे सति समगुणस्य समगुणः परिणामको भवति। अधिकगुणो हीनस्येति।

—Umāsvāti, Tattavārthādhigama, Chap. V).

The Nyāya-Vaiśeṣika chemical theory ; a Brief Summary.

I must content myself here with a brief and rapid sketch of the chemistry of the Nyāya-Vaiśeṣika, which I shall elaborate in connection with the mechanics and physics of the ancient Hindus in a separate paper.

The relation of the specific characters of molecules (and higher aggregates) to the original atomic qualities is reduced in the Nyāya-Vaiśeṣika to the following canons :—(a) कार्यगुणं कारणगुणपूर्वकम्। (b) समानजातीयसंयोगः द्रव्यारम्भकः न विजातीयसंयोगः (Here द्रव्य is used in a narrow technical sense, so as to exclude the quasi-compound substances). (c) अपाकजरूपरसगन्धस्पर्श परिमाणैकत्वैकपृथक्त्वगुरुत्वद्रवत्वस्नेहवेगाः कारणगुणपूर्वकाः। (d) रूपरसगन्धा-नुष्णस्पर्शशब्द-परिमाणैकत्वैकपृथक्त्वस्नेहाः। समानजात्यारम्भकाः। No separate explanation is necessary, as the cannos are embodied in the following exposition.

Theory of Atomic Combination

The four kinds of Atoms are Earth, Ap, Tejas and Vāyu atoms, possessed of characteristic mass, numerical unit, weight, fluidity (or its opposite), viscosity (Or its opposite), velocity (or quantity of impressed motion—Vega) ; also characteristic potential colour, taste, smell or touch, not produced by the chemical operation of heat अपाकजरूप-रसगन्धस्पर्शपरिमाणैकत्वैक-पृथकत्व-गुरुत्वद्रवत्वस्नेहवेगाः।) Ākāśa has no atomic structure (निरवयव), and is absolutely inert (निष्क्रिय) being posited only as the substratum of sound, which is supposed to travel wave-like in the manifesting medium or vehicle of Vāyu (air). Only the other four Bhūtas unite (or disunite) in atomic or molecular forms. The orthodox view is that, the presence of Earth-atoms is necessary whenever chemical transformation under the operation of heat. (पाकजोत्पत्ति) takes place.

Atoms cannot exist in an uncombined state in Creation (Śivāditya, Sapta-Padārthī–*vide* commentary, where, however, it is noted that still atmospheric air is believed to be monatomic in structure, *i.e.* to consist of masses of atoms in a loose uncombined state—(स्तिमितवायुस्तु परमाणु समूह एव अनारयधद्रव्यः।)

The atoms may combine in one or other of the following ways :—

I. One Earth-atom, by an original tendency, unites with another to form a binary molecule (द्व्यणुक). In the same way, binary molecules of the other Bhūtas are formed. The atoms are possessed of an inherent Parispanda (rotary or vibratory motion), and when they unite in pairs, *so long as there is no chemical operation under the action of heat corpuscles,* the original qualities of the atoms produce homogeneous qualities in the binary molecules.

The question as to the existence of a traid, a tetrad, a pentad etc. of atoms is one of the moot points of the Nyāya-Vaiśeṣika. The orthodox view is that, the primordial infinitesimal particles (atoms) state with an incessant vibratory motion अनवरतपरिस्पन्दमानापरिमितपवनादिपरमाणवः, Raghunātha Śiromaṇi—गतिशीलत्वात् पतत्रव्यपदेशाः पतन्तीति, Udayana. Kusumāñjali), and an inherent impulse that drives them to unite in pairs—a sort of 'monovalency', *as it were,* exhausted with the formation of a binary molecule. The binary molecules now combine by threes, fours, fives, etc. to form larger aggregates as well as the variety of elementary substances, the particular collocation any case being not only determined by physical cause, but also serving to satisfy the ends of the moral Law in creation. (अदृष्ट, कर्म), (द्व्यणुकैर्बहुभि-

रारभ्यते इत्यपि नियमो, न द्वाभ्यां। बहुषु त्वनियमः। कदाचित् त्रिभिरा रभ्यते इति। त्र्यणुकमित्युच्यते, कदाचिच्चतुर्भिरारभ्यते कदाचित् पञ्चभिरिति यथेष्टं कल्पना। अदृष्टवशात् तथा तथा तेषां व्यूहो यथा यथा तदारवधेषु अपरजातयो व्यज्यन्ते। अदृष्टकारिता सर्वभावानां सृष्टिः। Śrīdhara, Nyāya-Kandalī, पृथिवी-निरूपणम्।) *Cf.* Vācaspati's report, Bhāmatī, Chap. II., Pada. 11. Sūtra II. यदा चतुरणुकमारभते चतुर्णां द्व्यणुकानामारम्भकत्वात्). Praśastapāda appears to have originated this view (परमाणु द्व्यणुकेषुबहुत्वसंख्या-तैरारब्धे त्र्यणुकादिलक्षणे- Praśastapāda, परिमाणनिरूपणम्); but that another view was also maintained in the Vaiśeṣika school is evident from the brief summary of Kaṇāda's system given in Utpala's commentary on the Vṛhatsaṁhitā, and this indeed also follows from Śrīdhara's admissions in the Kandalī. On this view, also, atoms have an inherent tendency to unite, but some unite in pairs, others in traids, other in tetrads, etc. This may happen in two ways—either by the atoms falling into groups of threes, four etc. direct, or by the successive addition by one atom to each preceding aggregate.

A traid (Tryaṇuka), then, hold together three atoms (Aṇus), not there binary molecules (Dvyaṇukas) as on the orthodox hypothesis. Similiary with tetrads, pentads, etc. (चतुर्विधाः परमाणवः क्षितिजलाग्निवायूनाम्। द्वाभ्यां परमाणुभ्यां द्व्यणुकमारभ्यते त्रिभिः परमाणु भिस्त्र्यणुकमारभ्यते इति क्रमेण स्थूलकार्यद्रव्यस्योत्पत्तिः। Utapala, Chap. I. Śloka 7 ; *cf.* also Śrīdhara's admission, अथवा यदि परमाणवो द्व्यणुकमारभ्य तत्सहितास्त्र्यणुकमारभ्यन्ते त्र्यणुकसहितास्तु द्रव्यान्तरम् तथापि कुतो विश्वस्य अग्रहणम्). Śaṅkara seems to speak of two binary molecules in the Vaiśeṣika as forming a tetrad, (यदापि द्वे द्व्यणुके चतुरणुकमारेभेते–Śārīraka-Bhāṣya, Chap. II, Pada 20, Sūtra II, where are Bhāmatī gives a forced interpretation).

In Praśastapad is view these binary molecules are grouped by threes four, fives etc. (त्र्यणुक, चतुरणुक) to form different isomeric modifications. The variety of Earth-substances is due to differences in the arrangements of the molecules, (*e.g* their greater or less density, and, above all their grouping or collocation व्यूह, अवयवसन्निवेश), which account for the specific characters (अपरजाति) manifested by these isomeric substances. सा (पृथिवी) च स्थैर्याद्यवयवसन्निवेशविशिष्टा अपरजातिबहुत्वोपेता। Praśatapāda, पृथिवीनिरूपणम्। स्थैर्यं निड़ित्वम्। आदिशब्दात् प्रशिथिलत्वादिपरिग्रहः। परमाण्स्वादिषु अपरजात्यभावेऽपि अदृष्टवशात् तथा तथा तेषां व्यूहः यथा यथा तदारब्धेषु अपरजातयो व्यज्यन्ते। Śrīdhara, Kandalī *ibid.*

स्थैर्यं स्थिरता चिरकालावस्थायित्वमिति यावत्। आदिग्रहणाद्दिष्टम्मकत्वं जलादिव्यूहविरोधित्वञ्च। अवयवसन्निवेशाः तत्तत्सामान्यविशेषभिव्यञ्ज-कसंस्थानविशेषाः। न हि एतद् द्रव्यान्तरे सम्भवति। जलादीनां यत्किञ्चित्स्पर्शव-द्वेगवद्द्रव्यो पनिपातमात्रेणैव भङ्गुरत्वात्। Udayana, Kiraṇāvali, *ibid.* These original difference in molecular grouping leading to distinctions of genera and species, however mechanically or physically explained, come also under the operation of moral and metaphysical causes (अदृष्ट, कर्म), *i.e.*, of ideal ends in the moral government of the universe, which are superimposed upon the physical order, but which do not come within the scope of Natural philosophy. An elementary substance thus produced by primary atomic combination, may, however, suffer qualitative change under the influence of heat (पाकजोत्पत्ति). The process is as follows :—(1) the impact of heat corpuscles decomposes the binary (tertiary, or quanternary) molecules into homogeneous atoms possessing only the generic charcters of the Bhūta concerned ;—(2) the impacts of heat particles continue, and transform the characters of the atoms, determining them all in the same way ;—(3) the heat particles continue to impinge, and reunite the atoms so transformed to form binary (or rather) molecules, in different orders or arrangements, which account for the specific characters or qualities finally produced. The Vaiśeṣika holds that there is decomposition into homogeneous atoms, transformation of atomic qualities, and finally recombination, all under the influence of heat. The Nyāya on the other hand thinks that the molecules and larger aggregates assume the new characters under the influence of heat without decomposition into homogeneous atoms, or change of atomic characters (पिलुपाक or पिठरपाक).

तेषामनुमानेन विनाशः परिकल्पते। सर्वावयवेषु, अन्तर्वहिश्च पाकपूर्वकपूर्व-रूपादिविलक्षणगुणोपलब्धेरन्तः प्रवेशः कृशानोरनुमीयते तेन वेगवता वह्निद्रव्येण नोदनात् अभिघातात् वा नूनं घटाद्यारम्भकेषु अवयवेषु क्रिया जायते क्रियातो विभागः विभागात् द्रव्यारम्भकसंयोगविनाशः। तद्विनाशात् द्रव्यविनाशः। पक्वाश्च श्यामादिगुणानवजहतः रक्तादिगुणान्तरयोगमनुभवन्तः–अदृष्टप्रेर्यमाणाः परस्परं संयुज्य द्व्यणुकादिप्रक्रमेण तादृशमेव घटादिकार्यमारभन्ते। एवं तपनातपदृश्यमानेषु आम्रादिफलेषु एष एव न्यायः। शरीरे अपि उदर्येण तेजसा पच्यमानेषु अन्नपानादिषु रसमलधातुभावेन परिणाममुपगच्छत्सु प्रायेण प्रतिक्षणमुत्पादविनाशौ संभवत इति। (Jayanta. Nyāyamañjarī भूतचैतन्य-पूर्वपक्ष). This is the Vaiśeṣika view, but Jayanta himself inclines to the opposite view :—प्रकृतिसुषिरतयैव कार्यद्रव्यस्य घटादेः आरम्भात् अन्तस्तेजः कणानुप्रवेशकृतपाकोपपत्तेः अलं विनाशकल्पनया:। पिठरपाकपक्ष एव पेषलः। *ibid.* The Nyāya view :—ये वीजावयवास्ते पूर्वव्यूहपरित्यागेन व्यूहान्तरमापद्यन्ते व्यूहान्तरापत्तौ च पृथिवीधातुरब्धातुना संगृहीत आन्तरेण तेजसा पच्यमानो रसद्रव्यं निर्वर्त्तयति स रसः पूर्वावयवसहितोऽङ्कुरादिभावमापद्यते। परमाण्वस्थानि वीजानि भवन्तीत्येतन्न प्रतिपद्यामहे। यस्माच्छाल्यादिवीज-मुच्छूनावस्थामादिं कृत्वा यावदुपान्त्यं शालिवीजकार्यं तावन्न कदाचित् परमाणवस्थं भवति। यदि तु स्यात् कदाचित् नोपलभ्येत्। Udyotakara's Chap. II, Āhinka I, Sūtra 4.

II. Chemical combination (सम्भूयक्रिया, संहतक्रिया). Chemical combination takes place either between two or more substances which are isomeric modifications of the same Bhūta, or between substances which are mode of different Bhūtas.

A. ***Mono-Bhautic compounds***. The simplest compounds are Mono-Bhautic compounds, *i.e.* compounds of different substances which are isomeric modes of the same Bhūta.

(a) ***Mono-Bhautic compounds of the first order*** :—Under

the impact or impulse (अभिघात or नोदन) of heat corpuscles, the substances in chemical contact (आरम्भक संयोग) break up into their atoms. These atoms are homogeneous, possessing only the original physical and chemical characters of the Bhūta concerned. As the specific differences between isomeric substances arise from the arrangement or collocation of the atoms, the substances lose their distinctive marks on decomposition the latter. (न च परमाणुषु अपरजातिभेदो विद्यते न च यववीजशालिवीजपरमाणूनां कश्चिद्विशेषः। Udyotakara Chap. III, Āhnika I, Sūtra 4. न च शुक्रशोणितपरमाणूनां कश्चिद्विशेषः पार्थिवत्वविशेषात्। Śrīdhara, Nyāya-Kandalī, पृथिवीनिरूपणम्) । Under the continued impact (or, it may be, impulse) of heat particles (वेगवता वहिद्रव्येण नोदनात् अभिघातात् वा–Jayanta), these atoms take on new characters. It is heat and heat alone that can cause this transformation of the colours, taste, smells, etc., in these original Bhūta-atoms. What particular colours, testes, smells or physical characters will be produced in the atoms depends (1) on the nature of the constituent substances in contact; (2) the intensity or degree of the heat (खर, मृदु or मध्यम पाक); and (3) the species of Tejas-corpuscles that impinge on the atoms, or the nature of the impact (विलक्षणतेजः संयोग). (न ब्रूमोऽग्निसंयोगात् एकस्मात् रूपादय इति अपि तु पूर्वरूपादि-विशेषापेक्षात्। यद्द्रव्यं पच्यते अग्निसंयोगेन तस्य ये पूर्वरूपादयस्तेषां यः स्वगतो विशेषस्तमपेक्षमाणः अग्निसंयोगः उत्तरान् रूपादीन् विशिष्यानारभते। *Vide* also Udyotakara, III, 2, Sūtra 14. *Cf.* also Vācaspati, I, 1, Sūtra 4. अस्माकमभेदे अपि उपादानस्य पिठरस्य औष्ण्यापराख्यस्य च वह्निसंयोगस्य पूर्वरूपादिप्रध्वंसानां कारणानां भेदात् भिन्नजातीया जायन्त गन्धरूपरसस्पर्शा इति सिद्धान्तः)।

Now, when the atoms have all been determined in the same way, they begin to recombine again under the impact (or impulse) of the heat-particles in binary molecules, (or tertiary, etc.) and these in higher aggregates. It seems to be generally held that, at the final step one or more atoms of one constituent substance unite with one or more atoms of the other constituent substance or substances to form a molecular of the compound ; but the question is not of much significance for Mono-Bhautic compounds of the first order, as, in these cases, the atoms have before this all lost their distinctive characters and become homogeneously transfromed. The compound so prodcued will possess the new characters of the transformed atoms, so far as taste, colour, smell, etc. are concerned, but as the molecular arrangement or structure (व्यूह, अवयवसन्निवेश) may vary, different compound substance may result from the same components.

(b) ***Mono-Bhautic compounds of higher orders*** :— Again, Mono-Bhautic compounds of the first order may chemically combine to form higher compounds, and as the ultimate Bhūta substratum is the same, the process of decomposition and recomposition will be essentially the same as before. The only doubtful point is whether in this case the component compound substance are broken up only into their constituent molecules, into the original homogeneous Bhūta atoms. Some of the later Vaiśeṣika Scholiasts hold that the latter happend in every case of chemical composition, however complex, but the earlier Vaiśeṣika conceived that in the case of compounds of compounds, the decomposition does not proceed so far as the original Bhūta atoms ; but that it is the specifically determined atoms constituting the molecules of the component compounds that are transformed under the impact of the heat corpuscles; and then one such transformed atom (one or more according to another version) from the molecule of one component unites with one similarly transformed atom (one or more according to the other version) from the molecule of the other component. Praśastapāda, the great Vaiśeṣika Doctor, holds this view. When, for example, in the fertilised ovum, the germ and the sperm substances, which in the Vaiśeṣika view, are both isomeric modes of Earth (with accompaniments of other Bhūtas), unite, both are broken and into homogeneous Earth-atoms, and it is these that chemically combine under the animal heat (and bio-motor Energy, वायु) to form the germ-plasm (कलल). But next, when the germ-plasm develops, deriving its nutrition from the chyle (blood) of the mother, the animal heat breaks up the molecules of the germ-plasm into its constituent atoms (कललारम्भकपरमाणव:), *i.e.*, into atoms specifically determined, which by their grouping formed the germ-plasm, and then these germ-plasm atoms as radicals chemically combine with the atoms of the food constituents, and thus produce cells and tissues. (समुत्पन्नपाकजैः कललारम्भकपरमाणुभिः अदृष्टवशात् उपजातिक्रियैः आहारपरमाणुभिः सह सम्भूय शरीरान्तरमारभ्यते इत्येषा कल्पना। पितुः शुक्रं मातुः शोणितं तयोः सन्निपातानन्तरं जठरानलसम्बन्धात् शुक्रशोणितारम्भकेषु परमाणुषु पूर्वरूपादिविनाशे सति समानगुणान्तरोत्पत्तौ द्व्यणुकादि-प्रक्रमेण कललशरीरोत्पत्तिः।.....तत्र मातुराहाररसः मात्रया संक्रामति अदृष्टवशात् तत्र पुनर्जठरानलसम्बन्धात् कललारम्भकपरमाणुषु क्रियाविभागादिन्यायेन कलल शरीरे नष्टे समुत्पन्नपाकजैः कललारम्भक परमाणुभिः अदृष्टवशात् उपजातक्रियः आहारपरमाणुभिः सह सम्भूय शरीरान्तरमारभ्यते इत्येषा कल्पना। Śrīdhara, Kandalī, (पृथिवी-निरूपणम्।) In this hypothesis (कल्पना), it is assumed that the atoms are similarly transformed *i.e.*, become endowed with the colour, taste, smell etc. of the product (the cell or tissue), the moment before the chemical combination takes place. Similary, when milk is transformed into curd, one view is that the transformation take place (under internal heat) in the constituents atoms of the milk molecules, atoms specifically determined as milk, and not in the original atom of the Bhūta (or Bhūtas) entering into the composition of milk.

(*Cf.* एवं महादुग्धारम्भकैः परमाणुभिरेव दध्यारभ्यते। एकमहिम्नेव दध्यारम्भकैरेव परमाणुभिर्नवनीतारम्भः इति दिक्। Nyāyabodhinī, on Annam Bhaṭṭa's Tarkasaṅgraha.) Cf. Dinakarī, on the other hand—द्व्यणुकरूपदुग्धध्वंसजन्यत्वेऽपि तस्य (दध्नः)। In these cases, the atomic contact is called constituent contact (आरम्भकसंयोग) and all these the atoms are equally regarded as material causes (उपादानकारण or समवायिकारण) of the compound.

B. ***Hetero-Bhautic 'quasi-compounds'***.—The Nyāya-Vaiśeṣika maintains that in the case of bi-Bhautic (or poly-Bhautic) compounds, which are only quasi-compounds, there is another kind of contact between the heterogeneous atoms of the different Bhūtas, which may be called dynamic contact and is distinguished in its operations as Upaṣṭambha. Viṣṭambha or Avaṣṭambha (उपष्टम्भ, विष्टम्भ or अवष्टम्भ). In some cases, it so happens that the atoms of different isomeric modes of the same Bhūta do not chemically combine under the mere application of heat —they require to be surrounded (and 'excited', 'energised') by atoms of different Bhūtas. For example, in the case of the oils and fats as well as of plant saps and fruit juices the Earth-atoms must be dissolved in water (Ap), and it is only when the water atoms (Ap-atoms) congregate round the former that dynamic intra-atomic forces are set up, the Earth-atoms (with the water atoms in dynamic contact) now take on peculiar infra-sensible characters (colours, tastes, smells) under the impact of the heat corpuscles, and then, under further impact, fall into groupings or collocations (of a very peculiar nature, to be presently explained) which determine the nature of the composite substance thus produced. Here, it is the water atoms that are dynamic (उपष्टम्भक), and excite and Earth-atoms, and these substances, oils and fats (तैलवर्ग and घृतवर्ग), as well as acids (अम्लवर्ग) are, because of the Earth radicles regarded as Earth-compounds (or Earth-substances). पाकज तथाविधगन्ध-रसरूपादिमद्भिः परमाणुभिद्व्यणुकादिक्रमेण घृतादिद्रव्यमारभ्यते। तत्र च उपष्टम्भकतया निमित्ततामापन्नाः पानीयावयवाः। तेषां संयुक्तसमवायेन स्नेहस्तात्रोपलभ्यते। तैलक्षीरादिषु पार्थिवत्वसिद्धिः। तैलस्य भौमान-लेन्धनत्वात् घृतवत्। Udayana). In the above instances. Ap (water) acts as dynamic (Upaṣṭambhaka, (उपष्टम्भक), but Tejas and Vāyu can also act in the same way on Earth-particles. Conversely, Earth-particles may act dynimically on the atoms of the other Bhūtas. For example, in the case of mercury and the metals which are conceived in the Nyāya-Vaiśeṣika to be igneous bodies (in fact they are supposed to be formed under the subterranean heat (आकरज), the Tejas corpuscles are believed to form the radicles, and the Earth-particles are dynamic (उपष्टम्भक). (सुवर्णादि निरन्तरं ध्मायमानमपि न पूर्व रूपं जहाति। तेनैव द्रव्यान्तरेण प्रतिबद्धत्वात् उपष्टम्भकोऽपि पार्थिवभागः सदृशरूप एवानुवर्त्तते, यत्तु पुटपाकादिना रक्तसारता दृश्यते तन्मिश्रीभूताभिभावकतद्द्रव्योपगमात्। एतेन पारदादि व्याख्यातम्। Udayana, Kiraṇāvalī, तेजोनिरूपणम्। *Cf.* also योऽपि तैजसमिच्छति सुवर्णं तेनाप्यत्र पार्थिवोभाग उपष्टम्भक एष्टव्यः।)

It may be here noted Gaṅgeśa, the author of the Tattvacintāmaṇi, conjectures that even gold can be evaporated or made of dissappear by the application of intense heat ; अथ द्रुतद्रुततरादेः क्रमशः प्रतीतेः प्रबलानलसंयोगेन

सुवर्णनाशात् तद्द्रवता नश्यति। But Mathurānātha does notes here—भस्मप्राक्कालोच्छेदसमये। (सन्निकर्षवादरहस्य).

But while every Bhūta can act dynamically as उपष्टम्भक, 'energiser', 'exciter', it is the Earth-Bhūta alone which is capable of exercising the power of arrest of inhibition of molecular motion or the motion of particles due to gravity as in fluids (Viṣṭambha, विष्टम्भ), or the power of counteracting the tendency in a given set of atoms to all into a peculiar order or group (व्यूहविरोधित्वम्). (न च पार्थिववत् अम्भो द्रव्यान्तरं गच्छद्विष्टभाति पूर्व व्यूहं वा विरुणद्धि। Udayana, Kiraṇāvalī, पृथिवीनिरूपणम्। विष्टम्भकत्वं स्वाभाविकगुरुत्वाश्रयद्रव्यान्तरगति-प्रतिबन्धकत्वम्। व्यूहविरोधित्वं च स्वाभाविकगुरुत्वत्पतनप्रतिबन्धकत्वम्। Vardhamāna, Kiraṇāvalīprakāśa, *ibid.*)

Oils facts milks—Bi-Bhautic quasi-compounds with Ap as energiser

Oleaginous substances (आज्यतैलादयः स्नेहाः) are divided by Udayana into (1) oils, derived from vegetables, (2) butters derived from milk, and (3) fats derived from animals. The medical schools, as we have seen, recognise animal oils as distinguished from vegetable fats (*e.g.* विड़ङ्गघृत) are also mentioned. Vācaspati in the Tātparyaṭīkā contends that mustard oil has not the flavour and smell characteristic of the true oils (sesamum, inseed etc.) and is classesd with the latter by convention. Judged by the flavour test. Āmikṣā (the casein substance formed by mixing milk-curd with hot bolied milk) is to be classed with milk substances. So also Takra, whey,—but Vājina, the thin fluid that is left after the Āmikṣā (casein substance) is seperated, cannot be classed as milk. It may be added that the milks and curds as well as oils and fats derived from different species of vegetables or animals are supposed to differ in their ultimate structural arrangement and therefore in kind ; but Vallabha thinks that the ghees (clarified butters) prepared from different kinds of milk are of the same kind ; in other words, the milks and curds are 'polymeric'. the ghees (clarified butters), 'isomeric' using these terms as before, in a loose general sense....तैलघृतवसानां वीजक्षीरमांसयोनित्वात्। अतएव नारिकेलस्नेहस्यापि तैलत्वं रसादि सारूप्यादपितु-घृतव्यपदेशश्च। Udayana, Kiraṇāvalī, पृथिवीनिरूपणम्। Vallabha does that ghees do not differ in kind, as milks (and curds) do :— अतो महिषगव्यादिदुग्धेषु दधित्ववत् घृतादिषु जातिभेद एवास्तामिति चेन्न।– Vallabha, Līlāvatī. Compare Vācaspati, II, 2, 65 :—आज्यतैलादीनां जातिस्तु गन्धेन वा रसेन वा व्यज्यते। अतएव न सार्षपादीनां तैलत्वमस्ति। तद्वयञ्जकयोर्गन्धसयोरभावात्। भाक्तस्तु तैलशब्दयोगः। क्षीरजातिरपि रसव्यङ्ग्येव। अतएवामिक्षायाः क्षीरत्वम्। न तु वाजिनस्य तद्व्यञ्जकस्य रसभेदस्य याजिने अभावात् आभिक्षायां च भावात्।

III. Mixtures like Soups, Solution etc

A soup is a physical mixture of a peculiar kind, from which evaporation ordinarily sets the water free (*cf.* Udajayan's धनपीतयुक्त *infra*). When meat is boiled in water, there is the application of heat, with chemical changes in the meat, but the combination of meat-particles and water-particles in the soup is only physical combination, and not a chemical one. It is of course not a true compound, neither is it a quasi-compound, like milk (in which the water-particles are energisers of the Earth-particles). Milk, for example, retains its milky substance, when is coagulates or becomes solid, (this of course is also the case with mono-Bhautic substances whether elementary of compound, *e.g.* water, which becomes ice,) but the substance we call a soup or solution ceases to be a soup or solution, the moment it solidifies. Udyotakara notices gruel, baths, and lyes (alkaline solutions) as mixture of this class (एतेन पाककाञ्चिकविवेकावक्षारादयः प्रत्युक्ताः। Vārttika, Chap. I, Āhnika. I. Sūtra 14). यूवो हि नाम उत्पन्नपाकजानां द्रव्याणां कालविशेषानुग्रहे सति द्रव्यान्तरसंपृक्तानां पाकजोत्पत्तौ यः संयोगः स यूष इति। एतेन पाककाञ्चिकविवेकावक्षारादयः प्रत्युक्ताः। Udyotakara,

Vārttika, Chap. I. Āhanika I, Sūtra 14. *Vide* also Vācaspati's comment : उत्पन्नपाकजानां तैले सर्पिषि वा पिष्टमांसपिण्डावयवानां द्रव्यान्तरेण तोयेन संयुक्तानां, पाकजोत्पत्तौ सत्यां न चासौ सहसा इत्युक्तं कालविशेषानुग्रहे सतीति। स च संयोगभेदएव तोयमांसयोर्नतु अवयवी विजातीययोरनारम्भकत्वात्। नापि यूषजातीयं तोयसंयोगि द्रव्यं क्षीरजातीयमिवेति युक्तम्। तोयविरहे काठिन्ये क्षीरबुद्धिव्यपदेश- योस्तादवस्थ्यात्। इह तु काठिन्ये न यूषबुद्धिव्यपदेशौ इति सम्बन्धभेद एव तोयमांसयोर्यूष इति न्याय्यम्। स च अनुभवसिद्धः संयोगिव्यतिरिक्तः। Vācaspati, Tātparyyaṭīkā, *ibid.*, for salt and alkaline solutions, *vide* Kiraṇāvalī—यत् पुनरम्बुधिसिन्धुभेदेन क्षारादिरपि रसः पाथसे उपलभ्यते स दशमूलकषायस्येव पार्थिवद्रव्योपाधिकः। कथमन्यथा तस्यैव घनपीतमुक्तस्य रसो मधुर एव। Udayana, Kiraṇāvalī, अलनिरूपणाम्।

Chemical Action and Heat

The operation of heat is of course universally implied in chemical combination. Where the application of external heat is wanting, Vātsyāyana, the great Doctor of the Nyāya, points to the operation of internal heat. (*e.g.* पृथिवी धातुः अब्धातुना संगृहीतः आन्तरेण तेजसा पच्यमानः रसद्रव्यं निवर्त्तयति।–Vātsyāyana-Bhāṣya, Chap. IV, Āhnika I, Sūtra 47). In the case of combustion, we have seen Vijñāna-bhikṣu explain the heat as latent in the Earth-substance, the fuel, from which it breaks forth. Udayana points out that the solar heat is the source of all the stores of heat required for chemical change in the world. The change of colours in grasses, for example, is due to Tejas, in the form of latent (invisible) heat, not in the form of Agni ; and the cold in water cannot take away this store derived from the Sun. तृणादिविकारो हि यदि रूपादिरावृत्तिमात्रहेतुः स नूनमौष्ण्यापेक्षेण तेजसा कर्त्तव्यः। तादृश च पाके अनिमित्तं हिममिति। न किञ्चिदनिष्टमापद्यते। न हि सौरस्य तेजसः त्रैलोक्यपाकहेतोर्हिमादपगमः क्षमते। अथ विकारो भस्मादिरूपो विवक्षितः सोऽसिद्ध एव हिमहतेषु तृण्यादिषु क्व विरोधो बाधो वा। अथ रूपादिपरावृत्तिमात्रेणैव अग्निः साध्यते तदशक्यम्। तस्य दर्शनस्पर्शनग्राह्यस्य योग्यानुपलम्भवधितत्वात् अतादृशस्य तेजोमात्रस्य निवृत्तेरशक्यत्वात् अनिष्टत्वात् च। Udayana, Kiraṇvāvalī सृष्टिसंहारविधिनिरूपणम्।) Similary, it is under this solar heat that the unripe mango ripens, *i.e.* changes colour, taste, smell etc., showing that there is chemical transformation or subtile decomposition and recomposition going on ; and this is also the case with the resulting of the metals, which is a combustion due to the solar heat (सूर्यपाक) even as the conversion of food into chyle and of chyle into blood are instances of chemical action due to the internal animal heat (जठरानल and औदर्यं तेजः). But the kind of contact with heat-corpuscles, in other words, the kind of chemical action (पाक) which transforms colours is supposed to differ from that which transforms flavour (विलक्षणतेजः संयोग and पाक) and this last from that which produces a change of smell, tactile quality. (पाको नाम विजातीयतेजः संयोगः। स च नानाजातीयः। रूपजनको विजातीयतेजः संयोगस्तदपेक्षया रसजनको विजातीयः। एवं स्पर्शादौ अपि तथा। एवं प्रकारेण भिन्नभिन्नजातीयाः पाकाः कार्यवैलक्षण्येन कल्पनीयाः। तथाहि तृणपुञ्जनिक्षिप्ताऽम्रादौ उष्णलक्षणविजातीयतेजः संयोगात् पूर्वहरितरूपनाश रूपान्तरस्य पीतादेरुत्पत्तिः पूर्वरसस्य व्यक्तस्यैवानुभवात्। क्वचित् पूर्वहरितरूपसत्त्वेऽपि रसपरावृत्तिर्दृश्यते विजातीयतेजः संयोगरूपपाकवशात् पूर्वतनाम्लरसनाशे मधुररस्यानुभवात्। तस्माद्रूपजनकापेक्षया रसजनको विलक्षणा एवं गन्धजनको विलक्षण एवाङ्गीकार्य्यः रूपरसद्योरपरावृत्तौ अपि पूर्व गन्धनाशे विजातीयपाकवशात् सुरभिगन्धोपलब्धेः। एवं स्पर्शजनकोऽपि पाकवशात् कठिन्यस्पर्शनाशे मृदुस्पर्शानुभवात्। अतएव पार्थिवपरमाणूनामेकजातीयत्वेऽपि पाकमहिम्ना विजातीयद्रव्यान्तरानुभवः। यथा गोमुक्ततृणादीनां आपरमाण्वन्तं भङ्गे तृणारम्भकपरमाणुषु विजातीयतेजः संयोगवशात् पूर्वरूपादिचतुष्टयनाशे तदनन्तरं दुग्धे यादृशं रूपादिकं वर्त्तते तादृशरूपरसगन्धस्पर्श-जनकास्तेजः संयोगा जायन्ते। तदुत्तरं तादृशरूपादय उत्पद्यन्ते। तादृशरूपादिविशिष्ट-परमाणुभिर्दुग्धद्व्यणुकमारभ्यते। ततःत्र्यणुकादिक्रमेण महादुग्धारम्भ इति। Nyāyabodhinī or Annam Bhaṭṭa's Tarka-saṅgraha). Heat and light rays are supposed to consists of indefinitely small particles which dart forth or radiate in all directions rectilineally with a sort of conical dispersion and with inconceivable velocity. They may either (1) penetrate through inter-atomic (or inter-molecular) spaces as in cases of conduction of heat which when applied under the pot boils the water, or fries the paddy where there is no chemical action in the pot, no decomposition and recompostion of its atoms, no change in the molecular collocation ; or, as with light rays in case of translucency of transparency (स्वच्छता) penetrate through the inter-atomic spaces with Parispanda of the nature of deflection of refraction (तिर्यग्गमन, Udyotakara), in the same way as when fluids penetrate through porous bodies (तत्र परिस्पन्दः तिर्यग्गमनं परिस्रवः पात इति–Udyotakara, commenting on Vātsyāyana's परिस्पन्दपरिस्रवौ, Sūtra, 47 Āhnika 1, Chap. III.) or (2) impinge on the atoms, and rebound back—which explains reflection (मूर्च्छन, किरणविघट्ट–Varāhamihira, रश्मिपरावर्त्तन–Vātsyāyana) or otherwise be obstructed by the atoms in their path, which would explain degress of opacity, the casting of shadows, etc., all these operations being also physical, and unattended with decomposition and recomposition or alteration of molecular grouping, or (3) lastly, strike the atoms in a peculiar way, so as to break up their grouping, transform the physico-chemical characters of the atoms, and again recombine them, all by means of continual impact with inconceivable velocity, an operation which explains all cases of chemical combination. (अचिन्त्यो हि तेजसो लावघातिशयेन वेगातिशयः यत् प्राचीनाचलचूड़ावलम्बिनि एव भगवति मयूखमालिनि भवनोदरेषु आलोकइत्यभिमानो लौकिकानाम्। (Udayana, Kiraṇāvalī, तेजोनिरूपणाम्–taken from Vācaspati, Tātparyaṭīkā, प्रत्यक्षलक्षण-सूत्रम्।) *Cf.* also (चाक्षुषं तेजः वेगवता सावित्रेन तेजसा न प्रतिहन्यते। Vācaspati—वर्त्तिदेशे पिण्डितमपि तेजः प्रसर्पत् प्रासादोदरं व्याप्नोति। तत् कस्य हेतोः पृथ्व्यग्रत्वात्। स्वभावतः प्रसरदपि न स्वपरिमाणनुविधायिनं प्रत्यमाधत्ते किं तु विषयभेदानुविधायिनम्। *ibid, Cf.* स्फटिकाद्यन्तरितोपलब्धिरपि प्रसादस्वभावतया स्फटिकादीनां तेजोगतेरप्रतिबन्धकतया प्रदीपप्रभावदेवोपपन्ना। Udayana, तेजोनिरूपणम्, in reply to the objection—यदि हि प्राप्य गृह्णीयात् प्रतिघातिना स्फटिकद्रव्येण विष्टम्भादप्राप्तं प्रसर्पन्तृणादिकं नाददीत तस्मादप्राप्यकारि ततो न तैजसम्। (Udayana, *Ibid.*) Definition of स्वच्छताद्रव्यान्तरा-सम्पृक्तद्रव्यसमवायः स्वच्छता। दृष्टश्चाप्रतीघातः काचाभ्रपटलस्फटिकान्तरितोपलब्धैः।. ......स्थाल्यादिषु च पाचकस्य तेजसोऽप्रतीघातात्। Udyotakara, Chap. III, Āhnika I, Sūtra 38. आदित्यरश्मेः स्फटिकान्तरितेऽपि दाह्येऽविघातात्। Sūtra

47, where Udyotakara notes—कोऽयमविधातः-यस्य द्रव्यस्यावयवा न व्यूह्यन्ते तस्य अन्तरावयवैः अव्यूह्यमानस्य योऽभिसम्बन्धः सोऽविघात इति। Vācaspati explains यस्य द्रव्यस्य भर्जनकपालादेः अवयवा न व्यूह्यन्ते पूर्वोत्पन्नद्रव्यारम्भकंसयोगनाशेन द्रव्यान्तरकसंयोगोत्पादनं व्यूहनं तत्र क्रियन्ते तस्य द्रव्यस्य भर्जनकपालादेरव्यूह्यमानस्य अन्तरावयवैर्योऽभिसम्बन्धो वह्नेः सोऽप्रतिघातः। *Cf.* Vātsyāyana on Sūtra 47, Āhnika I, Chap. III. On the other hand in chemical combination, अन्तः प्रवेशः कृशानोरनुमीयते। तेन वेगवता वह्निद्रव्येण नोदनात् अभिघातात् वा अव यवेषु क्रिया, क्रियातो विभागः, विभागात् आरम्भकसंयोगविनाशः etc.—Jayanta, Mañjarī, भूतचैतन्यपूर्वपक्ष। For opacity, shadows, etc., *vide* छाया तु तेजः परमाणोरावरणात् मूर्त्तिमता परमाणुना तेजः परमाणुराव्रियते। यत्र च अस्य आवरणं तत्र च्छायेति। विरलतेजः सम्बन्धीनि द्रव्यगुणकर्माणि च्छाया इत्यभिधीयते। सर्वतो व्यावृत्ततेजः सम्बन्धीनि तु तानि तमः संज्ञकानि। Udyotakara, Chap. IV. Āhnika 2, Sūtra 25. For reflection, and its laws, I quote passages in my paper of Hindu Physics and Mechanics, to which the student of the history of Optics is referred.

Arrangement of Atomic in Space

The Nyāya conceives atomic magnitudes as Pārimāṇḍalya, a term which indicates a spherical shape. (नित्यं परिमण्डलम्-परिमण्डलमेव पारिमाण्डल्यम्-Śaṅkaramiśra). To conceive position in space. Vācaspati takes three axes, one proceeding from the point of sunrise in the horizon to that of sunset (on any particular day), (roughly speaking. from the East to the West) ;—a second bisecting this line at right angles on the horizontal plane, (roughly speaking, from the North to the South), and the third proceeding from the point of their section up to the meridian position of the sun on that day, (roughly speaking, up and down). The position of any point in Space, relatively to another point, may now be given by measuring distances along these three directions, *i.e.* by arranging in a numerical series the intervening points of contact, the less magnitude or distance being what which comes earlier in this series, and the greater that which comes later. The position of any single atom in space with reference to another may be indicated in this way with reference of the three axes. But this gives only a geometrical analysis of the conception of three-dimensioned Space, though it must be admitted in all fairness that by dint of clear thinking it anticipates in a rudimentary manner not only the foundations of solid (co-ordinate) geometry, but also of the geometry of position, and especially the conception of Space as a Manifold, which alone can serve as the basis of a generalisation comprehending all different possible kinds of geometry, Euclidean and non-Euclidean. (एकत्वेऽपि दिशः आदित्योदयदेशप्रत्यासन्नदेशसंयुक्तो यः स इतरस्माद्विप्रकृष्टप्रदेशसंयोगात् परमाणोः पूर्व एवमादित्यास्तमयदेशप्रत्यासन्नदेशसंयुक्तो यः स इतरस्माद्विप्रकृष्टदेशसंयोगात् परमाणोः पश्चिमः तौ च पूर्वपश्चिमौ परमाणू अपेक्ष्य यः सूर्यादयास्तमयदेश-विप्रकृष्टदेशसंयोगः स मध्यवर्त्ती। एवमेतयोर्यौ तिर्य्यग्देशसम्बन्धिनां मध्यस्य आर्जवेन व्यवस्थितौ पार्श्ववर्त्तिनौ तो दक्षिणोत्तरौ परमाणू। एवं मध्यन्दिनवर्त्तिसूर्यसन्निकर्षविप्रकर्षौ अपेक्ष्य उपर्यधोभावो द्रष्टव्यः। संयुक्तसंयोगाल्यत्वभूयस्त्वे च सन्निकर्षविप्रकर्षौ पूर्वसंख्यावच्छिन्नत्वं च अल्पत्वं परसंख्यावयच्छिन्नत्वं च भूयस्त्वम्। Vācaspati, Tātparyaṭikā, Chap. IV. Āhnika 2, Sūtra 25.

The original physical arrangement of atoms is also given. Each atom is in contact with six other atoms, which gives a cubical arrangement. एवं दिशोऽपि एकस्या अपि संयोगा एव भागाः सोऽयं परमाणोः षट्केन युगपद्योगे। मूर्त्तत्वमात्रप्रयुक्तः न सावयवत्वप्रयुक्तः। *ibid. Cf.* also the objection in the Buddhist Kārikā. षट्केन युगपद्योगे परमाणोः षड्ंशता। षणां समानदेशत्वे पिण्डः स्यादणुमात्रकः। This is the typical primordial arrangement, and variations in the collocation of atoms and molecules (व्यूह, अवयवसन्निवेश), as we have seen, were conceived to account for the variety of isomeric modes of the same Bhūta, as well as of mono-Bhautic and poly-Bhautic compounds.

The molecular arragement in the case of bi-Bhautic compounds is very peculiar. Two substances, say Earth and Ap (water), form a quasi-compounds, first, and each substance breaks up into atoms, one atom of Earth comes into contact with one of Ap. But the two do not form a binary molecule. Instead, this contact of heterogeneous atoms leads to a curious result. The atom of Earth combines with a neighbouring atom its own class, and forms a binary molecule. Simultaneously the atom of Ap combines with another Ap-atom, and forms a binary molecule. Now, the first binary molecule links on to the atom of Ap and similary the second binary molecule links on to the atom Earth. The moment after, the two binary molecules take on the physico-chemical characters of Earth and Ap respectively and simultaneously with the assumption of these physico-chemical characters, the binary molecules enter into complex contact (संयोगज संयोग). In all this process, work is done only in the first instant, in the contact of an atom of Earth with one of Ap—the resulting contacts, of atom with binary molecule and of the binary molecules with each other, involve no further expendition of Energy. Thus we get a particle holding two binary molecules (of Earth and Ap respectively) in complex contact, and such particles continue to be formed. In this way the particles of the two substances arrange themselves, and the peculiarity of this molecular arrangement explains the resulting mixed or compound qualities of this class of quasi-compounds. (संयोगजस्तु (संयोगः) उत्पन्नमात्रस्य चिरोत्पन्नस्य वा निष्क्रियस्य कारणसंयोग्रिकारणः कारणाकारण-संयोगपूर्वकः कार्याकार्यगतः संयोगः।.......एकस्माच्च द्वयोरुत्पत्तिः। कथम्। यदा पार्थिवाप्ययोरण्वोः संयोगे सति अन्येन पार्थिवेन पार्थिवस्य। अन्येन आप्येन च आप्यस्य युगपत्संयोगो भवतस्तदा ताभ्यां संयोगाभ्यां पार्थिवाप्ये द्व्यणुके युगपदारभ्येते। ततो यस्मिन् काले द्व्यणुकयोः कारणगुणपूर्वक्रमेण रूपाद्युत्पत्तिः तस्मिन्नेव काले इतरेतरकारणाकारणगतात् संयोगात् इतरेतरकार्याकार्यगतौ संयोगौ युगपदुत्पद्येते। किं कारणम्। कारणसंयोगिना हि अकारणेन कार्यं। अवश्यं संयुज्यते इति न्यायः। अतः पार्थिवं द्व्यणुकं कारणसंयोगिना आप्येन अणुना सम्बध्यते। आप्यमपि द्व्यणुकं कारणसंयोगिना पार्थिवेनेति (Praśastapāda-Bhāṣya, गुणग्रन्थे संयोग-निरूपणम्।)

The whole process may be graphically represented as follows :—

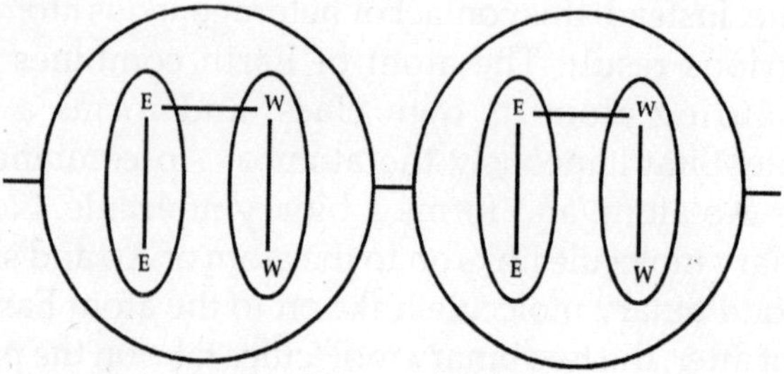

E=an atom of Earth.
W=an atom of water
E (Ap.)
|=a binary molecule
E of Earth.
W
|=a binary molecule of
W water (Ap).

Molecules of a bi-Bhautic quasi-compound;— graphic formula of complex contact.

I will conclude this account of ancient (and medieval) Hindu chemistry with a note on the conception of molecular (atomic) motion, Parispanda, and the different varieties of such motion which were conceived to account for the physical phenomena of sound light and heat. Any attempt to differentiate rigidly between Mechanics and physics on the one hand and Chemistry on the other at this primitive stage would be an idle affectation. My paper on Hindu Mechanics and Physics will give a detailed exposition in a separate treatise.

Conception of Molecular Motion (Parispanda)

Parispanda: Resolution of all Physical Action into Motion

Parispanda sometimes stands for motion molar as well

as molecular, but more often for the subtile motion of atoms or molecules. The radical meaning of the term is whirling or rotary motion, a circling motion, but it may also include simple harmonic motion (*e.g.* vibration). All action, operation, work (क्रिया, व्यापार) is ultimately traced to this form of subtile motion lodged in the atoms or in the matter-stuff. The Vedānta, for example, speaks of a cosmic vibratory motion (सर्वलोकपरिस्पन्दनम्–Śaṅkara).—Ākāśa, in the Vedānta, as we have seen, is the first, stadium in the evolution of Matter, which gives off Vāyu, which gives off Tejas, and so on ; but Ākāśa (Ether) itself passes through two stages before the emanation of the Sūkṣambhūta Vāyu :—(1) the motionless ubiquitous primordial matter-stuff (answering to the Sāṅkhya Bhūtādi) called Purāṇaṁ khaṁ (पुराणं खं); and (2) a subtile integration, the pure unquintuplicated Sūkṣmabhūta called Vāyuraṁ Khaṁ (वायुरं खं) (answering to the Sāṅkhya Tanmātra stage). It is this subtile Ākāśa, in its Tanmātric integration, *i.e.* in the derivative form,which is subject to an incessant Parispanda. The geaseous stage of matter (the Vedāntic Vāyu) is indeed matter in a state of Parispandic motion (वायोः परिस्पन्दात्मकत्वात्–Śaṅkara). So also the biomotor and sensori-motor principles apart from the directive Intelligence of the Self (प्राणस्य परिस्पन्दात्मकत्वादेव–यद्द्वैतम् स्थूलं सूक्ष्मञ्च ततसर्वं मनः स्पन्दितमात्रम्–Śaṅkara). The Sāṅkhya also conceives this Parispanda to characterise every process and phenomenon of cosmic evolution (व्यक्तं सक्रियं परिस्पन्दवत्– Vācaspati, Kaumudī), Bhūtas, organisms, mental organs, as modes of Prakṛti (considered apart from the Inteligence of Puruṣa) are all subject to this Parispanda (बुद्ध्यादयो देहं त्यजन्ते देहान्तरं उपाददते इति तेषां परिस्पन्दः। शरीरपृथिव्यादीनां च परिस्पन्दः प्रसिद्ध एव।–Vācaspati on Kārikā 10). On the other hand, Prakṛti as the Avyakta, the a-cosmic, the un-manifest ground, with resolution only of like to like (सदृशपरिणाम). is devoid of all Parispandic motion (यद्यपि अव्यक्तस्यापि परिणामलक्षणा क्रिया, तथापि परिस्पन्दो नास्ति। *ibid*. On

Kārikā 10). The Nyāya-Vaiśeṣika finds Parispanda in all forms of matter, except Ākāśa which, in that system is non-atomic and incapable of any change or activity (निष्क्रिय). But all atoms from those of Vāyu downwards, are in incessant motion. The world at bottom is an infinitude of continually whirling (or vibratory) particles (अनवरतपरिस्पन्दमानापरिमित-पवनादिपरमाणवः-Raghunātha ;—compare also Udayana-Kusumāñjalī, Stavaka V.—परमाणवः हि गतिशीलत्वात् पतत्रव्यपदेशाः पतन्तीति।). All physical action consists in motion. The Nyāya-Vaiśeṣika rejects, force, power, operation (शक्ति) except as modes of motion. Jayanta indeed states : we do not acknowledge any mysterious power or operation, which the senses do not and cannot report to us. But this denial or Force (शक्ति) and of unperceived and unperceivable operation (अतीन्द्रियव्यापार) is put forward as a philosophical (epstemo-logical and metaphysical) proposition to justify the Nyāya analysis of the causal *nexus* into mere invariable and unconditional antecedence among phenomena without productive power or efficiency (अन्यथासिद्धिशून्यस्य नियतपूर्ववर्त्तिता-Bhāṣā-Pariccheda). It is not of course intended to question the existence of Parispanda, which is of the nature of motion, and which, though subtile and therefore infra-sensible (सूक्ष्म and अनुद्भूतरूप, not अतीन्द्रिय), is the ultimate form of all physical activity. (परिस्पन्द एव भौतिको व्यापारः करोत्यर्थः। अतीन्द्रियस्तु व्यापारः नास्तीति ब्रूमहे।........तस्मात् कारकचक्रेण चलता जन्यते फलम्। न पुनश्चलनादन्यो व्यापार उपलभ्यते। Jayanta, Nyāya-Mañjarī, Āhnika I). The effect (no less than the action) is, in all cases of material causation, the resultant of the combined motions of the various (material and efficient) causes involved (*e.g.* in the case of पाक-समुदितदेवदत्तादिसकल-कारकनिकरपरिस्पन्द एव विशिष्टफलावच्छिन्नः पाक इत्युच्यते।...अथ व्यापार एवैषः सर्वः संभूय साध्यते। किं फलेनापराद्धं वः तद्धि संभूय साध्यताम्।-Jayanta, Nyāya-mañjarī, Āhnika I).

But, in the Nyāya-Vaiśeṣika, though all action of matter

on matter is thus resolved into motion, conscious activity is sharply distinguished from all forms of motion, as against the Sāṅkhya-Vedānta, which, as we have seen, considered every thing other than Inteliigence, the Puruṣa or the transcendental Self, to arise in the course of cosmic Evolution, and therefore to be subject to Parispandic motion. (क्रियाविशेष एवायं व्यापारो ज्ञातुरान्तरः। स्पन्दात्मकवहिर्भूतक्रियालक्षणविलक्षणः।–Quoted in Jayanta's Nyāya-mañjarī, Āhnika IV).

Santāna, *i.e.* Gati-santāna (including wave motion and current motion or convection); Kampa santāna, Spandana, (vibration);—Caraka notes three kinds of santānas, serial motions, *viz.*, those of water, sound and light (जलसन्तान, शब्दसन्तान and अर्चिःसन्तान). Cakrapāṇi points out that a wave of sound travels more rapidly than a wave of water, and much less repidly than a ray of light. In Hindu Mechanics, a current of water (downward flow, स्यन्दन) is conceived to consists of particles moving is an uninterrupted series under the action of gravity and fludity (गुरुत्व and द्रवत्व). A ray of light implies the rectilinear propagation of indenfinitely minute corpuscles, in all directions, with inconceivable velocity, and a short of concial disperssion (अचिन्त्यो हि वेगातिशयः तेजः प्रसर्पत् व्याप्नोति पृथ्वग्रत्वात् Udyotakara—Vācaspati). A wave of water (योचितरङ्ग) implies the transmission of vibratory motion in the water particles. (*Cf.* Jayanta—पतद्घनपयोबिन्दु-संदोहस्पन्दनक्रमात्, Āhnika II) A wave of sound is conceived by some on the analogy of a wave of water (वीचितरङ्गन्याय), only the air-waves (*cf.* the Mīmāṁsā) or the sound-waves in and through the vehicle of air-waves (*cf.* the Nyāya-Vaiśeṣika) travel by concentric circles not in one but in all planes . (N.B. this assums transverse waves). Others hold that the air-waves (*cf.* the Vākyapadīya) or the sound-and-air waves (*cf.* Udyotakara) are propagated by the transmission of the vibration in all directions, leading to conjunction and disjunction of air-particles, so that the wave may be said to expand by alternate concentric spherical layers or rarefaction and condensation (प्रचय) (N.B. this assumes longitudinal waves),—(कदम्बगोलकन्याय–*cf.* संयोगविभागा नैरन्तर्येण क्रियमाणाः शब्दमभिव्यञ्जन्तो नादशब्दवाच्याः (Śavara-Bhāṣya 1-2-17) अभिघातेन हि प्रेरिता वायवः स्तिमितानि वायुवन्तराणि प्रति धावमानाः सर्वतोदिक्कान् संयोगविभागान् उत्पादयन्ति।–*ibid*. 1-1-13).- The Vākya-padīya describes articulate sounds (Varṇas) and indeed all sounds (śabdas) as only forms of air set in motion. with rarefaction and condensation (प्रचय), and capable of variations of velocity and configuration (स्थानेष्वभिहतो वायुः शब्दत्वं प्रतिपद्यते। तस्य कारणसामर्थ्यात् वेगप्रचयधर्मिणः। संनिपातात् विभज्यन्ते सारवत्योऽपि मूर्त्तयः–Vākya-padīya. Kāṇḍa I. Śloka 109) (*vide* my paper on Hindu Mechanics and Physics). (For the Hindu doctrine of scientific Method, *Vide* Appendix ; for certain interesting recipes of chemical technology).

I had intended to conclude this survey of Hindu Physico-chemical science of scientific ideas in the culture-history (kultur-geschichte) of the Chinese, the Greeks and the Arabs, as an Essay in the historico-comparative method of investigation (*vide* the introduction to my comparative studies in Vaiṣṇavism and Christianity for a correction of this method), but space forbids, and the reader too, I fancy.

The Date of Rasaratnasamuccaya

While the present volume was about to issue from the press, Mr. T.G. Kāla, Editor of the Marhattā Journal "Samālocaka". sent us a critical notices of R.R.S. As there are some important historical facts thought to light and as the date of this work arrived at from quite independent sources tallies with that assinged by us (Vol. 1. Intro. LXXXIX), we make no apology for reproducing its substance in a condensed form.

Carpaṭi is Carpaṭinātha of the Nātha school is mentioned in the R.R.S., (see VI, 58, Poona edition) as also king Siṅghaṇa.

The Navanātha Śaktisāra, a Marhattā work by Narahari

Māla gives some legendary information about this Carpaṭinātha and speaks of him as a contemporary of Matsyendranātha. Below has been given the geneology of the pupils of मत्स्येन्द्रनाथ. The last, ज्ञानदेव or ज्ञानेश्वर, was the celebrated Marhattā Saint and author of a commentary on the भगवद्गीता called ज्ञानेश्वरी.[1] It was written in Śaka 1212, *i.e.,* A.D. 1290. So मत्स्येन्द्र and Carpaṭinātha must have lived at the *beginning of the thirteenth century A.D.*

मत्स्येन्द्रनाथ
↓
गोरक्षनाथ
↓
गैनीनाथ
↓
निवृत्तिनाथ
↓
ज्ञानदेव

The R.R.S.which mentions the Siddha Carpaṭi must be therefore later than the first half of the thirteenth century A.D.

सेवणचन्द्र II (A.D. 1069
↓
सिंघण I
↓
मल्लुगि
↓
भिल्लम (A.D. 1191)
↓
जैभपाल (A.D. 1191-1210)
↓
सिंघण (A.D. 1210-1247)

Among the Yādava kings of Devagiri or Daulatābād, there were two kinds by the name of Siṅghaṇa. Taking the Siṅghaṇa. Taking the Siṅghaṇa mentioned by R.R.S. to be the second, we are required to place the composition of the work in the latter half of the 13th century. On the whole. the R.R.S, may be safely taken to be a work written *about 1300 A.D.*

1. See the concluding portion of ज्ञानेश्वरी, a Marhatt'a work.

The Weight of Air

(By Principal B. N. Seal)

Experiments were of course conducted for purposes of chemical operation in relation to the arts and manufactures, *e.g.*, metallurgy, dyeing, pharmacy, perfumery, cosmetics horticulture and the making of glass (lenses and mirrors of various kinds are mentioned, the spherical, oval—वृत्त, and वर्त्तुल, being well-known—Pliny indeed mentions that the best glass over made was Indian glass). But of Experiment as an independent method of Proof of Discovery, the instances are rare. I may note one Interesting example in Udayana's Kiraṇāvalī, relating to the weight of air. Udayana argues that air must be a distinct and independent Bhūta, for if air were made of the Earth-Bhūta, it would have weight, and it has none. To prove the absence of weight, he refers to an experiment. A small bladder made of thin membrance, filled with air, will not cause a greater descent in the scale than the same bladder weighed empty. Hence, the air possesses no weight. Then Udayana makes in interesting statement. It may be objected, he says by one who accepts the weight of air—that this argument is inconclusive. For a counter-experiment may be suggested. The balloon filled with smoke (or gas, धूम) rises in the air, whereas the air-filled ballon comes down. This would go to show that air has weight. Udayana replies that this would only show that both smoke (gas, धूम) and air have no weight. The Hindus appear to air have been ignorant of the principle of Archimedes, at least as applied to gases. Vallabhācārya in the Līlāvatī speaks of peculiar resistance to sinking (or gravity) exercised by water, which explains the tendency in certain objects to float or to come up to the surface of the water but the description show that he had no clear ideas on the subject. *Cf.* Udayana, Kiraṇāvalī, वायुनिरूपणम्। *Cf.* also Vallabhācārya, Līlāvatī.

APPENDIX

Scientific Method of the Hindus

(By Principal Brajendranath Seal)

The Doctrine of Scientific Method

A study of the Hindu Methodology of Science is absolutely essential to a right understanding of Hindus positive Science, its strength and its weakness, its range and its limitations. Apart from this rigorous Scientific Method, Hindu Chemistry, such as it is, would be all practical recipe, or all unverified speculation. This, however, would be a very inadequate and indeed erroneous view of this early achievement of the human mind. That the whole movement was genuinely and *positively* scientific, though arrested at an early stage, will apear from the following brief synopsis of the Hindu Methodology of Science.

Criterion of Test of Truth (after the Buddhist)

The ultimate Criterion of Truth is found, not in mere cognitive presentation, but in the correspondence between the cognitive and the practical activity of the self, which together are supposed to form the circuit of consciousness. That knowledge is valid which prompts an activity ending in fruition. (*Cf.* the distinction between सम्वादिज्ञान and विसम्वादिज्ञान। Also, compare प्रमाणतोऽर्थप्रतिपत्तौ प्रवृत्तिसामर्थ्यात् अर्थवत् प्रमाणम्–Vātsyāyana). Truth the Buddhists contented is *not* self-evidence, *not* the agreement between idea, *nor* the agreement of the idea with the reality beyond if any, for this cannot be attained direct, but the harmony of experience (सम्वाद), which is implied when the volitional reaction, that is prompted by a cognition and that completes the circuit of consicousness, meets with fruition, *i.e.*, realises its immediate end (*cf.* Śrīharṣa, Khandana Khanda Khādya on the relation of प्रमा to लोकव्यवहार). This is the material aspect to Truth. The formal aspect is given in a principle which governs all presentations in consciousness, and which combines the three moments of Identity, non-Contradiction and Excluded Middle in every individual cognitive operaion [तदुक्तं तत् परिच्छिनत्ति (identity) अन्यद् व्यवच्छिनत्ति (non-contradiction) तृतीयप्रकारभावं च सूचयति (excluded middle) इति एकप्रमाणव्यापारः–Jayanta, Nyāya-Mañjarī, प्रमाणद्वैविध्यखण्डनम्].

Perception

The conditions of perception, and its range and limits, were carefully studied, The minima sensible (*e.g.*, the minimum visible, the Trasareṇu, the just perceptible mote in the, slanting sunbeam), the infra-sensible (अनुद्भूतरूप सूक्ष्म, sometimes termed अतीन्द्रिय), the obscured (अभिभूत, *e.g.*, a meteor in the mind-day blaze), and the potential (अनुद्भूतवृत्ति), are distinguished ; but finer instruments of measurement were wanting, and this was a principal cause of arrested progress. It may be noted that the measurement of the relative pitch of musical tones was remarkably accurate and original (*vide* my Paper on Hindu Mechanics and Physics).

Observation (दर्शन–Vācaspati nad Udayana)

The entire apparatus of scientific method proceeded on the basis of observed instances carefully analysed and sifted. This was the source of the physico-chemical theories and classifications, but in Anatomy, the Hindus went one step further ; they practised dissection on dead bodies for purposes of demonstration. Ingenious directions are given *e.g.*, the body must be first disembowlled and wrapped round with the kuśa and other grasses, then kept immersed in still water for seven days, after which the medical student should proceed to remove the layers of the skin with a carefully prepared brush made of the fresh elastic fibres of green bamboos ;—which will enable the tissue, vessels and ducts to be observed. *Post-mortem* operations as well as major operations in obstetric surgery (the extraction of the foetus, etc.) were availed of for embryological observations (*e.g.*, it is stated as a resulf of observation that the rudiments of the head and the limbs begin to appear in the foetus in the third month, and are developed in the fourth ; the bones, ligaments, nails, hair, etc. becoming distinct in the sixth);— and also embryological theories, *e.g.*, the indication of sexual character in the second month by the shape of the foetus, the shape of a round joint indicating the male sex, and an elongated shape as of a muscle the female sex (*cf.* Caraka, Sūtrasthāna, Chap, IV,—द्वितीये मासि घनः सम्पद्यते पिण्डः पेश्यर्वुदं वा। तत्र घनः पिण्डः पुरुषः स्त्री पेशी अर्वुदं नपुंसकम्। Cakrapāṇi notes घनः कठिनः। पिण्डो ग्रन्थ्याकारः। पेशी दीर्घमांसपेश्याकारा। अर्वुदं वर्त्तुलोन्नतम् *loc. cit.*). In Phonetics (as in the Prātiśākhyas, *circa* 600 B.C.), in Descriptive and Analytical Grammer (as in Pāṇini), and in some important respects in Comparative Grammar (as in Chanda's and Hemachandra's Grammars of the Prākṛta Dialects), the observation was precise, minute and thoroughly scientific, This was also the case in Materia Medica, and in Therapeutics, especially the symptomology of diseases. In Meteorology, the Hindus used the rain-guage in their weather forecasts for the year, made careful observations of the different kinds of clouds and other atmospheric phenomena (*e.g.*, they give the heights of the clouds, the distance from which lightning is ordinary visible, or the thunder is heard, the area of distrubance of different earth-quakes, the height to which the terrestrial atmosphere extends, etc. *vide* Varāhmihira Śrīpati, and the authorities quoted by Utpala). In Astronomy, the observation was, generally speaking, very defective as in the determination of the solar and the planetary elements, and this was probably due to the lack of practical interest, but the determination of the lunar constants entering into the calculation of lunar periods and eclipses, matters in which the Hindus had a practical ceremonial interest, reached a remarkable degree of approximation (much above Graeco-Arab computations) to the figures in Laplace's Tables, which can only be explained by the circumstance that in the case of these constants the Hindus carried out for more than a thousand years a systematic process of verification and correction by comparison of the computed with the observed results (like the navigator's correction of the course of the ship at sea), a process which was termed दृग्गणितैक्य. In Zoology, the enumeration of the species of Vermes, Insecta, Reptilia, Batrachia, Aves, etc., makes a fair begining, but the classification proceeds on external characters and habits of life, and not on an anatomical basis. In Botany, the observatoin was mainly in the interests of Materia Medica, and the classfication was as superficial as possible. (*Vide* my paper on the Hindu Classification of Plants and Animals).

Experiments

Experiments were of course conducted for purposes of chemical operations in relation to the arts and manufactures *e.g.*, Metallurgy, Pharmacy, Dyeing, Perfumery and Cosmetics, Horticulture, the making and polishing of glass (lenses and mirrors of various kinds are mentioned, the spherical and oval वृत्त and वर्तुल, being well-known—Pliny indeed mentions that the best glass ever made was Indian glass). And the results of such experiments were freely drawn upon for building up scientific hypotheses and generalisations. But of experiment as an *independent* method of proof or discovery, the instances recorded in books are rare. I may note one interesting example in Udayana's Kiraṇāvalī, relating to the weight of air. Udayana argues that air must be a distinct and independent Bhūta, for if air were a form of the Earth-Bhūta, it would have weight, and it has none. To prove the absence of weight, he refers to an experiment. A small bladder made of a thin membrane, filled with air, will not cause a greater descent in the scale than the same bladder weighed empty. Hence, the air possesses no weight. Then Udayana makes an interesting statement. It may be objected, he says, by one who accepts the weight of air—that this argument is inconclusive. For a counter-experiment may be suggested. The balloon filled with smoke (or gas, धूम) rises in the air, whereas the air-filled ballon comes down. This would go to show that air has weight. Udayana replies that this would only show that both smoke (or gas, धूम) and air have no weight. The Hindus appear to have been ignorant of the principle of Archimedes. Vallabhācārya in the Līlāvatī, it is true, speaks of a peculiar resistance to sinking (or gravity) exercised by water, which explains the tendency in certain objects to float or to come up to the surface of the water, but the description show that he had no clear ideas on the subject.

Mathurānātha, again, states that the determination of the degree of purity (the carat) of gold by rubbing against the assaying stone and observing the character of the yellowish streak against the black smooth background, is only an indirect means of ascertaining weight (गुरुत्ववैलक्षण्य *lit.* specific gravity)—which seems to suggest that there was a more direct means of arriving at the latter. Probably this refers to the common Indian method of comparing the lengths and weights of wires of uniform thickness than can be formed by drawing different pieces of gold through the same diamond bore, I think it may be regarded as fairly certain, that the Hindus were ignorant of Archimedes's discovery, an ignorance which, at any rate, they could not have well borrowed from the Greeks, no more than they could have thus borrowed their knowledge of things unknown to the Greeks themselves. [*Cf.* Udayana, Kiraṇāvalī]

वायुनिरूपणम्, कि च (वायोः) पृथिवीत्वे गुरुत्वमपि स्यात्। न च पवनापूरितस्य चर्मपुटकादे अपूरणदशातः अधिकं अवनमनम्। उन्नीयमानस्य धूमापूरितेन इदम् अनैकान्तिकं इति चेत्। न। अलं धूमदशायां तत्रापि गरिमविशेषहेतोरवनतिविशेषस्य अप्रतीतेः ।-*cf.* Vallabhācārya, Līlāvatī—गुरुत्वविशेषेऽपि पाषाणलावुप्रभृतिषु मज्जनोन्मज्जन नियमवत् अदृष्टहेतुकता साम्येऽपि पतनोत्पतन-नियमो भविष्यतीति। इति चेत् न। जलाधोगमनं जलेन धारणं पतनप्रतिबन्ध उन्मज्जनं एतच्च जलस्य योगस्य कस्यचिदेव पतनप्रतिबन्धसामर्थ्यात्। दृष्टहेतुकमेव मज्जनञ्च उन्मज्जनञ्च]-*cf.* Mathurānātha, सन्निकर्षवादरहस्य on Gaṅgeśa's Tattvacintāmaṇi—तस्य (निकषोपलस्य) न सुवर्णपीताभिव्यञ्जकत्वं किन्तु पीतद्वारा गुरुत्ववैलक्षण्यव्यञ्जकत्वमेव तस्य।

Fallacies of Observation-Mal-observation and Non-observation

These are carefully studied in relation to errors of observation, and Hallucination (भ्रम, अध्यास, आरोप),-which were ascribed to three causes :—(a) Doṣa, दोष, defect of sense organ, as of the eye in jaundice, or of the skin in certain forms of leprosy (leading to tactile insensibility *cf.* Suśruta), or defect of necessary stimulus, *e.g.*, too faint light, or undue distance or nearness, in vision ; (b) Samprayoga, सम्प्रयोग, presentation of a part or an aspect instead of the whole; and (c) Saṁskāra, संस्कार, the disturbing influence of mental predisposition, *e.g.* expectation, memory, habit, prejudice, etc.

The Doctrine of Inference

Anumāna (Inference) is the process of ascertaining, not by perception or direct observation, but through the instrumentality or medium of a mark, that a thing possesses a certain character. Inference is therefore based on the establishment of an invariable concomitance (Vyāpti, व्याप्ति) between the mark, and the character inferred. The Hindu Inference (Anumāna) is therefore neither merely formal nor merely material, but a combined Formal-Material Deductive-

Inductive Process. It is neither the Aristotelian Syllogism (Formal-Duductive Process), nor Mill's Induction (Material-Inductive Process), but the real inference which must combine formal validity with material truth, inductive generalisation with deductive particularisation.

An inference admits of a rigorous formal statement—in the shape of five propositions, for dailectical purposes (*i.e.* in demonstrating to others),—or of three propositions when the inference is for onself (स्वार्थानुमान):–(1) the probandum, the statement of the proposition to be established (प्रतिज्ञा, साध्यनिर्देश, *e.g.*, yonder mountain is fiery (say, an active volcano) ; (2) the reason, the ascription of the mark (हेतु, लिङ्गव्यपदेश), *e.g.*, for it smokes ; (3) now, the general proposition, stating the invariable concomitance which is the ground of the inference, clenched by an example bringing home the responsibility of the reasoner to establish a real relation, *e.g.*, whatever smokes is fiery, as an oven ; (उदाहरण);–(4) next, the application, the ascertainment of the existence of the mark in the present case (उपनय) *e.g.*, younder mountain smokes ; (5) finally, the conclusion, the probandum proved (निगमन), *e.g.*, younder mountain is fiery.

1. Yonder mountain is fiery.
2. For it smokes.
3. Whatever smokes is fiery, as an oven.
4. Yonder mountain does smoke.
5. Therefore, yonder mountain is fiery.

For inference for oneself, only the first three or the last three propositions are held to be sufficient.

The Hindu Anumāna, it will be, seen, anticipates J. S. Mill's analysis of the syllogism as a material inference, but is more comprehensive ; for the Hindu Udāharaṇa, the third or general proposition with an example, combines and harmonises Mill's view of the major premise as a brief memorandum of like instances already observed, fortified by a recommendation to extend its application to unobserved cases, with the Aristotelian view of it as a universal proposition which is the formal ground of the inference. This Formal-Material Deductive-Inductive process thus turns on one thing—the establishment of the invarible concomitance (व्याप्ति) between the mark and the character inferred, in other words, an inductive generalisation. The question is—what is our warrant for taking the leap from the observed to unobserved cases ? Under what conditions are we justified to assert a Universal Real Proposition on the basis of our necesarily limited observation ?

The Cārvāka View

Among the Cārvākas there were two classes, the cruder school of materialists who accepted perception (प्रत्यक्ष) as a valid source of knoledge, as well the reality of Natural Law (स्वभाव) and the finer school of sceptics, who impugned all kinds of knowledge, immediate as well as mediate, and all evidence, Perception as well as Inference, *vide* Jayanta's reference in the Nyāya-Mañjarī to सुशिक्षितचार्वाकाः ; also चार्वाकधूर्त्तस्तु अथातस्तत्वं व्यख्यास्याम इति प्रतिज्ञाय प्रमाणमेयसंख्या-लक्षणनियमाशक्यकरणीयत्वमेव व्याख्यातवान्–Āhnika. I, Mañjarī.

The Cārvākas hold that the principle of causality, which the Buddhists assume to be a ground of an induction (व्याप्ति) is itself an induction (a case of Vyāpti), which amounts to reasoning in a circle (चक्रक); that every inference is based on an unconditional invariable concomitance which itself must be inferred, as universal propositions cannot be established by our limited perceptions, and thus there is a regressus ad infinitum (अनादि परम्परा); and that the nexus between cause and effect, or between the sign and the thing signified (*e.g.*, smoke and fire) is only a mental step or subjective association based on former perception, a mental step which by accident is found justified by the result in a number of cases.

The Buddhist : their Analysis and Vindication on Inference

The Buddhists, however, take their stand on the principle of the Uniformity of Nature (प्रतिबन्ध, स्वभाव प्रतिबन्ध, Nyāya-Vindu). This uniformity, for scientific purposes, has to be divided into two different relations—(1) the uniformity of succession in the relation of cause and effect *e.g.*, of smoke to fire (कार्यकारणभाव, तदुपत्ति, Nyāya-Vindu); (2) the uniformity of co-existence (in the first form of co-inherence in the same substance) in the relation of genus and species, *e.g.*, the relation of invariable concomitance expressed in the proposition—all Sinsapās are trees, which is not a relation of causality, but of co-existence or co-inherence in the same substrate (*i.e.*, the co-inherence of the generic qualities of a treee with the specific characters of a Sisu tree, in this particular individual before me, a Śiśu tree)—a relation which may be termed essential identity (तादात्म्य, स्वभाव–Nyāya-Vindu). To these two, the Buddhist add a third ground of inference, non-perception of the perceptible (अनुपलब्धि, दृश्यानुपलब्धि), which is employed in inferring the absence (प्रतिषेध) of a thing from the non-perception of something else. In all cases of inference based on the Uniformity of Nature, the relation is that of inseperableness or non-disjunction between the mark and the character inferred. The question is—how is this inseparableness (अविनाभाव) ascertained, and what is the warrant of our belief in it, in these cases ?

Ascertainment of Inseparableness or Non-disjunction: Buddhist Account

First take the case of causation. The cause is the invariable antecedent of the effect. What is meant is that the specific effect (with the all distinctive and relavant accompaniments, कार्यविशेष) is invariably preceded by a specific cause (कारण-विशेष). It is not that clouds always lead to rain, or that floods in the river valley always imply rain in the hills higher up. But this particular conjuction of antecedent circumstances (*e.g.* the appearance of a particular kinds of clouds accompanied with flashes of lightning, the roll of thunder and flights of Valākās—driven by the wind from a particular quarter of the horizon, and ascending in black masses, etc.) is as a rule the precursor of a particular assemblage of rain effects (rain with particular accompaniments). Again, this particular kind of flood (over flowing of the river banks accompanied with muddy discoloration of the water, rapid currents, the bearing down of tree trunks, etc.) is always preceded by rain in the hills higher up (though, no doubt, other cases of floods in a river may be due to a breach in an embankment of the melting of the snows). In other words, the Buddhist (and the earlier Nyāya Schools) avoid the difficulty arising from the plurality of causes by taking into consideration the accompanying phenomena, which, if properly marked, would always point to a specific cause of a specific effect.

I quote Nyāya authorities, but this device to obviate the plurality of causes is common to the early Nyāya and the Buddhist systems.

पूर्ववत्। यत्र कारणेन कार्यमनुमीयते। उदाहरणम्–"

मेघोन्नत्या भविष्यति वृष्टिरिति कारणेन कार्यानुमानम्। कथं पुनरस्य प्रयोगः। वृष्टिमन्तः एते मेघाः गम्भीरध्वनिवत्वे सति बहुवलाकावत्वे सति अचिरप्रभावत्वे सति उन्नतिमत्वात् वृष्टिमन्मेघवत्।

शेषवत् (यत्र कार्येन कारणमनुमीयत)–उदाहरणम्–

उपरिबृष्टिमद्देशसम्बन्धिनी नदीस्रोतः शीघ्रत्वे सति पूर्णफलकाष्ठादिवहनवत्वे सति पूर्णत्वात् पूर्णवृष्टिमन्नदीवदिति। (Udyotakara, Chap I. Āhnika 1, Sūtra 5. Vācaspati puts this clearly :—यद्यपि कारणमात्र व्यभिचरति कार्योत्पादं तथापि यादृशं न व्यभिचरति तत निपुणेन प्रतिपत्रा भवितव्यम्। अन्यथा धूममात्रमपि वह्निसत्तां व्यभिचरतीति न धूमविशेषो गमको भवेत्। In other words, a single condition called a cause is not invariably succeeded by the effect, nor does the effect phenomenon in general point to any particular cause as antecedent, for there may be a plurality of causes of a general effect. The skilful observer will therefore select the full complement of causal conditions, which is invariably succeeded by the effect—and also the specific effect (*e.g.*, धूमविशेष) which points to a specific causal antecedent. Compare also Jayanta: —we infer an effect from a specific assemblage of causes—न च कारणमात्रस्य हेतुत्वं ब्रूमो येनास्य व्यभिचारः स्यात्। अपि च विशिष्टमेव कारणं हेतुः। न च कारणविशेषो दुरवगमः। गम्भीरगर्ज्जितारम्भनिर्भिन्नगिरिगह्वराः रोलम्बगवलव्यालतमालमलिनत्त्विषः.....वृष्टिं व्यभिचरन्तीह नैव प्रायः पयोमुचः। अनभ्युपगमे चैवमनुमानस्य जीवितम्। न स्याद्धर्मविशेषाणामपि वोद्धुमशक्तिः। Similarly, we infer a specific cause from a specific assemblage of effects :—शेषवदिति यत्र कार्येण कारणमणुमीयते यथा नदीपूरेण उपरितने देशे वृष्टिरिति। वृष्टिमतपृष्ठदेशसंसृष्टा इयं नदी फेनिलकलुषत्वादिविशिष्टपूरोपेतत्वात्। यत्तु सेतुभङ्गहिमविलयनादिनापि नदीपूरोपपत्ति र्दृष्टा इति तत्रापि उच्यते–

प्रमामातु रपराधोऽयं विशेषं यो न पश्यति।
नानुमानस्य दोषोऽस्ति प्रमेयाव्यभिचारिणः।।

(Nyāya-Mañjarī, Āhnika 2, on Gotama Sūtra 5, Āhnika, 1, Chap I.)

A specific assemblage of causes, therefore, has only one specific assemblage of effects, and *vice versa*. Of course, the observer is to find out the essential or relevant features (as distinguished from the irrelevant ones) which, being included, will enable him to specify the particular cause of the particular effect.

Now this being premised to be the exact meaning of the inseparableness or non-disjunction in the case of cause and effect, we come to the question with which we started—how is this relation to be ascertained or established between two phenomena or assemblages of phenomena ? Obviously, mere observation of their agreement in presence (अन्वय) and their agreement in absence (व्यतिरेक) is no help in the matter. Take a concrete example. The ass is customarily employed to bring the fuel with which fire is lighted. In a hundred cases you have observed the ass among the antecedents of smoke. In a hundred cases you may have observed that when there is no ass, there is no smoke. This is no warrant for concluding a relation of cause and effect between an ass and smoke. It may be that you happen to have never observed smoke without an antecedent ass, or an ass without smoke following. Even this is of no avail. It is not agreement (unbroken and uniform though it be) in presence, or in absence, or in both, that can settle the matter. There is one and only one way of ascertaining the causal relation. Suppose *A* with certain accompaniments is found to precede B immediately. Now if *A* disappearing, *B* disappears, even though all other anteceedents remain and there is no other change in the case, then and then only can the causal relation be ascertained. It is not a mere table of positive instances or negative instances (अदर्शनान्न न दर्शनात्);–it is this method which we may term the Method of Subtraction (the Method of Difference in its negative aspect) that is the only exact and rigorous scientific. Method. Such was the statement of the earlier Buddhist (*cf*. Udyotakara's and Vācaspati's report of the Buddhist Doctrine of Inference—स हि प्रतिबन्धो न दर्शनमात्रा-वसेयः।

तस्मात् तादात्म्यतदुत्पत्तिनिबन्धनः एव प्रतिबन्धः। यदाह कार्यकारण भावाद्वा स्वभाबाद्वा नियामकात्। अविनाभावनियमोऽदर्शनान्न न दर्शनात् —a Buddhist Kārikā quoted in Vācaspati, Udayana, Śrīharṣa, Mādhava, etc.)–कार्यकारणभावश्च इदम्–अस्मिन् सति भवति–सत्सु अपि तदन्येषु अस्मिन्नसति न भवति–एवमाकारः। न अन्वयव्यतिरेकाभ्यामतिरिच्यते।...(एवं तादात्म्यमपि विपर्यये बाधकप्रमाणोपपत्ता निश्चेतव्यम्।...तस्मात् तादात्म्यतदुत्पत्तिभ्यामेव प्रतिबन्धो नान्यतः। Vācaspati, Tātparyaṭīkā, Chap I, Āhnika 1, Sūtra 5, अनुमानलक्षणम्–व्याप्तिग्रहोपायः)

But the canon is this form is not sufficiently safe-guarded against possible abuse. Two points have to be empahasised:–

(1) It must be carefully observed that no other condition is changed, (2) that the appearance and disappearance of *A* must immediately precede the appearance and disappearance of *B*. The definition of cause is based on two fundamental characters (1) the unconditional invariableness of the antecedence, and (2) the immediateness of the antecedence. The canon of the Method of Difference must therefore be stated in such a form as to amphasise each of these aspects. And one main difficulty in the practical application of the canon is that along with the introduction or sublation of an antecedent, some other phenomeon may be introduced or sublated unobserved. As a safe-guard against this radical vitiation of the Mehtod, the later Buddhist formulated the canon of a Method, termed the Pañcakāraṇī, a joint Method, of Difference, which combines the positive and the negative Methods of Difference (the Method of Addition and the Method of Subtraction) in a series of five steps, and which equally emphsises the unconditionality and the immediateness of the antecedence as essential moments of the causal relation. This is neither agreements, in presence, nor agreements in presence as well as absence (the foundation of J.S. Mill's joint Method of Agreement), but the Joint Method of Difference. The Pañcakāraṇī runs thus :—

The following changes being observed, everything else remaining constant, the relation of cause and effect is rigorously established:—

First step—The 'cause' and the 'effect' phenomena are both unperceived.

Second step—Then the 'cause' phenomena is perceived.

Third step—Then in immediate succession, the 'effect' phenomenon is perceived.

Fourth step—Then the 'cause' phenomenon is sublated or disappears.

Fifth step—Then in immediate succession, the 'effect' phenomenon disappears.

Throughout, of course it is assumed that the other circumstances remain the same (as least the relevant or material circumstances).

This Pañcakāraṇī, the joint Method of Difference, has some advantage over J. S. Mill's Method of Difference, or what is identical therewith, the earlier Buddhist Method ; and the form of the canon bringing out in prominent relief the unconditionality and the immediateness of the antecedence, is as superior from a theoretical point of view to J. S. Mill's canon, and is as much more consonant that the latter to the practice of every experimenter, as the Hindu analysis of Anumāna as a Formal-Material Deductive-Inductive Inference is more comprehensive and more scientific than Aristole's or Mill's analysis of the Syllogism (or Mediate Inference).

But even the Pañcakāraṇī is no sufficient answer to the question with which we started. The Pañcakāraṇī is only a method ; it shows only how in a particular case the relation of cause and effect is to be established (प्रतिबन्धग्रहोपाय). But we want more than this—we require a warrant for the process. The Buddhist therefore supply the following proof of the Method:—Doubt is legitimate, but there is a limit to doubt. When doubt lands you in an unsettlement of a fundamental ground of practice, and would thus annul all practical exercise of the will, the doubt must cease ; else the doubt would be suicidal or sophistical. In this particular case, when the Pañcakāraṇī is satisfied, the antecedent in question must be the cause, for there is no other antecedent to serve as cause ; the proof is indirect but rigid. If this be not the cause, there is no cause of the phenomenon. It was not, and it begins to be, without a cause ; which would be a contradiction of the rational ground of all practice, for all volitional activity proceeds by implication on the principle of causality. If things could happen without a cause, all our motives to action would be baffled. The link between a presentation and the instinctive volitional reaction would snap, and the circuit of consciousness would be left incomplete. In fact, the Buddhist go further ; they hold causal efficiency (अर्थक्रिया) to be of the essence of empirical (relative) Reality. The proof of the join Method of Difference, then, lies in a strict application of the principle of causality in its negative form (*viz.*, there can be no phenomenon without a cause) and the truth of this last is guaranteed by the same ultimate criterion of empirical (relative). Reality as the truth of Perception itself, *viz.*, the correspondence between the rational and the practical activity of the self.

But invariable concomitance (or non-disjunction), the Buddhist argue, has another form *e.g.*, the relation of the genus to the species. We may have perceived a hundred instances of the association of certian characters with certain others ; we may also have never perceived the former when the latter are absent ; but this would not enable us to generalise and establish invariable and unconditional co-existence. We must be first satisfied that there is indentity of essence (तादात्म्य स्वभाव). It is only when we perceive that the characters of a Siṁsapā are co-inherent. with the generic characters of a tree in the same individual object (a Siṁsapā tree before me), and when we further perceive that the characters, are held together by the relation of identity of essence, that we can say that all Siṁsapās are trees. For as there is identity of essence, a Siṁsapā would not be a Siṁsapā, if it were not a tree. It would lose its self-indentity which is a contradiction. Hence, the relation of identity of essence (तादात्म्य स्वभाव), as in the relation of the species to the genus, is the sole ground for establishing uniformity of co-existence (प्रतिबन्ध, स्वभावप्रतिबन्ध).

For the Buddhist Method of Induction, in its later form, the Pañcakāraṇī, *vide* Sarvadarśana Saṁgraha—Buddhist reply to the Cārvāka attach on Inference :—

यदभ्यधायि अविनाभावो दुर्वोध इति तदसाधीयः। तादात्म्यतदुत्पत्ति-

भ्यामविनाभावस्य सुज्ञानत्वात्। तदुक्तं–कार्यकारणभावात् वा स्वभावात् या नियामकात्। अविनाभावनियमोऽदर्शनान्न न दर्शनात्। As for the Nyāya view अन्वयव्यतिरेकौ अविनाभावनिश्चायकौ इति।–the Buddhist objects—ननू पक्षे साध्यसाधनयोरव्यभिचारो दुरवधारणः भवेत्। भूते, भविष्यति, वर्त्तमाने अनुपलभ्यमाने च व्यभिचारशङ्काया अनिवारणात्। the Nyāya retorts ननु तथाविधस्थले तावकेऽपि मते व्यभिचारशङ्का दुष्परिहरा इति।–the Buddhist answers—मैवं बीचः, विनापि कारणं कार्य्यमुत्पद्यताम् इत्येवं विधायाः शङ्कायाः व्याघातावधिकतया निवृत्तत्वात्। तदेव हि आशद्धयेत यस्मिन्नाशङ्क्यमाने व्याघातादयः नावतरेयुः। तदुक्तं व्याघाताबधिराशङ्का इति। तस्मात् तदत्पत्तिनिश्चयेन अविनाभावो निश्चीयते। तदुत्पत्तिनिश्चयश्च कार्यहेत्वोः प्रत्यक्षोपलम्भानुपलम्भपञ्चकनिबन्धनः। कार्यस्य उत्पत्तेः प्रागनुपलम्भः, कारणोपलम्भे सति उपलम्भः, उपलब्धस्य पश्चात् कारणानुलम्भात् अनुपलम्भः इति पञ्चकारण्या धूम-धूमध्वजयो कार्यकारणभावः निश्चीयते। तथा तदात्म्यनिश्चयेन अपि अविनाभावः निश्चीयते। यदि शिंशपाः वृक्षत्वम् अतिपतेत् स्वात्मानमेव जह्यात् इति विपक्षे बाधकप्रवृत्तेः। अप्रवृत्ते तु बाधके भूयः सहभावोपलम्भेऽपि व्यभिचारशङ्कायाः को निवारयिता। शिंशपावृक्षयोश्च तादात्म्यनिश्चयः वृक्षोऽयं शिंशपा इति सामानाधिकरण्यवलात् उपपद्यते। तस्मात् कार्यात्मानौ कारणम् आत्मानम अनुमापयत इति सिद्धम्। Sarvadarśana Saṅgraha, Buddha-Darśanam. *Vide* also Śrīdhara Kandalī.

The Nyāya Doctrine of Inference

The Nyāya easily demolishes the Buddhist contention about identity of essence. The Nyāya writers, being realists, do not impugn the reality of the genus (जाति) like the nominalists of the nominalisitic conceptualists; but they point out that the inseparableness (or non-disjunction) in such cases can only be established by the experience of unbroken uniformity (अव्यभिचारित्व *i.e*, by अन्वयव्यतिरेकौ, दर्शनादर्शने). Uniform agreement in presence with uniform agreement in absence—not the mysterious identity of essence irresistibly perceived in any individual case or cases—is the only basis for constituting genera and species in Natural Classification. Indeed some of the later Nyāya writers point out that individuals do not always possess in Nature all the characters that go to form the definition of the class to which they are referred.

Similarly, as regards the relation of cause and effect, *nexus* is sometimes fancied to be perceived, a power in the cause to produce the effect (शक्ति), or an ultimate form (जाति सोक्ष्म्य), which is supposed to be present, whenever the effect (quality or substance) is produced (*cf.* Bacon's view of the 'Forms' of Simple Qualities). all this is neither a matter of observation nor of legitimate hypothesis. There is nothing except the invariable time-relation (antecedence and sequence) between the cause and the effect. But the mere invariableness of an antecedent does not suffice to constitute it the cause of what succeeds, it must be an unconditional antecedent as well (अन्यथासिद्धिशून्यस्य नियतपूर्ववर्त्तिता, being the definition of कार्यकारणभाव). For example, the essential or adventitious accompaniments of an invariable antecedent, may also be invariable antecedents, but they are not unconditional but only collateral and indirect, in other words their antecedence is conditional on something else (न स्वातन्त्र्येण). The potter's stick is an unconditional invariable antecedent of the jar, but the colour of a stick, or its texture of size or any other accompaniment or accident, which does not contribute to the work done (so far as we are considering it) is not an unconditional antecedent, and must not therefore be regarded as a cause. Similarly, the co-effects of the invariable antecedents, or what enters into the production of these co-effects, may themselves be invariable antecedents but they are not unconditional, being themselves conditioned by those of the antecedents of which they are effects. For example, the sound produced by the stick, or by the potter's wheel, invariably precedes the jar, but it is a co-effect,—and Ākāśa (ether) as the substrate, and Vāyu (air) as the vehicle, of the sound, enter into the production of this co-effect, but these are not 'unconditional' antecedents, and must therefore be rejected in an enumeration of conditions or causes of the jar.

Again, the conditions of the conditions, the invariable antecedents of the invariable antecedents, are not unconditional. The potter's father is an invariable antecedent of the potter who is an invariable antecedents of the jar, but the potter's father does not stand in a causal relation to the potter's handiwork. In fact, the antecedence must not only be unconditionally invariable, but must also be immediate (अव्यवहितपूर्वकालावाच्छेदेन कार्यदेशे सत्त्वम्). Finally, all seemingly unvariable antecedents which may be dispensed with or left out, are *ipso facto* not unconditional, and cannot therefore be regarded as causal conditions ; in short, nothing that is unnecessary is unconditional ; for this class, *vide* Viśvanātha:–नियतावश्यकपूर्वभाविनः अवश्यक्लृप्तनियतपूर्ववर्त्तिन एव कार्यसम्भवे। तद्भिन्नम् (अतिरिक्तं) अन्यथासिद्धं– Viśvanātha, Siddhānta Muktāvalī, on Śloka 20. For example it is the custom to point to spatial position or direction with the fingers, but finger-pointing, though invaribly present, is not causally related to the perception of direction or spatial position, because we can imaging such perception without finger-pointing, (अन्यथासिद्धतया अङ्गुल्या निर्देशस्य दिक्प्रत्यक्षत्वेन न स्वाभाविकः सम्बन्धः, Vācaspati, Tātparya-Ṭīkā, Chap. I. Āhnika 1, sūtra 5,—this shows that the doctrine of अन्यथासिद्धिशून्य was long anterior or Gaṅgeśa).

[Viśvanātha in the Bhāṣa-Pariccheda mentions five kinds of अन्यथासिद्ध, conditional antecedents—(1) येन सहपूर्वभावः, (2) यस्य वा कारणमादाय (पूर्वभावः), (3) अन्यपति पूर्वभावे ज्ञाते यत्पूर्वभावविज्ञानम्, (4) जनकं प्रति पूर्ववर्त्तिताम् अपरिज्ञाय यस्य (पूर्ववर्त्तिता) न गृह्यते, and (5) अथापि यत् नियतावश्यकपूर्वभाविनः अतिरिक्तम् भवेत् (एतेषु पञ्चसु आवश्यकस्त्वसौ–Ślokas 19, and 20), यस्य स्वातन्त्र्येण अन्वयव्यतिरेकौ न स्तः किन्तु स्वकारणमादायैव अन्वयव्यतिरेकौ गृह्येते तदन्यथासिद्धम्, Siddhānta Mutāvalī, *loc. cit.*—The Dinakarī points out that the first two cases are comprehended under the formula इतरान्वयव्यतिरेक प्रयुक्तान्वयव्यतिरेकशालित्वम्। There are several classification of these irrelevant antecedents (अन्यथासिद्ध); I quote one of the best known].

The unconditional (अन्यथासिद्धिशून्य) as interpreted in this comprehensive sense is a far more fruitful conception than Mill's, and is well adapted to its work—the elimination of the irrelevant factors in the situation. In the end, discrimination of what is necessary to complete the sum of causes, from what is dependent, collateral, secondary, superfluous, or inert, (*i.e.* of the relavant from the irrelevant factors) must depend on the test of expenditure of energy. This test the Nyāya would accept only in the sense of an operation analysable into molar or molecular motion (परिस्पन्द एव भौतिकी व्यापारः करोत्यर्थः–अतीन्द्रियस्तु व्यापारो नास्ति, Jayanta, Mañjarī, Āhnika 1); but would emphatically reject, if it is advanced in support of the notion of a mysterious causal power or efficiency (शक्ति). This is a peculiarity of the Nyāya—its insisting that the effect is only the sum or resultant of the operations of the different causal conditions—that these operations are kinetic, being of the nature of motion, in other words, holding firmly to the view that causation is a case of expenditure of energy, in the kinetic form,—but at the same time absolutely repudiating the Sāṁkhya conception of power or productive efficiency as metaphysical or transcendental (अतीन्द्रिय) and finding nothing in the cause other than an unconditional invariable complement of operative conditions (कारणसामग्री) and nothing in the effect other than the consequent phenomenon which results from the joint operations of the antecedent conditions (समुदितसकलकारकनिकर परिस्पन्द एव–Jayanta—सामग्री कार्यं– सामग्रीकार्यत्वे तु कार्यमस्तु किमन्तरालवर्त्तिन्या शक्त्या–Jayanta, *ibid.*, Āhnika, 1, शक्तिनिराकरणम्–it may be noted that the Nyāya, while repudiating transcendental power (Śakti) in the mechanism of nature and natural causation, does not deny the existence of metaphysical conditions like merit (धर्म), which constitute a system of moral ends that fufil themselves in and through the mechanical system and order of Nature—*vide* Jayanta, अतीन्द्रियं किमपि कारणं कल्पितमेव धर्मादि–Āhnika 1, शक्तिनिराकरणम्).

The causal relation, then, like the relation of genus to species, is a natural relation of concomitance (व्याप्तिः स्वाभाविक सम्बन्धः–Vācaspati) which can be ascertained only by the uniform and uninterrupted experience of agreement it presence and agreement in absence, and not by deduction from a certain *a priori* principle like that a Causality or Identity of Essence.

Nyāya objection to the Buddhist Method of Difference as a means of ascertainning causality.

Take for example the Buddhist deduction of Causality in any particular conjunction by means of the negative Method of Difference, or of the Pañcakāraṇī. The ascertainment of the causal relation by these Methods is open to the following objections:—(1) The unconditionality of the antecedent cannot possibly by ascertained. As the Cārvāka rightly points out, the Methods enable you to eliminate irrelevant antecedents that are or can be perceived ; but the introduction or sublation of latent or undetected antecedents can be imagined against which the Method of Difference is powerless. In the case of the production of smoke, for example, by fire,—what it I say that an invisible demon intervenes in every case between the fire and the smoke, that this demon (पिशाच) is the immediate antecedent and real cause of the latter, and that the fire is an accident which, in every such case, is brought about by its own causal antecedents ; in saying this, I do not go counter to the principle of causality and am landed in no contradiction (व्याघात) such as strikes at the very roots of all practice, or baffles the completed, circuit of consciousness, however much I may violate probability.

(2) In the second place, even supposing that the fire, in this particular case (which satisfies the Method of Difference rigidly) is ascertained to be the cause of the smoke, how can I know that fire is the cause in other cases, or that there is no other cause. You will perhaps argue that if there were an *indefinite* number of cause of the same specific phenomenal effect, it would violate the principle that phenomena are all conditioned, *i.e.*, exist only under certain conditions (कादाचित्कत्व), which is more comprehensive than the principle of causality, and the contradiction of which equally overthrows all rational practical activity. Yes, I accept the conditionality of phenomena, but this is not violated by supposing that one specific assemblage of phenomena has more than one cause. It is true that if you suppose plurality of causes, you cannot establish the invariableness of the particular conjunction (green-wood fire and smoke) which your Method of Difference fixes upon as a case of cause and effect ; in other words, with your special principle of Causality so restricted, and without any general principle of Uniformity of Nature to fall back upon, you cannot ascertain, from the present case, or from any number of similar cases that you may have observed, that all green-wood fires are followed by smoke ; *i.e.*, you are helpness in demonstrating with apodictic certainty (or ascertaining indubitably) the relation of cause and effect. But this is an objection against your own position, not mine. Why not admit at once that certain phenomena are naturally connected (as invariable concomitants or antecedents) with other phenomena, and take your stand on observed concomitance (uniform and uninterrupted experience of agreement is presence as well as absence) without assuming causality as an *a priori* principle and making deductions therefrom, and without the trouble of ascertaining the relation of cause and effect in every individual case. I am free to admit that theoretical objections irresistible force (like those of the Cārvāka Sceptics) can be urged against this ascertainment of universal invariable and

unconditional concomitance (व्याप्तिग्रहोपाय) on these basis of mere observation. Doubts of this kind can no more be laid by my view of the matter than by your canons of causality and essential indentity (तदुत्पत्ति and तादात्म्य). Ultimately we all have to fall back on the rational practice of thinking persons (प्रेक्षावन्तः), and such persons are always content to act on practical certitude instead of hankering after an unattainable apodictic certainty in the affairs of life (प्रामाणिकलोकयात्रामनुपालयता यथादर्शनं शङ्कनीयम्). This same practical certitude is also the ultimate warrant of the Deductive-Inductive Inference by which we ascertain the characters of things without direct perception and through the medium or instrumentality of a mark.

[To the earlier Buddhist canon of the Method of Substraction, *i.e.*, the negative Method of Difference, Udyotakara and Vācaspati of the Nyāya school pertinently and acutely objects as follows :—

सत्यं यत् किञ्चित् क्वचित् दृष्टं तस्य यत्र प्रतिबन्धः तद्विदः तस्य तत् गमकं तत्र इति अनुजानीमः। स एव तु प्रतिबन्धो न तावत् तदुत्पत्त्या सम्भवति। का पुनरियं तदुत्पत्तिः। धूमस्य हि वह्न्यनन्तरं भावः। स तादृशः अस्ति रासभस्यापि। तत्प्रतिवद्धोऽपि धूमः स्यात्। अथ तदनन्तरमेव भाव, न च रासभानन्तरं भवन्नपि तदनन्तरमेव भवति, तस्मिन् सति अपि असति अग्नो तदभावात्। असति अपि तस्मिन् सत्यार्द्रेन्धनवति वह्नौ तद्भावात्। अथ यद्यपि धूमस्य वह्निभावानुविधानं तत्रोपलब्धं, तथापि देशान्तरादिषु तद्भावोऽन्य कुतस्त्यः। तथा हि भूयो भूयो रासभे दृष्टे धूमोदृष्टः, तदभावे च अदृष्ठः, न स तत्कार्यः। तज्जातीयस्यैव धूमस्य रासभं विना सति वह्नौ भावात्, एवं सत्यपि अग्नौ पिशाचेन जनितो धूमः कचिद्देशादौ तज्जातीय एव रासभवदवह्न्यभावेऽपि पिशाचादेव भविष्यति, इति अवश्यं शङ्क्या भाव्यं नियामकं अपश्यताम्। न च सति भावमात्रं नियामकं, तस्य रासभादिषु अविशेषात्। तदनन्तरमेव इति च अवधारणस्य शङ्कायामभावात्। अवधारणेन तु शङ्कापनयने परस्पराश्रयप्रसङ्गात्।

Buddhist reply:—स्यादेतत्। यो यो धूमोदृष्टः स सर्वस्तावत् आद्रेन्धनसहित वह्न्यनन्तरमेव, न पिशाचानन्तरं। स च कादाचित्कतया निमित्तमपेक्षमाणो यदनन्तरमेव गम्यते तदेव अप्रतीतव्यभिचारं निमित्तं करोति, न तु प्रतीतव्यभिचारं रासभादि। नापि

सर्वथाऽनुपलब्धपिशाचादिः। यदि न तन्निमित्तं कस्मात् विनाऽपि वह्निं क्वचित् धूमो न उपलभ्यते। अथ असौ (पिशाचः) सर्वथा वह्निसहितः तथा सति आर्द्रेन्धनवत् कथं वह्निरपि न कारणं। कारणं चेत् कथं तमन्तरेण धूमभाव शङ्का। Now, the phenomenon that is contingent (कादाचित्क) cannot be uncaused—अकारणस्य हि कार्यस्य नित्यं सत्त्वं असत्त्वं वा स्यात्, अनपेक्षत्वात्, कादाचित्कत्त्वम्। And with the limitation of a specific effect, there must be only one specific cause, for an unrestricted plurality of cause would amount to the denial of uniformity in causality (*i.e.*, of the *unconditional invariable* antecedence)—

नापि अनेकं कारणं, अकारणत्वप्रसङ्गादेव। वह्न्यनन्तरमेव भाव इति हि धूमस्य वह्निकार्यत्वम्। स चेत् अवह्नेरपि अनन्तरं, न एवकारार्थः स्यात्, इति न वह्नेः कार्यम्। एवम् अन्यस्यापि न कार्यं। न हि अन्यानन्तरमेव भवति, वह्नेरपि अन्तरं भावात्, ततश्च अहेतुको धूमः स्यात्। तथा च कादाचित्कत्वव्याहतिः।

Nyāya rejoinder:—सत्यम्। यद्यपि विना वह्निं नोपलब्धो धूमः, यद्यपि च पिशाचान्तरं नोपलभ्यते, तेषामनुपलब्धेः तथाऽपि पिशाचकार्य्य एव धूमः। तत्र तत्र वह्निः कुतश्चित्, स्वहेतोः उपनिपतितः, रासभ इव, न धूमस्य जनकः। तेन तदभावेऽपि तज्जातीय एव कारणभेदजन्मा कदाचित्, कादाचित्को धूमः स्यात्। इति अनिवृत्तिरेव शङ्कायाः। न च दृष्टसम्भवे नाददृष्टं कल्पयितुं युक्तम् इति शक्यं भवद्भिर्वक्तुम्। अनुपलब्धिलक्षण प्राप्तस्य अशक्यनिराकरणत्वात् (*vide* Nyāya Vindu)। न च अनुपलब्ध्यन्तराणि अपि तन्निषेधे प्रभवन्ति। तस्मात् अनन्तरमेव इति अवधारणाभावात् नैवं कार्यकारणभावावधारणं युक्तम्। न च यदन्यसहितानन्तरं उपलब्धं तत्, तदन्यरहितात्, तस्मात् भवत् भिन्नजातीयं भवति। उक्तं हि रासभसहितात् वह्नेः यादृशो धूमः तादृश एव तद्रहितात् वह्ने रिति। तस्मादेवंविद्या शङ्कापिशाची स्वनिवारकं तदुत्पत्तिनिश्चयमास्कन्दती न शक्या निवारयितुम्।

The Nyāya proceeds to point out that what is contingent need not necessarily be produced by a cause. It is enough if there is natural connection with something else, a relation of antecedence and succesion (or concomitance), without any element of productive or causation.

अपि चास्तु तदुत्पत्तिनिश्चयः तथाऽपि कस्मात् कारणमन्तरेण न कार्यं भवति।

तथा च सति अनपेक्षतया कादाचित्कत्वविहतिः इति चेत्। अस्तु तर्हि सम्बन्धः स्वाभाविकतया अन्यानपेक्षः अव्यभिचारी गमकत्वाङ्गम्। स च यो वा स वा भवतु। कृतं कार्यकारणभावावधारणायासेन। यथा चैतत् तथा अग्रे दर्शयिष्यामः। Vācaspati, Tātparya-Ṭīkā, Chap. I, Āhnika 1, Sūtra 5, व्याप्तिग्रहोपायविचारः।

The Nyāya analysis of the causal relation continued—Co-effects:—In the enumeration of different varieties of irrelevant antecedents (अन्यथासिद्ध), we have already noticed that co-effects of the same cause are apt to be confounded as cause and effect. In some cases, the co-effects may be simultaneous, *e.g.*, the case of the ascending and the descending scale in a balance, which are co-effects of gravity (*vide* Pārthasārathi Miśra on Kumārila, Śloka-Vārttika—तुलानमनोन्नमनयोस्तु मिथः कार्यकारणत्वम्, उभययोरेककारणत्वात्,–Śloka, 157, Śūnya-Vāda). In other cases, the co-effects may be successive effects of the same cause, and here the risk is great of mistaking the antecedent co-effect to be the cause of the succeeding co-effect to be the cause of the succeeding co-effect ; *e.g.*, the case of ants moving in a line to carry their eggs upward, which is observed before the summer rains;—where the movement of ants and the rains are not cause and effect, but successive effects of the same cause, *viz.*, the heat (उष्मा), which disturbs the elements, *viz.*, the earth and the atmosphere (महाभूतक्षोभ) ; the ants being affected by this heat earlier than the atmospheric movements which bring the clouds and the rain.

(Udyotakara, and Vācaspati—न च पिपीलिकाण्ड सञ्चरणं वर्षस्य कारणमनुपलब्धसामर्थ्यात्। असत्यपि तस्मिन् वर्षस्योत्पत्तेः। वर्षमूलकारणस्य तु महाभूतक्षोभस्य पिपीलिकाण्डसञ्चरणं पूर्वकार्यम्। क्वथ्यमानाः खलु पिपीलिका भौमेनोष्मणा स्वानि अण्डानि भूमिष्ठानि उपरिष्ठात् नयन्ति–Tātparya–Tīkā II, Sūtra 37).

Synchronousness of Cause and Effect

This is resolved into a case of simulataneous co-effect of

the same ultimate cause, *e.g.*, the ascent of one scale and the descent of the other in the balance, which are not related as cause and effect, but are simultaneous effects of gravity. In other cases, the synchronousness is only apparent, the interval between the antecedent and the consequent being too small (सूक्ष्मकाल) to be apprehended (यौगपद्याभिमान) *e.g.*, in the case of the needle piercing a hundred soft lotus petals laid one upon another, where the steps are really successive; or the illumination of the whole room by the light of a lamp, where the succession is unperceived owing to the inconceivable velocity of light (*cf.* Kumārila, Śloka-Vārtika,— हेतुहेतुमतां सिद्धं यौगपद्यनिदर्शनम्,–यत् प्रदीपप्रभाद्युक्तं, सूक्ष्मकालोऽस्ति तत्र नः। दुर्लक्षस्तु यथा वेधः पद्मपत्रशते तथा। Śūnyavāda, Śloka 156-157.—I quote Mīmāṁsā authorities, but the view is common to the Mīmāṁsā and the Nyāya-Vaiśeṣika).

The Time-relation in a Chain of Causes and Effects

A careful study in the time-relation in a chain of causes and effects is a peculiarity of the Vaiśeṣika system (and the later Nyāya). A moment (ultimate unit of time, Kṣaṇa, क्षण) is defined to be the time-interval between the completion of the sum of conditions and the appearance of the effect. The Vaiśeṣika conceives the unit to be determined by reference to the division of one atom from another (विभागानुत्पाद्यविभाग प्रागभावसम्बद्धकर्मावच्छिन्नः कालः क्षणः–Sapta Padārthī, Śivāditya ;— *i.e.*, the ultimate unit of time is the time during which motion exists in an atom prior to its division from another atom, in a case of division due to motion). The Sāṅkhya, we have already seen, determines this ultimate unit by reference to the motion of a Tanmātra.

The number of such units will determine the time-interval between a given set of physical conditions and a particular effect, for between a so-called sum of causes and a so-called sum of effects, there intervenes a series of atomic (or molecular) motions, with conjunctions add disjunctions which form the casual chain. However, crude in the practical application, the fundamental idea is, in connection with the principle of work and energy (for which both the Sāṅkhya and the Nyāya-Vaiśeṣika furnish a rudimentary basis), immensely suggestive of a possible Time Calculus.

Plurality of Causes—This will be discussed when we consider. The relation of Vyāpti to the principle of Causality.

The Nyāya Ground of Inference

Vyāpti नियतसाहचर्य- Vācaspati Inference, then, in the Nyāya, depends on the ascertainment, not of the causal relation, not of the relation of genus of species, but of a natural relation, between two phenomena, of invariable and unconditional concomitance (उपाधिविधुरः स्वाभाविकः सम्बन्धः व्याप्तिः-Udyotakara and Vācaspati). Of the two phenomena so connected, one is called the Vyāpya or Gamaka (the sign, mark or indicator), and the other Vyāpaka or Gamya (the thing signified, marked, or indicated). In the relation of fire and smoke, for example, smoke is the Vyāpya or Gamaka (sign or mark) ; and fire, the Vyāpaka or Gamya (the thing signified or marked). Now, the relation of Vyāpti between *A* and *B* may be either unequal of equipollent (विषमव्याप्ति or समव्याप्ति). When *A* is the sign of *B*, but *B* is not the sign of *A*, the Vyāpti is one-sided or unequal, and here a Vyapti is said to exist between *A* and *B*, but not between *B* and *A*. For example, smoke is a sign of fire, but fire is not universally a sign of smoke. When, therefore, the relation of Vyāpti is an unequal one, as between smoke and fire, it is expressed in the proposition :—whenever the Vyāpya (sign or mark *e.g.*, smoke) exists, the Vyāpaka (the thing signified or marked, *e.g.*, fire) also exists. From this it follows by necessary implication (a sort of अर्थापत्ति) that whenever the Vyāpaka (*e.g.*, fire) is absent, the Vyāpya (*e.g.*, smoke) is also absent (व्यापकाभावे व्याप्याभावः). Again, the Vyāpti may be a mutual or equipollent one, *i.e.*, *A* and *B* may be signs of each other, *e.g.*

green wood fire and smoke. Here each in turn is Vyāpya and Vyāpaka, and this is expressed in the two propositions:—(1) Whenever there is smoke, there is green-wood fire, and (2) whenever there is green-wood fire, there is smoke. By necessary implication it follows—(1) where there is no green-wood fire, there is no smoke ; (2) where there is no smoke, there is no green-wood fire. We have seen that a Vyāpti exists between smoke and fire, for whenever there is smoke, there is fire. But we cannot say that a Vyāpti exists between fire and smoke, for we cannot say that wherever there is a fire, there is smoke. The combustion of an iron-ball (अयोगोलक), for example, is a case of fire without smoke. But it would be correct to say that a Vyāpti exists between green-wood fire and smoke, as well as between smoke and green-wood fire. The question, therefore, is—What is the relation between fire and smoke ? The relation between fire and smoke is a conditional relation; *i.e.*, on condition that the fire is green-wood fire, it would be a sign of smoke. In other words, a Vyāpti implies unconditional invariable concomitance, and the relation between fire and smoke is not therefore a Vyāpti (natural unconditional concomitance), for fire requires a 'condition,' Upādhi, *viz.*, green-wood to be followed by smoke. Smoke, on the other hand, requires no 'condition' to indicate fire. For the purposes of Inference, therefore, relations between phenomena may be considered as of two kinds :—(1) Contingent conditional relations, holding good on the fulfilment or a certain condition or Upādhi, and (2) Vyāpti, or unconditional invariable relation, between a mark and that which it marks, a relation without any Upādhi or determining condition (उपाधिविधुरः सम्बन्धः). It is this latter kind of relation that serves as the ground of Inference. If we can ascertain that a Vyāpti exists, between *A* and *B*, then *A* is a sign of *B*, and an inference of the presence of *B* from the presence of *A*, and of the absence of *A* from the absence of *B*, would be warranted. The question, therefore is—how to ascertain the relation of Vyāpti between two phenomena.

Ascertainment of Vyāpti according to the early Nyāya (व्याप्तिग्रहोपाय)

Briefly speaking, the observation of agreement in presence (अन्वय) as well as agreement in absence (व्यतिरेक), between two phenomena, with the non-observation of the contrary (व्यभिचारादर्शनम्), is the foundation of our knowledge of Vyāpti (दर्शनादर्शने–Vācaspati). This suggests a natural relation (स्वाभाविकसम्बन्ध–Vācaspati) of invariable concomitance (नियतसाहचर्य–Vācaspati) between the phenomena, which is fortified by our non-observation (अदर्शन) of the contrary (व्यभिचार). But this does not establish the unconditionally of the concomitance (उपाधिविधुरसम्बन्ध), which is essential to a Vyāpti. We have therefore to examine the cases carefully to see if there is any determining condition (Upādhi—*i.e.*, some hidden undetected but really operative or indispensable accompaniment). Which conditions the relation between the supposed sign of mark (Gamaka) and the supposed signate (thing signified, Gamya). Now let us consider what constitutes an Upādhi. It is a circumstance which always accompanies, and is always accompanied by, the supposed singnate (the thing signified, Gamya), but does not invariably accompany. the supposed sign or mark (Gamaka). If, therefore, in the set of positive instances where both the sign and the signate are present, nothing else is constantly present there can be no Upādhi. Or, again, if in the set of negative instances where both the sign and the signate are absent, no other material circumstance is constantly absent, there is no Upādhi. This follows from the very definition of an Upādhi. It is impracticable to fulfil these requirements rigorously. Still, every one of the accompanying circumstances (of course the likely ones) may be taken successively, and it may be shown that the concomitance continues even when the suspected Upādhi (शङ्कितोपाधि) is absent, and therefore it cannot be the Upādhi. And this is to be fortified by the observance of uniform and uninterrupted agreement in absence (Vyatireka) between the two concomitant phenomena. In this way, when we have disproved all suspected. Upādhis, we conclude by establishing the Vyāpti. It is true that we may still go on doubting. But doubt has a certain limit for the 'experimenter' and the thinking person (परीक्षक, प्रेक्षावान्). When doubt overthrows the foundation of all rational practice (प्रामाणिकलोकयात्रा), or leads to a stoppage or arrest of all practical activity (लोकव्यवहरा), it stand *ipso facto* condemned, and must be abandoned (व्याघातावधिराशङ्का) Śrīharṣa and Udayana—व्याघात is mentioned by Vācaspati). Thus it is that Vyāpti is ascertained. In this way, we observe innumerable instances of Vyāpti. Now by means of repeated observations of this kind (भूयोदर्शन), we have established the principle of the Uniformity of Nature (स्वभावप्रतिबन्ध), and also of Causality; and these two principles thus ascertained may be made use of in their turn as the basis of an argumentation or deduction (Tarka, Uha, तर्क) to confirm a particular Vyāpti in a particular case. Tarka or Uha, then, is the Verification and vindication of particular inductions by the application of the general principles of Uniformity of Nature and of Causality, principles which are themselves based on repeated observation (भूयोदर्शन) and the ascertainment of innumerable particular inductions of uniformity of causality (भूयोदर्शन जनितसंस्कारसहितम् इन्द्रियमेव स्वाभाविकसम्बन्धग्राहि–Vācaspati). Thus Tarka, also helps in dispelling doubt (सन्देह). Śrīharṣa, however, question the validity of this verification,—*cf.* the well-known couplet ending तर्कः शङ्कावधिः कुतः।

It will be seen that the process of disproving all suspected Upādhis (उपाधिशङ्कानिरास), in the early Nyāya, answers exactly *as a process* to Mill's Method of Agreement. In fact, the disproof of a suspected Upādhi by pointing to instances of Agreement in presence (अन्वय) even in the absence of the Upādhi, fortified as this is by the instances of agreement in absence (व्यतिरेक), virtually amounts to Mill's joint Method of Agreement. But the Fundamental difference is this:—Mill's Method of Agreement is formulated in view of the phenomena of causation (including co-effects, etc.) and, as usually enunciated, confessedly breaks down in dealing with cases of Uniformities of Co-existence unconnected with Causation ; the Nyāya Method based on the disproof of suspected Upādhis is a more daring and original attempt, and is for more comprehensive in scope, being applicable to all Uniformities of Co-existence and of Causation alike. And this the Nyāya successfully accomplished, by introducing the mark of unconditionally (उपाधिविधुरता) into the relation of Vyāpti (Concomitance), even as the same mark of unconditionality (अन्यथासिद्धिशून्यता) has been previously introduced into the definition of Causality (कार्यकारणभाव). The difference between the early Nyāya and the Buddhist systems may be briefly put thus : The former relied on empirical induction based on uniform and uninterrupted agreement in Nature, and accordingly regarded the Method of Agreement as the fundamental Method of Scientific Induction, founding Inference on Vyāpti to which they subordinated Causality in the doctrine of Method ; the latter assumed two *a priori* principles, *viz.*, causality and indentity of essence, deduced the canon of the Method of Difference by an indirect proof from the principle of causality, and made this Method the foundation of all scientific Induction of Causality just as they based all natural classification of Genera and Species on their *a priori* principle of Identity of Essence.

Texts from the early Nyāya

Method of Agreement and the joint Method without the device of the Upādhi :—

Cf. Jayanta on साहचर्य्य-तस्मिन् सत्येव भवने न विना भवनं ततः। अयमेवाविनाभावो नियमः सहचारिता। तस्मिन् सत्येव 'only this remaining throughout', while others change—implying the Method of

Agreement. The set of positive instances, in which this antecedent alone is constant, must be supplemented by a set of negative instances (agreement in absence) :— व्यतिरेकनिश्चयमन्तरेण प्रतिबन्धग्रहणानुपपत्तेः। नियमश्चायं उच्यते यत् तस्मिन् सति भवनं ततो विना न भवनं इति भूयोदर्शनम्। तच्च तस्मिन् सति भवनं इत्यन्वयमात्रपरिच्छेदात् अर्द्धगृहीतो नियमः स्यात्। ततो विना न भवन इत्यस्य अर्थस्य अपरिच्छेदात् इति Jayanta, Āhnika II, व्यतिरेकव्याप्तिनिश्चयावश्यकत्वम्।

Doctrine of the Upādhi :—Unconditional concomitance distinguished from conditional :—

यो वा स वा अस्तु सम्बन्धः यस्य असौ स्वाभाविको नियतः स एव गमकः, गम्यश्च इतरः सम्बन्धतीति युज्यते। तथा हि धूमादीनां वह्न्यादिसम्बन्धः स्वाभाविकः। नतु वह्न्यादीनां धूमादिभिः। उपलभ्यते ते हि विनापि धूमादिभिः। यदातु आर्द्रेन्धनादिसम्बन्धम् अनुभवन्ति तदा धूमादिभिः सह संवध्यन्ते। तस्मात् वह्न्यादीनाम् आर्द्रेन्धनाद्युपाधिकृतः सम्बन्धो न स्वाभाविकः। ततो न नियतः। स्वाभाविकस्तु धूमादीनां वह्न्यादिसम्बन्ध उपाधेः अनुपलभ्यमानात् क्वचित् व्यभिचारस्य अदर्शनात् अनुपलभ्यमानस्यापि कल्पनानुपपत्तेः। अतो नियतः सम्बन्धः अनुमानाङ्गम्।

Concomitance sufficient without causal *nexus* :—

The Buddhist objects—In Nature, everything is connected with everything else. Hence, if there were no *nexus* of causality between antecedent and consequent, everything might follow from everything else. The Nyāya replies—You admit uniformities (of co-existence, etc.) other than causal;— so you confess that a natural fixed order can exists without the causal *nexus*.....

An Upādhi, how established and how disproved :

The Nyāya then proceeds to show how an Upādhi is established, or how disproved by observation :— तत्र वह्नेः अनार्द्रेन्धनस्य विना धूमम् अयोगोलकादौ दर्शनात् आर्द्रेन्धनोपाधिः अस्य धूमेन सम्बन्धः न तु स्वाभाविक इति निश्चीयते। धूमविशेषस्य तु विना वह्निम् अनुपलम्भात्, उपाधिभेदस्य च अदृश्यमानस्य कल्पनायां प्रमाणाभावात्, विशेषस्मृत्यपेक्षस्य च संशयस्य

अनुपलब्धपूर्वे अनुत् पादात्, उत्पादे च अतिप्रसङ्गात् प्रेक्षावत् प्रवृत्युच्छेदात्, स्वाभाविकः सम्बन्धः अवधार्यते।

General Method of Induction by exhaustion of the Upādhi, more comprehensive than Mill's Joint Method :—

For the definition of an Upādhi, vide Śrīharṣa and Udayana : अव्याप्तसाधनो यः साध्यसमव्याप्ति उच्यते स उपाधिरिति। Hence to avoid an Upādhi (which is साध्यसमव्याप्तिः), the constant presence of anything relevant other than the sign and the signate in the positive instances (of agreement in presence, अन्वय), and the constant absence of any such thing in the negative instances (of agreement is absence, व्यतिरेक) must be safe-guarded against. This amounts to Mill's joint Method.

Suspicion of non-perceptible Upādhi—Limits of legitimate Doubt :—

न च अदृश्यमानोऽपि दर्शनानर्हत्तया साधकवाधकप्रमाणाभावेन संदिह्यमान उपाधिः स्वाभाविकत्वं प्रतिवध्नाति इति साम्प्रतम्। अवश्यं शङ्कया भवितव्यं नियामेकाभावादिति दत्तावकाशा खलु ईयम् प्रमाणमर्यादातिक्रमेन शङ्का पिशाची लब्धप्रसरा न क्वचित् नास्तीति नेयं क्वचित् प्रवर्त्तेते। सर्वत्रैव कस्यचित् कथञ्चित् अनर्थस्य शङ्कास्पदत्वात्। अनर्थशङ्कायाश्च प्रेक्षावतां निवृत्त्यङ्गत्वात्। अन्ततः स्निग्धान्नपानीपयोगेषु अपि मरणादिदर्शनात्। तस्मात् प्रामाणिकलोकयात्राम् अनुपालयता यथादर्शनं शङ्कनीयम्।. .....तस्मादुपाधिं प्रयत्नेन अण्विष्यन्तोऽनुपलभमाना नास्तीति अवगम्य स्वाभाविकत्वं सम्बन्धस्य निश्चिनुमः।-

Tarka, ūha,—Deductive verification of particular inductions by applying the two fundamental inductions of Uniformity of Nature, and Causality :—Doubt finally dispelled :—

Now innumerable particular uniformities of this kind (Vyāptis) are observed, and as a result of this repeated observation, a belief in the Uniformity of Nature (स्वभाव प्रतिबन्ध), as well as in the principle of Causality (कार्यकारणभाव) is generated in the mind, a belief which has evidential value and validity. It is not intuition (न मानसम्) but a mental predisposition based on uniform and uninterrupted experience (भूयोदर्शनजनितसंस्कारसहितम् इन्द्रियमेव स्वाभाविकसम्बन्धग्राहि,–अत्र भूयोदर्शनम् असाधारणम् इति प्रमाणान्तरं जातम्). Then armed with these new resources, the belief in uniformity and in causality as general principles, we proceed to fortify our particular inductions (Vyāptis), whether of uniformity of nature or of causality, by indirect deduction from these general principles —We argue if, under these observed circumstances, *A* were not the mark of *B*, the principle of uniformity of nature would be violated—Nature would not be unifrom or, if under these observed circumstances *A* were not the cause of *B*, the principle of causality would be violated, the phenomenon *B* would be without a cause ; and such indirect proof (तर्क, ऊह) gives us the overwhelming probability which we call practical certitude, and on which every reasonable man (every thinking and judging person) proceeds to act in due natural course—स्वभावतश्च प्रतिवद्धा हेतवः स्वसाध्येन, यदि साध्यमन्तरेण भवेयुः, स्वभावादेव प्रच्यबेरन्निति तर्कसहायेः अनिरस्तसाध्यव्यतिरेकवृत्तिसंदेहा यत्र दृष्याः, तत्र स्वसाध्यम् उपस्थापयन्त्येव। Vācaspati I, 1, Sūtra 5 ; *cf.* also, I, 1, 40—ऊहः कारणस्योपपत्त्या संभवेन ज्ञायते। कारणासम्भवे कार्यस्य असत्वात्।]

Instances of Vyāpti (uniformity) not comprehended under Causality, or the relation of Genus and Species —The Nyāya points out that the relations of cause and effect and of genus and species do not exhaust the grounds of Inference. There are cases of Inference based on Vyāpti (*i.e.*, on invariable and unconditional concomitance) which come neither under Causality (तदुत्पत्ति) nor under Indentity of Essence (तादात्म्य). Vācaspati notes that to-day's sunrise and yesterday's sunrise, the rise of the moon and the tide in the ocean, the relative positions of the stellar constellations, are instances of Vyāpti (invariable concomitance) between phenomena which are neither related as cause and effect, nor as genus and species. Jayanta adds the conjunction of sun-set with the appearance of the stars, of ants moving in procession (with their eggs) with the approach of the rains—of the rising of the constellation Agastya (Canopus) above the horizon with the drying up of rivers ; of the spring tide with the full moon ; and dismisses as sophistical and far-fetched the Buddhist attempt to explain all these cases by means of causality. We have seen that the Nyāya and the Mīmāṁsā reduce most of these conjunctions to cases of co-effects of the same cause, co-effects which may be either simultaneous or successive.

अपि च अद्यतनस्य सवितुरुदयस्य ह्यस्तनेन सवितुरुदयेन, चन्द्रोदयस्य च समानकालेन समुद्रवृद्ध्या....न कार्यकारणभावः तादात्म्यं वा। अथच दृष्टोगम्यगमकभावः। Vācaspati, Tātparya-Ṭīkā I, 1, Sūtra 5 ; *cf.* also Jayanta, Nyāya Mañjarī—अन्येषामपि हेतूनां भूम्नां जगति दर्शनात्। सूर्यास्तमयमालोक्य कल्पते तारकोदयः। पूर्णचन्द्रोदयाद्वृद्धिरम्बुधेरवगम्यते। उदितेनानुमीयन्ते सरितः कुम्भयोगिनः। शुष्यत्पुलिनपर्य्यन्त विश्रान्तखगपंक्तयः। पिपीलिकाण्डसञ्चारचेष्टानुमितवृष्टयः। भवन्ति पथिकाः पर्णकुटीरकरणोद्यताः।

Vyāpti between Cause and Effect:—Relation of Causality to Vyāpti

On the Buddhist (and early Nyāya) view that one specific assemblage of 'effect' phenomena has one specific assemblage of causal conditions, there would be two aggregates, the sum of causal conditions (कारणसामग्री), and the sum of effects (कार्यसामग्री). For example, fire requires green-wood to complete the sum of causal conditions to give rise to smoke with some particular marks (धूमविशेष, बहुलपाण्डुतादिधर्म विशिष्ट, Jayanta II ; compare Gaṅgeśa, कृष्णागुरुप्रभववह्निमानयं कटुकासुरभिपाण्डरधूमत्वात्।–Cintāmaṇi, Anumāna, Viruddha Siddhānta). Here, between an effect and a single condition (termed a cause) there is a relation of Vyāpti. The effect is Vyāpya or Gamaka (the sign or mark) ; the cause (or condition) is Vyāpaka or Gamya (the thing signified). In other words the presence of the effect indicates the presence of the causal condition, and the absence of the causal condition will by implication indicate the absence of the effect. Smoke of this particular kind is supposed to be an effect of which there is one and only assemblage of causal conditions (fire and green-wood) ; hence where there is smoke, there is fire ; and when there is no fire, there is no smoke.

Now Introduce the Complication of the Plurality of Causes

Fire, for example, is the effect of several assemblages, *e.g.*, (1) blowing on heated grass, (2) focussing rays through a lens on a combustible like paper or straw, (3) friction with the fire drill, etc. Here each assemblage is regarded as a sum of causes. But in this case there is no Vyāpti between the effect 'fire' and any particular assemblage of causal conditions, say, of the lens on the fire drill. For the presence of fire does not indicate the presence of the lens or the fire drill assemblage, nor does the absence or either of the latter in particular, indicate the absence of fire. Indeed in such a case, the effect 'fire' is not a mark or sign (Gamaka or Vyāpya) of any one in particular of the different possible causal assemblages, though each of these particular assemblages of causal conditions is a mark or sign (Gamaka or Vyāpya) of fire.

The plurality of causes requires a further consideration in the light of the definition of the causal relation. A cause is defined to be the unconditional invariable antecedent. From the unconditionally, it follows that the entire sum of conditions, and not one single condition, is, properly speaking, the cause. In view of the plurality of causes, an invariable antecedent must be taken to mean that any particular cause (*i.e.*, assemblage of causal conditions) is invariably followed by the effect—not that the effect is invariably preceded by any particular cause.

Popularly, a single condition, say the lens or the fire-drill, is said to be a cause of fire ; but, in view of the plurality of causes, this is apt to be misleading, as there is no Vyāpti in this case ; the lens or the fire-drill is no more a mark of fire than fire is a mark of the lens or the fire-drill.

The plurality of causes strains the definition of a cause, and undermines the relation of Vyāpti between an effect and a cause. Any particular cause (causal agreegate) still indicates the effect, but not *vice versa*. The earlier Nyāya (down to Vācaspati and Jayanta) obviated the plurality, as we have seen, by introducing distinctive marks in the effect such as would indicate a single specific cause. (कार्यवैलक्षण्य or कार्यभेद indicating कारणवैलक्षण्य or कारणभेद). Some Indeed went further and held that when the antecedent causal assemblages differ in kind, the effect phenomena, though apparently the same,do really differ specifically (or in kind)—कारणवैजात्यात् कार्यवैजात्यम्। But the Nyāya discards this hypothesis ; the fire is the same, though the possible causes (or causal aggregates) differ, *e.g.*, the lens, the drill, etc. But the effect phenomenon to which we attand is not the only effect ; in the case of plurality of causes, we must carefully examine the accompaniments of the effect, *i.e.*, the sum of effects, and the examination will shew some distinctive or specific circumstances or accompaniment which will enable us to definitely determine the particular assemblage of causal conditions that must have preceded in the case under examination. This is the device of the earlier Nyāya as well as of the Buddhists, as we have seen ; but the later Nyāya doubts the practicability as well as the theoretical validity of such a step on an unrestricted assumption of the plurality of causes, and feels troubled by the circumstance that no effect for which more than one cause (or causal aggregate) can be assigned, can be regarded as a mark or sign (Gamaka or Vyāpya) of any one of the causes in particular. Accordingly, some adherents of the later Nyāya advanced the proposition that when more than one causal aggregate can be supposed for any effect, the latter is a mark or sign (Gamaka or Vyāpya), not of any one of the causal aggregates in particular, but of one or other of them ; and the absence, not of one such cause, but of each and every one of them, alone indicates the absence of the effect. A cause therefore should be defined to mean *one or other* of the possible alternative aggregates which, being given, the effect follows invariably and unconditionally. If we ask what is the defining mark (or quiddity) of the cause (कारणतावच्छेदक), we are told that it is *one-or-otherness* (अन्यतमत्व), and nothing else : others cut the Gordian knot by assuming that the different possible causes of the same effect possess a common power or efficieny (अतिरिक्तशक्ति), or a common 'form' (जातिसौक्ष्म्य), which accounts for the production of a common effect. The latter is therefore a sign or mark of this power (शक्ति), or this form (जातिसौक्ष्म्य), which is a manifested by each of the causal aggregates. This hypothesis (कल्पना), they hold, is simpler and more plausible than the hypothesis of specific differences latent in the apparently idential effect of a plurality of causes (कारणवैजात्यात् कार्यवैजात्यम्).

(*Cf.* Dinakarī on the Siddhānta Muktāvlī—केचित्तु वह्निं प्रति तृणफुत्कार-संयोगादीनां तृणफुत्कारसंयोगत्वादिरूपेण कारणतया व्यभिचारेण असम्भवात् अतिरिक्तशक्तिसिद्धिः। न च तृणफुत्कारयोः अरणिनिर्मन्थेयाः मणितरणिकिरणयोश्च सम्बन्धस्य जन्यतावच्छेदकं वह्निवृत्तिबैजात्यत्रयं कल्प्यमिति न व्यभिचार इति वाच्यम्। तज्जन्यतावच्छेदकवैजात्यत्रयकल्पनामपेक्ष्य तत्तत्सम्बन्धानां एकशक्तिमत्वेन कारणत्वकल्पनाया एव लघुत्वेन न्याय्यत्वात् इत्याहुः। तत्र।..... अन्यतमत्वेन कारणतासम्भवात्....परे तु तृणादिसम्बन्धकालीनवायसंयोगादीनां एकशक्तिमत्वेन हेतुतामादाय विनिगमनाविरहात् न शक्तिसिद्धिः इत्याहुः। Śloka 2, *cf.* also, तृणारणिमण्यन्यतमत्वं कारणस्य विनिगमकम्।

Cf. Udayana, Kusumāñjalī, Stakvaka I—कथं तर्हि तृणारणिमणिभ्यो भवन्नाशुशुक्षणिरेकजातीयः। एकशक्तिमत्वात् इति चेत् न। यदि हि विजातीयेषु अपि एकजातीयकार्यकारण-शक्तिः समवेयात्, न कार्य्यात् कारणविशेष

क्वचित् अनुमीयेत। कारणव्यावृत्त्या च न तज्जातीयस्यैव कार्यस्य व्यावृत्तिरवसीयेत।. ...एतेन सूक्ष्मजातीयमिति निरस्तम्। अवह्लेरपि तत् सौक्ष्म्यात् धूमोतपत्त्यापत्तेः।

The Scientific Methods already noticed, the joint Method of Difference (the Pañcakāraṇī), and the joint Method of Agreement (Vyāptigraha with Upādhiśaṅkā-nirasa and Tarka), are not the only methods of ascertaining causality or concomitance, or establishing a theory (सिद्धान्त); nor are these Methods always practicable. Very often, we reach the explanation of a fact (उपपत्ति) by means of a Hypothesis (कल्पना) properly tasted and varified (निर्णीत). A legitimate Hypothesis must satisfy the following conditions:—(1), the hypothesis must explain the facts (दृष्टसिद्धि, or उपपत्ति); (2) the hypothesis must not be in conffict with any observed facts or established generalisations (दृष्टसिद्धये हि अदृष्टं कल्प्यते, न दृष्टविघाताय–Jayanta, Nyāya-Mañjarī, Āhnika 1) ; (3) no unobserved agent must be assumed, where it is possible to explain the facts satisfactorily or observed agencies (यदि अदृष्टमन्तरेण दृष्टं न सिद्ध्यति काममदृष्टं कल्प्यताम्, अन्यथाऽपि तदुपपत्तौ किं तदुपकल्पनेन, *ibid.*) ; (4) when two rival hypotheses are in the field, a crucial fact or test (विनिगमक, *ratio sufficiens*) is necesary; the absence of such a test (विनिगमनाविरह) is fatal to the establishment of either, (5) of two rival hypotheses,the simpler, *i.e.*, that which assumes less, is to be preferred (*ceteris paribus*) (कल्पनालाघव *versus* कल्पनागौरव); (6) of two rival hypotheses, that which is immediate or relevant to the subject-matter is to be preferred to that which is alien or remote (प्रथमोपस्थितत्व) ; (7) a hypothesis that satisfies the above conditions must be capable of verification (निर्णय) before it can be established as a theory (सिद्धान्त). The process of verification of a hypothesis consists in showing that it can be deduced as a corollary from (or is involved by implication in) some more general proposition which is already well-established (*cf.* Vātsyāyana's exposition and illustration of

Verification, निर्णय, including both the Deductive Method and, Colligation).

This doctrine of Scientific Method, in Hindu Logic, is only a subsidiary discipline, being comprehended under the wider comception of Methodology, which aims at the ascertainment of Truth whether scientific (Vijñāna) or philosophical (Jñāna) (मोक्षे धीर्ज्ञानम् स्यात् विज्ञानं शिल्पशास्त्रयो:, Amara-Koṣa) ;—the latter being the ulterior aim. In the investigation of any subject, Hindu Methodology adopts the following procedure : (1) the proposition (or enumeration) of the subject-matter (Uddeśa), (2) the ascertainment of the essential characters, or marks, by Perception, Inference, the Inductive Methods, etc. resulting in definitions (by लक्षण) or descriptions (by उपलक्षण); and (3) Examination and Verification (परीक्षा) and (निर्णय). Ordinarily the first step. Uddeśa, is held to include not mere Enumeration of topics, but Classification or Division proper (विभाग: उद्देशरूपानपायात्तु उद्देश एवासौ। सामान्यसंज्ञया कीर्त्तनमुद्देश: प्रकारभेदसंज्ञया कीर्तनं विभाग इति–Jayanta, Mañjarī); but a few recognise the latter as a separate procedure coming after Definition or Description. Any truth established by this three-fold (or four-fold) procedure is called a Siddhānta (an established theory). Now, the various Pramāṇas Proofs, *i.e.*, sources of valid knowledge) in Hindu L'ogic, *viz.*, Perception, Inference, Testimony. Mathematical Reasoning (सम्भव, including Probability in one view) are only operations subsidiary to the ascertainment of Truth (तत्त्वनिर्णय). And the Scientific Methods are merely ancillary to these Pramāṇas themselves.

I have explained the principles of the Hindu doctrine of Scientific Method, avoiding the technicalities of Logic as far as possible; and I cannot here enter upon the logical terminology or the logical apparatus and machinery, which would require a separate volume to themselves. For these, I would refer the reader to my paper on Hindu Logic,—as also for an account of the later Nyāya (नव्यन्याय), which, in spite of its arid dialectics, possesses a three-fold significance in the history of thought : (1) logical, in its conceptions of Avacchedaka and Pratiyogi, being an attempt to introduce quantification on a connotative basis, in other words, to introduce quantitative notions of Universal and Particular, in both an affirmative and a negative aspect, into the Hindu theory of Inference and Proposition regarded connotatively as the establishment of relations among attributes or marks; (2) scientific, in its investigation of the varieties of Vyāpti and Upādhi, (and of अन्यथासिद्ध) being an elaboration of Scientific Method, in the attempt to eliminate the irrelevant; and (3) ontological and epistemological, in its classification and precise determinations of the various relations of Knowledge and Being, with even greater rigidity and minuteness than in Hegel's Logic of Being and Essence. I will conclude with a few observations on Applied Logic, *i.e.*, the logic of the special sciences, which is such a characteristic feature of Hindu scientific investigation. What is characteristic of the Hindu scientific mind is that without being content with the general concepts of Science and a general Methodology, it elaborated the fundamental categories and concepts of such of the special sciences as it cultivated with assiduity, and systematically adapted the general principles of Scientific Method to the requirements of the subject-matter in each case. The most signal example of applied logic (or Scientific Method) worked out with systematic carefulness is the Logic of Therapeutics in Caraka, which a Logic which adapts the general concepts of cause, effect, energy, operation, etc., and the general methodology of science, to the special problems presented in the study of diseases, their causes, symptoms and remedies (*vide* Caraka, Vimānasthāna, chap IV,— also Sūtra Sthāna,—*vide* my Paper on Hindu Logic). Here I will give an illustration of Applied Methodology from the Science of Analytical and Descriptive

Grammar. Patañjali, in the Mahābhāṣya, *Circa.* 150 B.C. is very careful as regards Methodology. I take no note now of the philosophical presuppositions of this philology (*vide* my Paper on the Hindu Science of Language,) but will confine myself to his presentation of the Applied Logic of Descriptive and Analytical Grammar. The sentence is the unit of speech, as every Hindu philologer contends, but the first besiness of Analytical Grammar is to analyse the sentance into its significant parts and their coherent relations to one another. Assuming that articulate sounds are significant, the question is : how is the sentence, which is the unit of articulate speech, broken up into significant words and their mutual relations ? Patañjali answers that this is done by an instinctive use (*cf.* Vācaspati's भूयो दर्शनजनितसंस्कारम् इन्द्रियम्) of the Joint Method of Difference (combined Addition and Subtraction), fortified by the Joint Method of Agreement (अन्वय व्यतिरेकौ). Patañjali starts with a simple case. Take the two sentence-units :— Paṭhati (he reads) and Pacati (he cooks). Suppose you start with the assumption that these sounds are significant, and *that separate elements have separate meanings.* Then you hear Paṭhati, and, at the same time, a man reading is pointed at. Then the assemblage of sounds Paṭhati (*i.e.*, Paṭh+ati) stands for the assemblage—'one reads,' (*i.e.* the action reading+an individual agent). Similarly by finger-pointing or other indication, you find that the assemblage Pacati (*i.e.* Pac+ati) stands for the assemblage 'one cooks' (*i.e.* the action cooking+an individual agent, say, the same agent as before).

Now, look at the groupings —

Paṭh+ati=reading+one agent.

Pac+ati=cooking+one agent.

From repeated observation of similar groupings, one is led to conclude that Paṭh is the invariable concomitant of the action 'reading', because the disappearance of the former (the other element remaining) leads to the disappearance of the latter (a rough Method of Difference by substraction—

हीयते); that Pac is the invariable concomitant of the action 'cooking' because the introduction of Pac (with nothing else added) leads to the introduction of the action 'cooking' (a rough Method of Difference by addition,—उपजायते); and that Ati, which is the only 'common antecedent' (अन्वयी) is the invariable concomitant of 'one agent' which is the only common. 'consequent' (अन्वयी). In fact, the last should come first by the Method of Agreement, then the other two either by Residue, or by joint Agreement in presence and absence (अन्वयव्यतिरेकौ). So far all is plain sailing, though only very simple and very rough application of the Methods are given. But—and this is the point—through out the argument, it is assumed that one should is the concomitant of one idea, in other words, there is no plurality of causes to vitiate the application of the Method of Agreement in the above example. And now the objection is advanced that this basal assumption is untrue. The same sound is not the concomitant of the same idea. Different words (sounds) may have the same meaning. and the same world (sound) may have different meanings. We cannot therefore by the Joint Method of Agreement (अन्वयव्यतिरकौ) determine the meanings of words, or the separate functions of roots (stems) and inflections, at least not so simply as is pretended above. Patanjali states this difficulty in the way of applying the Scientific Methods to the problems presented by Analytical Grammar, a difficulty arising from the plurality of causes, but does not state the solution. The solution, however, depends on the Method of Subtraction and Addition (Patañjali's हीयते and उपजायते). For it will be found by extensive observation that the number of meanings of which a word (sound) may be capable is limited. So also is the number of words (sounds) expressive of a given meaning. Hence, by the method of Difference etc., the causal relations of words and meanings may be determined. Patañjali thus establishes the doctrine of Prakṛti and Pratyaya (roots and inflections), with their separate significance, which is of course a necessary postulate in the case of an inflectional language like Sanskrit. Isolating and Agglutinative languages offer less difficulty. Whereas the difficulties are in some respects enhanced in the case of languages with a polysynthetic or incorporating (incapsulating) morphological structure.

कथं पुनर्ज्ञायते अयं प्रकृत्यर्थः अयं प्रत्ययाथ ईति। अन्वयव्यतिरेन्काभ्याम्। कोऽसौ अन्वयो व्यतिरेको वा। इह पचतीत्युक्ते कश्चित् शब्दः श्रूयते, पठतीत्युक्ते कश्चित् शब्दो हीयते (पच्),-कश्चित् उपजायते (पठ्),-कश्चित् अन्वयी (अति);-अर्थोऽपि कश्चित् हीयते (विक्लित्तिः),-कश्चित् उपजायते (पठिक्रिया),-कश्चित् अन्वयी (कर्त्तृत्वं चैकत्वंच)। तेन मन्यामहे यः शब्दो हीयते, तस्य असौ अथः योऽर्थो हीयते। यः शब्द उपजायते तस्य असौ अर्थः योऽथं उपजायते। यः शब्दोऽन्वयो तस्य असौ अर्थः योहर्थोऽन्वयी।

विषम उपन्यासः-वहवो हि शब्दा एकार्था भवन्ति। तद्यथा इन्द्रः शक्रः पुरन्दरः। एकश्च शब्दो वह्वर्थः। तद्यथा यक्षाः पादा माषा इति। अय किं न साधीयोऽर्थवत्ता सिद्धा भवति। नापि व्रूमोऽर्थवत्ता न सिध्यतीति। वर्णिता हि अर्थवत्ता अन्वव्यतिरेकाभ्यामेव। तत्र कुत एतदयं प्रकृत्यर्थः अयं प्रत्ययार्थ इति। (Mahābhāṣya, 1—3, 9.)

Patañjali not only applies the Scientific Methods to the foundations of Grammatical Analysis, but also *by their means* establishes and elaborates the fundamental categories and concepts of Grammatical (and Philological) Science, *e.g.*, the concepts of action (क्रिया), agent (कर्त्ता) instrumental cause करणसाधकतमम्), end (निमित्त), origin (हेतु), limit (अवधि), substance (द्रव्य), quality (गुण), and genus (जाति);-also of the foundamental relations (at the bottom of all thought and speech)—the relations, of Time (कालिकसम्बन्ध), of space (दिग्देशसम्बन्ध), of Causality (कार्यकारणभाव), of inherence (समवाय), of co-inherence सामानाधिकरण्य of substance and attribute (विशेषणविशेष्यभाव), of the sign and the signate (संज्ञासंज्ञिभाव), of mutual dependence (इतरेतराश्रय);-an entire grammatical (and philological) apparatus, which will serve as a *'pointappui'* for generations of philologists and gramerians to come.

ADDENDA

(By Principal Brijendranath Seal)

Empirical Recipes from Varāhamihira (circa 550 A.D.) relating to chemical Technology

A. Searing of heards rocks to enable them to be cut (or pulverised) (शिलादारण).

Sprinkle on the rock taken red-hot from the fire of Palāśa and Tinduka wood (Butea Frondosa and Diospyros Embryopteris) :—(a) diluted milk, or (b) a solution of wood ashes (the ashes of the *Mokṣaka* mixed with those of reeds), or (c) a decoction of (the fruit) of the jujube (Zizyphus Jujuba) kept standing for seven nights in a mixture of whey, vinegar and spirits, in which Kulattha (Dolichos Uniflorus or Biflorus) has been steeped, or (d) a solution of the ashes of the Neema bark and leaves (Azadirachta Indica), the seasame pod, the resinous fruit of the Diospyros Embryopteris, and the Guduchi Tinospora Cordifolia), with cow's urine. Repeat the process seven times (in the last case six times).

(Varāhamihira, Vṛhat-Saṁhitā, Chap. 33, Ślokas 112-117).

B. Hardening of steel (शस्त्रपान).

(1) Plunge the steel red-hot into a solution of plantain-ashes in whey, kept standing for 24 hours ; then sharpen on the lathe.

(2) Make a paste with the juice of the Arka (Calotropis Gigantea), the gelatine from the horn of the sheep, and the dung of the pigeon and the mouse ; apply it

T